AF775302

Selected Titles in This Series

54 **Erik Winfree and David K. Gifford, Editors,** DNA Based Computers V

53 **Nathaniel Dean, D. Frank Hsu, and R. Ravi, Editors,** Robust Communication Networks: Interconnection and Survivability

52 **Sanguthevar Rajasekaran, Panos Pardalos, and D. Frank Hsu, Editors,** Mobile Networks and Computing

51 **Pierre Hansen, Patrick Fowler, and Maolin Zheng, Editors,** Discrete Mathematical Chemistry

50 **James M. Abello and Jeffrey Scott Vitter, Editors,** External Memory Algorithms

49 **Ronald L. Graham, Jan Kratochvíl, Jaroslav Nešetřil, and Fred S. Roberts, Editors,** Contemporary Trends in Discrete Mathematics

48 **Harvey Rubin and David Harlan Wood, Editors,** DNA Based Computers III

47 **Martin Farach-Colton, Fred S. Roberts, Martin Vingron, and Michael Waterman, Editors,** Mathematical Support for Molecular Biology

46 **Peng-Jun Wan, Ding-Zhu Du, and Panos M. Pardalos, Editors,** Multichannel Optical Networks: Theory and Practice

45 **Marios Mavronicolas, Michael Merritt, and Nir Shavit, Editors,** Networks in Distributed Computing

44 **Laura F. Landweber and Eric B. Baum, Editors,** DNA Based Computers II

43 **Panos Pardalos, Sanguthevar Rajasekaran, and José Rolim, Editors,** Randomization Methods in Algorithm Design

42 **Ding-Zhu Du and Frank K. Hwang, Editors,** Advances in Switching Networks

41 **David Aldous and James Propp, Editors,** Microsurveys in Discrete Probability

40 **Panos M. Pardalos and Dingzhu Du, Editors,** Network Design: Connectivity and Facilities Location

39 **Paul W. Beame and Samuel R Buss, Editors,** Proof Complexity and Feasible Arithmetics

38 **Rebecca N. Wright and Peter G. Neumann, Editors,** Network Threats

37 **Boris Mirkin, F. R. McMorris, Fred S. Roberts, and Andrey Rzhetsky, Editors,** Mathematical Hierarchies and Biology

36 **Joseph G. Rosenstein, Deborah S. Franzblau, and Fred S. Roberts, Editors,** Discrete Mathematics in the Schools

35 **Dingzhu Du, Jun Gu, and Panos M. Pardalos, Editors,** Satisfiability Problem: Theory and Applications

34 **Nathaniel Dean, Editor,** African Americans in Mathematics

33 **Ravi B. Boppana and James F. Lynch, Editors,** Logic and Random Structures

32 **Jean-Charles Grégoire, Gerard J. Holzmann, and Doron A. Peled, Editors,** The SPIN Verification System

31 **Neil Immerman and Phokion G. Kolaitis, Editors,** Descriptive Complexity and Finite Models

30 **Sandeep N. Bhatt, Editor,** Parallel Algorithms: Third DIMACS Implementation Challenge

29 **Doron A. Peled, Vaughan R. Pratt, and Gerard J. Holzmann, Editors,** Partial Order Methods in Verification

28 **Larry Finkelstein and William M. Kantor, Editors,** Groups and Computation II

27 **Richard J. Lipton and Eric B. Baum, Editors,** DNA Based Computers

26 **David S. Johnson and Michael A. Trick, Editors,** Cliques, Coloring, and Satisfiability: Second DIMACS Implementation Challenge

For a complete list of titles in this series, visit the AMS Bookstore at **www.ams.org/bookstore/**.

DIMACS

Series in Discrete Mathematics
and Theoretical Computer Science

Volume 54

DNA Based Computers V

DIMACS Workshop
DNA Based Computers V
June 14–15, 1999
Massachusetts Institute of Technology

Erik Winfree
David K. Gifford
Editors

NSF Science and Technology Center
in Discrete Mathematics and Theoretical Computer Science
A consortium of Rutgers University, Princeton University,
AT&T Labs–Research, Bell Labs (Lucent Technologies),
Telcordia Technologies, and NEC Research Institute

American Mathematical Society

This DIMACS volume contains papers from a DIMACS workshop entitled DNA Based Computers V, held at the Massachusetts Institute of Technology, Cambridge, MA, June 14–15, 1999.

2000 *Mathematics Subject Classification.* Primary 68–06, 68Q05, 68Q10, 68M99, 92–06, 92B05, 92B20, 92C05.

Library of Congress Cataloging-in-Publication Data

DIMACS Workshop on DNA Based Computers (5th : 1999 : Massachusetts Institute of Technology)
DNA Based Computers V : DIMACS Workshop DNA Based Computers, June 14–15, 1999, Massachusetts Institute of Technology / Erik Winfree, David K. Gifford, editors.
p. cm. — (DIMACS series in discrete mathematics and theoretical computer science ; v. 54)
Includes bibliographical references.
ISBN 0-8218-2053-2 (alk. paper)
1. Molecular computers—Congresses. I. Winfree, Erik, 1969– II. Gifford, David K., 1954– III. Series.
QA76.887.D5 1999
511.3—dc21 00-040621

Copying and reprinting. Material in this book may be reproduced by any means for educational and scientific purposes without fee or permission with the exception of reproduction by services that collect fees for delivery of documents and provided that the customary acknowledgment of the source is given. This consent does not extend to other kinds of copying for general distribution, for advertising or promotional purposes, or for resale. Requests for permission for commercial use of material should be addressed to the Assistant to the Publisher, American Mathematical Society, P. O. Box 6248, Providence, Rhode Island 02940-6248. Requests can also be made by e-mail to reprint-permission@ams.org.

Excluded from these provisions is material in articles for which the author holds copyright. In such cases, requests for permission to use or reprint should be addressed directly to the author(s). (Copyright ownership is indicated in the notice in the lower right-hand corner of the first page of each article.)

© 2000 by the American Mathematical Society. All rights reserved.
The American Mathematical Society retains all rights
except those granted to the United States Government.
Printed in the United States of America.

∞ The paper used in this book is acid-free and falls within the guidelines
established to ensure permanence and durability.
Visit the AMS home page at URL: http://www.ams.org/

10 9 8 7 6 5 4 3 2 1 05 04 03 02 01 00

Contents

Foreword

The Fifth International Meeting on DNA Based Computers was held at MIT on June 14–15, 1999. We would like to express our appreciation to Erik Winfree and David Gifford for their efforts to organize and plan this successful conference.

The workshop was part of the Special Focus on DNA Computing. We extend our thanks to Laura Landweber and Richard Lipton for their work as special focus organizers.

The conference brought together theoreticians and practitioners working on DNA computing from a variety of perspectives, with an emphasis on DNA computing based on combinatorial search and on DNA computing based on finite-state machines.

DIMACS gratefully acknowledges the generous support that makes these programs possible. The National Science Foundation, through its Science and Technology Centers program, the New Jersey Commission on Science and Technology, and DIMACS's partners at Rutgers, Princeton, AT&T Labs-Research, Bell Labs, NEC Research Institute, and Telcordia Technologies generously supported the special focus.

Fred S. Roberts
Director

Robert Sedgewick
Co-Director for Princeton

Introduction

The Fifth International Meeting on DNA Based Computers was held at MIT June 14-15, 1999 and presented papers that explored two distinct styles of DNA computing:

1. DNA computing based on combinatorial search, where randomly created DNA strands are used to encode potential solutions to a problem, and constraints induced by the problem are used to identify DNA strands that are solution witnesses.
2. DNA computing based on finite-state machines, where the state of a computation is encoded in DNA, which controls the biochemical steps that advance the DNA-based machine from state to state.

Early results in DNA computing focused on idealized combinatorial search. The first result by Leonard Adleman in *Science* titled "Molecular Computation of Solutions to Combinatorial Problems" showed how to solve a small instance of the directed hamiltonian path problem. Pure combinatorial search proposes to use exponential amounts of DNA instead of the exponential amount of time used by traditional approaches. Lipton, Boneh, and colleagues presented molecular algorithms for solving formula and circuit satisfiability problems, and several other groups showed that universal Turing machines can be simulated in parallel, with a single DNA molecule encoding the tape and head state of each machine. However, despite some very real experimental advances, laboratory demonstrations of DNA computing are still firmly in Toyland. The question on everyone's mind during the meeting was, "Can we get out of Toyland?"

The meeting was organized into four half-day sessions: Satisfying Experiments, Design and Simulation, Self-Assembly and Nanomachines, and Theoretical Perspectives. After each session, a panel discussion was held to bring into the open questions about where the research is going and what needs to be done next.

The first session opened with three back-to-back talks each describing the successful solution of a small Formula Satisfiability problem using a different technique. The paper by **Dirk Faulhammer, Anthony Cukras, Richard Lipton, and Laura Landweber** describes their work using RNA. The Princeton group's work is notable not only because it is the largest SAT problem yet attempted, with 9 input variables, but perhaps more importantly because they have made a considerable effort to characterize the error rates and error mechanisms in individual reactions.

Hiroshi Yoshida and Akira Suyama describe their work developing methods for solving 3-SAT by breadth-first search. The University of Tokyo researchers' goal is to find more efficient algorithms than the 'classical' brute-force approach wherein all possible solutions are generated and tested. In the breadth-first algorithm, solutions are constructed recursively, first finding all 1-variable solutions, then expanding those to all 2-variable solutions, and so forth. Empirical trials of the algorithm suggest that the number of different DNA strands required for 'typical' problems scales as $2^{\frac{n}{2}}$ instead of 2^n – a dramatic improvement (but not uncommon in computer science) that leads the authors to think that 100-variable problems might be solved by this method. It remains an open question whether the most difficult 3-SAT problems will succumb to the breadth-first approach, and whether breadth-first algorithms for formula-SAT or circuit-SAT can be devised. The algorithm presented here is particularly elegant, with all key steps experimentally prototyped on a 4-variable formula; importantly, error rates were also measured.

The talk by **Liman Wang** described the Wisconsin group's recent success doing DNA computing on surfaces. Their effort works to bring to DNA computing the lesson of automated DNA synthesis: that it is easy to subject a population of DNA strands to a sequence of chemical reagents when the DNA is attached to a solid support. They have now developed the technology to fully implement their algorithm on a 4-variable 3-SAT problem. An ancillary, but also important, accomplishment of the Wisconsin group is that, at least for small problems, DNA chips can be used to read-out the entire set of DNA sequences at any point in the algorithm, allowing them to measure the sequence-dependence of error rates – knowledge that will be critical for scaling up all DNA computing approaches. Unfortunately, their paper will not appear in this volume.

The discussion panel brought up many tough questions about the feasibility of solving significant SAT problems with DNA. A critical question, if SAT is to remain a major target for DNA based computers, is whether there are any problems in the size range DNA based computers might be able to solve that would pose any real difficulty for electronic computers. It was pointed out that modern heuristic algorithms such as Walk-SAT can routinely solve "typical" thousand-variable 3-SAT problems as well as the hardest randomly-generated 250-variable problems, and there are other 3-SAT algorithms for which $O(1.61^n)$ bounds have been proved. But are there any hard 100 variable problems, of the scale that DNA computers can potentially handle?

Going one step beyond Yoshida and Suyama on the road from rigorous algorithms to heuristics, **Junghuei Chen, Eugene Antipov, Bertrand Lemieux, Walter Cedeño, and David Wood** described preliminary results applying the logic of evolutionary computation to DNA-based computation. Although their experimental protocol is designed to solve a trivially simple problem, it is one that has been the subject of many theoretical studies of genetic algorithm performance. It will be interesting, in the future, to experimentally evaluate the effectiveness of genetic algorithms on large DNA population. It would be a welcome turn of events if the synthesis of evolutionary computation and DNA based computers also leads to better ways to evaluate the effectiveness of traditional in-vitro evolution and finds methods to improve them.

As hinted above, understanding and controlling error rates are central to solving large problems with DNA based computers. If at each step, a strand has even a 10% chance of being misprocessed or accidentally destroyed, after just 100 steps only 1 in 40,000 strands will have been processed correctly. Addressing this issue, **Kevin Chen and Erik Winfree** analyzed error-correcting algorithms for both separation errors and for strand loss errors, presuming that sequence independence of error rates can be achieved by biochemical means. However, errors are not the only potential bottleneck to scaling up DNA based computations to significant problems. The limiting factors could equally well be time if many steps are required, or strand length, or test tube complexity. It behooves us to figure out which are the critical bottlenecks. New implementations that allow more computation to occur in a single test tube, such as the use of liposomes to compartmentalize reactions as proposed by **Brian Bloom and Carter Bancroft** or the self-assembly approaches mentioned below, may make applications to larger problem instances feasible. Overall, things look pretty good. I will even venture to bet that a 20-variable SAT problem will be solved within the year. However, we are still very far from doing solid engineering, where we built systems with reliable performance by using well-characterized components; we do not yet have the intellectual framework for characterizing and predicting the behavior of molecular systems to the required level of accuracy.

To make matters more complicated, the set of available molecular tools is ever increasing. In a captivating invited talk, **Eric Kool** described three molecular technologies he has been developing. Closest to the atomic scale is a new pair of artificial DNA "bases" that can be incorporated by DNA polymerases. They are remarkable because one is essentially nothing (an abasic site) and the side group on the other is a large four-benzene-ring structure, pyrene, and thus specificity is not mediated by hydrogen bonds. At a slightly larger scale are activated DNA oligonucleotides, with modified 3′ and 5′ ends, that can easily be ligated without the need for enzymes or special reagents. His last example made novel use of an enzyme by asking RNA polymerase to transcribe from small circular oligonucleotides – somewhat like teaching an elephant to balance on a circus ball. Indeed, in this reaction, "rolling circle transcription," the polymerase will go around and around the template, producting long transcripts – which, if they contain the appropriate ribozyme motif, in turn can self-cleave to produce a well-defined product.

The second session on Design and Simulation focussed on how to design DNA computers within the limits set by physics. The paper by **Allen Mills, Bernard Yurke, and Phil Platzman** describes how parallel DNA hybridization reactions could implement vector algebra operations, including neural networks. After analysing the cycle time and statistical limits for this form of computation, he concluded that vector spaces as large as 10^7 dimensions are conceivable. An unsolved problem is how to find a set of 10^7 oligonucleotides with no or very little cross-hybridization. **Amit Marathe and Anne Condon** addressed this problem by defining several criteria related to cross-hybridization, and finding bounds on the size of sets of strings that satisfy the criteria. **Max Garzon, Russell Deaton, John Rose, and Donald Franceschetti** discuss a different set of criteria for the same problem, which have been implemented in the software package `Edna`. **Masahito Yamamoto et al.** have their own criteria for designing non-interacting DNA sequences; their work is notable for having an experimental investigation of the sequences they designed. **Alexander Hartemink, Tarjei Mikkelsen, and David Gifford** describe modular software for simulating biological reactions; submodules include prediction of hybridization and the action of enzymes such as polymerase and ligase. The package has been used to predict the products of the PCR-based unary counter that their laboratory has been investigating experimentally. Such a tool can be useful not only for prediction, but also for identifying hypotheses for unexpected experimental results.

Panel discussion identified several challenges for simulation and design software. The first is how to assess validity and scope – that is, how to compare the simulation results to real experiments and understand the limits for accurate prediction of new experiments. There was discussion of what the fundamental data structures should be for such as simulation; should modelling be done at an abstract logical level, at a quantitative thermodynamic level, at the atomic or quantum level, or do all these levels of abstractions need to be combined to accurately model molecular systems? As a practical matter, there was general agreement that algorithms and code should be made available on the Web, to promote communication between labs; it is a bit frustrating that every lab has their own DNA design criteria, and no one can convincingly argue that one is unambiguously better than another.

The third session on Self-Assembly and Nanomachines explored some alternative approaches to and alternative uses for molecular computation. **Thomas LaBean, Erik Winfree, and John Reif** report experimental progress toward implementing the String Tile model, which holds promise using a single self-assembly reaction to construct a molecular lookup table for addition. **Michail Lagoudakis and Thomas LaBean** give a theoretical example of performing more complex computations by 2D self-assembly: they show how to construct a set of tiles that solve the satisfiability problem in a single self-assembly reaction. **Takashi Yokomori** proposes a new model of computation that involves a particularly simple experimental reaction: the self-assembly of linear chains of pre-formed components, followed by melting and re-hybridization into a low-energy structure. Remarkably, this model is Turing universal.

Andrew Turberfield, Bernard Yurke, and Allen Mills describe a series of fiendishly clever, yet remarkably simple, experiments where DNA oligonucleotides are used as catalysts to control the hybridization of other DNA molecules. This careful and quantitative work is sure to serve as the groundwork for more elaborate DNA mechanisms. In fact, the group have already constructed their first "machine", a jointed DNA molecule that can be opened and closed like tweezers. A paper by **Michael Robertson, Jay Hesselberth, Colin Cox, and Andrew Ellington** describes the group's work using *in vitro* evolution and rational modification to discover ribozymes that can ligate DNA under control of a complex switch that can depend on, so far, up to three inputs, such as a DNA oligonucleotide, adenosine, and theophyline. Combining these switches into interacting molecular systems will be a fascinating endeavour.

It is interesting to speculate on where this work is going. DNA and RNA can perform logical computations during self-assembly; it can catalyze and perform mechanical work on other DNA molecules; it can undergo conformational switching in response to effector molecules; it can catalyze chemical reactions; are these not sufficient building blocks for constructing arbitrarily sophisticated biochemical systems and molecular devices? Might we be creating a "DNA world" as a stage in molecular engineering, regardless of whether such a stage existed in the evolution of life on Earth? These were all topics of the panel discussion.

What theoretical constructs will be necessary for molecular computation and complex biochemical systems? What theoretical goals can help use guide experimental investigations into promising areas? This was the topic of the fourth and final session, Theoretical Perspectives. The paper by **Andrzej Ehrenfeucht, Hendrik Jan Hoogeboom, Grzegorz Rozenberg, and Nikè van Vugt** gives an elegant mathematical analysis of sets generated by simultaneous enforcing and forbidding conditions. An advantage of this formalism is that it does not depend explicitly on the order of elementary events – a key property for describing molecular systems where reactions occur asynchronously. **Lila Kari and Laura Landweber** ask whether concepts from computer science can help us understand biological phenomena, such as the mysterious gene unscrambling found in single-cell ciliates. **Gheorghe Păun and Takashi Yokomori** present new results for a family of models inspired after the subcellular compartments found in Eukaryotic cells. Going ever smaller, **Ehud Shapiro** presented his model of computation in which a hypothetical holoenzyme – like the ribosome but tailored for computation – acts on a DNA-like polymer to simulate a Turing Machine, possibly producing molecules as output. His paper will appear elsewhere. The paper by **Ashish Gehani, Thomas Labean, and John Reif** analyses a new and unexpected direction for DNA-based cryptography: rather than try to break electronic cryptosystems, use DNA to encrypt and transmit secret messages. It seems there is no shortage of surprises in the molecular world.

During the panel discussion it was noted that in fields such as computer science and quantum physics, theory preceeds experiment, which only goes so far as to find a way to obtain what the theory predicts; whereas in fields such as biology and solid-state physics, experiment preceeds theory, which only manages to put the pieces together and explain them in a more understandable form. So far, DNA based computing has been driven by theory. But it is quite possible, as unexpected molecular computing components are discovered, as continually occurs in Ellington's lab, that the reverse trend will become dominant.

So, the future of DNA computing looks bright. But don't reach for your sunglasses. It is still before the dawn, and in the dark shadows I just stubbed my toe on something.

Erik Winfree

David Gifford

Organizers & Editors

DIMACS Series in Discrete Mathematics
and Theoretical Computer Science
Volume **54**, 2000

When the Knight Falls: On Constructing an RNA Computer

Dirk Faulhammer, Anthony R. Cukras, Richard J. Lipton,
and Laura F. Landweber

ABSTRACT. Recently, we introduced and developed RNA based computing as a general approach for the solution of 'satisfiability' (SAT) problems [**1**]. As an example, we chose to solve a 9-bit instance of the "Knight Problem," which asks generally how many knights and what configurations can one place on an $n \times n$ chess board such that no knight is attacking any other knight on the board. By introducing a destructive algorithm of sequential ribonuclease (RNase) digestions to manipulate strands of a 10-bit binary RNA library (representing all positions on a 3×3 chessboard), we recovered a set of 42 out of 43 molecules that describe solutions to this problem [**2**] . Here, we present experimental data on several controls and highlight our solutions to possible pitfalls encountered on the way to a successful algorithm.

1. Introduction

Chess problems create a class of SAT problems that are flexible in both scale and complexity. For the "Knight Problem," a 'true' or '1' variable represents the presence of a knight at positions a–i, while a 'false' or '0' variable represents the absence of a knight at that position. Though not NP-complete for any defined board, this is a particular instance of a class of NP-complete SAT problems. For a 3×3 board

a	b	c
d	e	f
g	h	i

we can represent the knight problem in the following, simplified way ($\wedge$ = 'AND', $\vee$ = 'OR'):

$$((\neg h \wedge \neg f) \vee \neg a) \wedge ((\neg g \wedge \neg i) \vee \neg b) \wedge ((\neg d \wedge \neg h) \vee \neg c)$$
$$\wedge ((\neg c \wedge \neg i) \vee \neg d) \wedge ((\neg a \wedge \neg g) \vee \neg f)$$

2. Implementing the destructive algorithm

Because of its $2'$ hydroxyl group, RNA is well suited for a destructive algorithm, as we described earlier [**1**]. By using complementary DNA oligonucleotides and RNase H, unique 'bit' sequences in a binary RNA library [**2**] can be easily destroyed and removed from the library. The algorithm consists of four repetitive steps, followed by a fifth, readout, step. We describe the approach as follows:

Step 1: Set bits to 1 or 0. Split the RNA pool into 2 tubes and anneal DNA oligonucleotides complementary to 'bit' positions to be digested. (For example, to set bit position a to 1 in clause 1 , anneal the oligonucleotide complementary to a set to 0. Correspondingly, in this test-tube bit positions f and h must be set to 0. In the other test-tube, set bit position a to 0.)

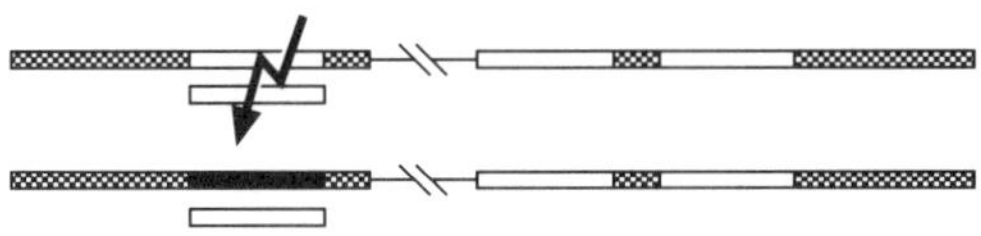

Step 2: RNase H digestion sets bit positions.

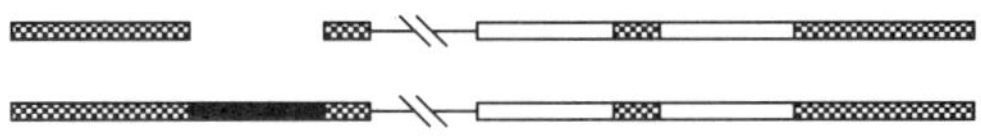

Step 3: Remove DNA oligonucleotides and short digested RNA products by denaturing polyacrylamide gel electrophoresis; reamplify RNA by reverse transcription (RT), polymerase chain reaction (PCR) and *in vitro* transcription.

Step 4: Combine the contents of the two test-tubes and go to steps 1-4 for the next clause.

Step 5: After the last digestion, perform RT-PCR and clone representative molecules from the enriched pool.

The construction of the library is decribed elsewhere [**1**]. Sequences of the 'bits' are as follows [**1**]:

Bit	Knight Absent Code Set to 0	Knight Present Code Set to 1	Spacer Code
a	CTCTTACTCAATTCT	TCCTCACATTACTTA	TCTAC
b	CATATCAACATCTTA	ACTTCCTTTATATCC	ATAAC
c	ATCCTCCACTTCACA	TTATAACAAACATCC	CTTAA
d	TTAAAATCTTCCCTC	ACATAACCCTCTTCA	TTTAC
e	CTATTTATCCACACC	ACCTTACTTTCCATA	TACAA
f	GCTTCAAACAATTCC	GTACATTCTCCCTAC	TCCTT
g	AACTCTCAAATTCAA	CATAATCTTATATTC	TCAAT
h	CTAACCTTTACTTCA	ATAATCACATACTTC	TCCAA
i	CATTCCTTATCCCAC	TCCACCAACTACCTA	ACACA
j	CACCCTTTCTCCTCT	TTTTAAATTTCACAA	SUFFIX

3. Results and conclusions

3.1. Digestion of the 10-bit RNA library by RNase H. To test the efficiency of RNase H digestion on the library, we digested the $5'$ end radiolabeled 10-bit RNA library under standard conditions [**1**] in the presence of three complementary bit oligonucleotides. We separated the reaction products on a 12% polyacrylamide gel and quantified the bands of the undigested precursor and the major cleavage products (often appearing as double bands probably due to the exonuclease activity of RNase H [**3**]) by using a phosphorimager (Molecular Dynamics) and Molecular Dynamics Imagequant software. FIGURE 1 shows a typical gel of the 5′ end radiolabeled RNA library after digestion with RNase H in the presence of three bit oligonucleotides as indicated ("-" indicates a negative control) where $\overline{a_0}$ is the complement to the sequence of bit position a set to 0, etc. Digestion of the library in the presence of $\overline{a_0}, \overline{f_1}, \overline{g_1}$ and $\overline{g_0}, \overline{b_1}, \overline{f_1}$, respectively, resulted in extra bands (marked with arrows) probably due to mis-hybridization of $\overline{f_0}$ to the RNA.

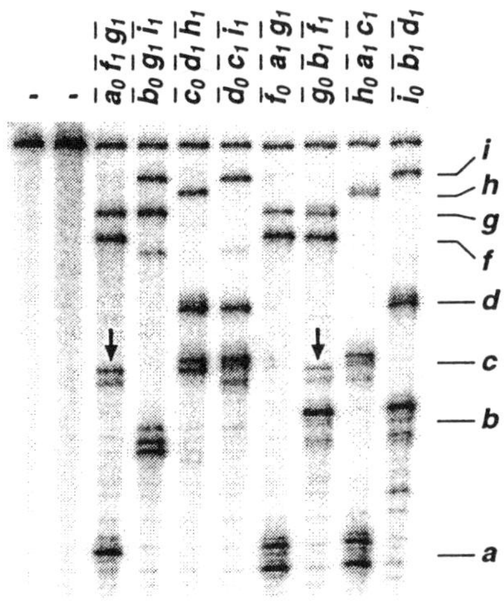

FIGURE 1

The sum of the intensities of the bands in a lane (full-length RNA and shorter product bands) was used as a measure of the total RNA library digested (including strong extra bands). We calculated the relative strength of a single band by dividing its intensity by this sum. Relative intensities (in %) were as follows[1]:

Bit complements	Precursor	Product 1	Product 2	Product 3
$\overline{a_0}, \overline{f_1}, \overline{g_1}$	12.1	11.4	20.3	42.8
$\overline{b_0}, \overline{g_1}, \overline{i_1}$	10.3	11.7	17.4	56.8
$\overline{c_0}, \overline{d_1}, \overline{h_1}$	12.1	10.2	26.4	50
$\overline{d_0}, \overline{c_1}, \overline{i_1}$	7.5	8.3	13.5	69
$\overline{f_0}, \overline{a_1}, \overline{g_1}$	7.1	6.2	17	70
$\overline{g_0}, \overline{b_1}, \overline{f_1}$	12.3	11.6	18.7	48.5
$\overline{h_0}, \overline{a_1}, \overline{c_1}$	7.1	6.5	23.2	63.2
$\overline{i_0}, \overline{b_1}, \overline{d_1}$	9.1	7.7	27.5	56

The expected relative intensities of the reaction products are: undigested precursor 12.5% (1/8 of the pool), product 1 12.5% (1/8), product 2 25% (1/4), and product 3 50% (1/2). The values found in our experiment are either smaller (precursor, products 1 and 2) or larger (product 3) than the expected values. This result is probably due either to overdigestion or to an unequal representation of bit positions in the library. However, our results suggest that the digestion efficiency was sufficient under these conditons.

[1]Product 1 is the longest, and product 3 is the shortest digestion product. Because of extra bands, the sum of relative intensities is not 100% in all lanes. See also FIGURE 1.

Digestion of single RNA clones under the same conditions was almost complete ($\geq$ 98% in all cases but 96% for i_0, b_1, d_1) as shown below.

Bit positions	Undigested RNA (%)	clone
b_0, g_1, i_1	0.46	100 000 101
c_0, d_1, h_1	0.64	010 110 010
d_0, c_1, i_1	0.97	101 011 011
f_0, a_1, g_1	0.67	100 000 101
g_0, b_1, f_1	1.26	011 111 001
h_0, a_1, c_1	1.33	111 101 101
i_0, b_1, d_1	3.63	111 101 101

3.2. Readout by multiplex primer extension. We developed a streamlined approach to readout all bit settings at once. After colony PCR [**1**], cloned strands were first amplified for 20 cycles of linear PCR [**4**] in the presence of the primer complementary to the 5′ end only; then, an equimolar mixture of 5′ end radiolabeled '0'- or '1'-complementary DNA oligonucleotides ($\overline{a_0}$,...,$\overline{i_0}$ or $\overline{a_1}$,...,$\overline{i_1}$) was added to the same tube and the reaction continued for 5 more cycles. FIGURE 2 shows a schematic view of the experiment (on the right) and a typical 'bar-coding' gel (on the left). Clone A corresponds to the sequence 010 011 010 and clone B 001 011 000 (shown in the schematic).

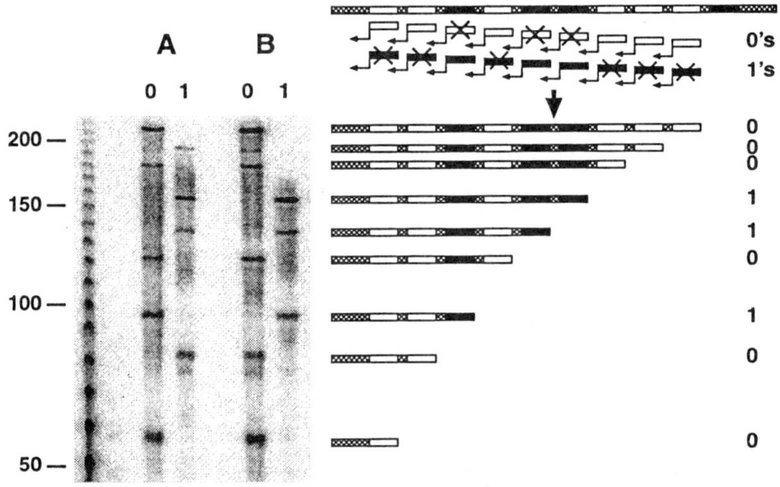

FIGURE 2

3.3. Gel elution and reverse transcription. To test the efficiency of gel elution, three cloned RNA molecules (b9 = 001 011 000, b10 = 010 011 010, and d2 = 001 000 001) were loaded on a 6% polyacrylamide gel, eluted after running the gel (500 mM ammonium acetate, 1 mM EDTA, 0.1% SDS), ethanol precipitated [**5**] and resuspended in 25 μl H_2O. The optical density (OD_{260}) of the RNA was measured before and after gel purification.[2] The efficiency of reverse transcription

[2]The RNA concentration (in μg/ml) was calculated by [RNA] = (OD_{260}) $\times$ (dilution factor) $\times$ (40 μg/ml) [**5**] and the total amount of RNA (pmol) in n μl by (pmol) = ([RNA]/1000) $\times$ n/(330 g/mol).

(RT) was tested using the eluted RNA and 100 pmol 5′ end radiolabeled primer in 80 μl [**2**]. Reaction products were separated on a 12% polyacrylamide gel and quantified by using a phosphorimager. We calculated the efficiency of the reaction by dividing the amount of full length cDNA by the total amount of radioactive products and primer in a lane (considering that one RNA molecule gives rise to one cDNA molecule). The results are shown in the table. Although 50% efficiency of gel elution seems to be quite satisfactory, it is very important to have more than one copy of each strand available in the test-tube. Otherwise, solutions to the problem will be inevitably lost. This is especially true for the reverse transcription step, where efficiencies are as low as 22%.

Sample	Input Amount (pmol)	Output Amount (pmol)	RT Efficiency
b9	52.8	27.5 [52%]	48%
b10	31.6	16.3 [52%]	22%
d2	36.8	27.5 [75%]	25%

3.4. Pitfall 'bit swapping' by PCR. Recombination events are likely to occur during PCR amplification of heterogeneous target sequences [**6**], especially as the strands in our 10-bit library share several stretches of sequence identity constrained by the bit structure of the library. FIGURE 3 shows two possible ways of 'bit swapping' between two DNA strands (`wt-I 011 111 001` and `wt-II 010 110 010`) during PCR. 'Bit swapping' happens either by annealing and extension of incompletely extended strands to heterologous target sequences or by annealing of non-complementary full length DNA sequences. (In heteroduplexes only one strand would be repaired by the bacterial cell, giving rise to false readouts after cloning [**8**].) Three bit positions (c, f, and h) were analyzed by readout PCR.

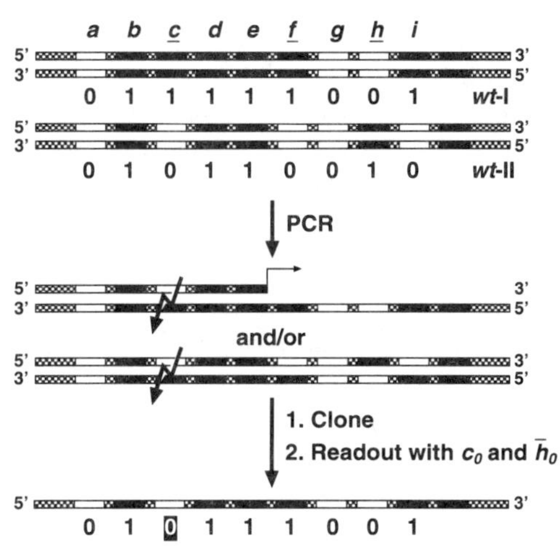

FIGURE 3

To measure this effect, the PCR products derived from two clones (`wt-I 011 111 001` and `wt-II 010 110 010`, $a, ..., i$) were combined and amplified for 25 cycles of PCR. The resulting PCR products were cloned, and 20 clones were randomly chosen and analyzed by multiplex colony PCR readout at positions c, f, and h. Eight out of 20 resulting clones (40%) could be explained as the product of 'bit swapping'. This indicates that 'bit swapping' poses a serious threat to successful execution of the algorithm or any manipulation of a binary nucleic acid library using PCR; however, when only 15 instead of 25 PCR cycles were used, there was no evidence of 'bit swapping' in 20 randomly chosen clones. This suggests that 'bit swapping' in our system is mostly due to a high proportion of incompletely extended strands annealing to heterologous target sequences, and it suggests that we can essentially eliminate this problem by limiting the number of PCR cycles. Consequently, no more than five cycles of PCR were performed between any two digestion steps and before cloning. Instead, we exploited RNA polymerase as an amplifier. The chance of 'bit swapping' during RNA synthesis caused by initiation of transcription by hybridization of RNA to single-stranded DNA template or to RNA itself [**7**] is reduced in our system where double-stranded DNA serves as a template and intra- and intermolecular interactions of RNA molecules are minimized by pool design [**1**].

3.5. RNA Solutions. Taking into account these possible sources of error, we executed the successful molecular algorithm as follows: The RNA library was reamplified (only 2–5 cycles of PCR) after each digestion step and at least 200 pmol ($\sim$17 μg) of RNA were used in each digestion reaction. As an additional control, we also cloned the library after each digestion step and performed a readout on 12 clones, which were legal. We subjected the RNA library to the algorithm outlined above. Of 43 unambiguously screened clones chosen at random, 42 represented solutions to the "Knight problem". There were nine occurrences of solutions which appeared two or more times in our random sampling. The 32 different unambiguous 'boards' are shown in FIGURE 4. The overall distribution of the complete 42 clones was not different from a random sampling of all 94 possible solutions. One clone (`010 111 010`) represented a maximal solution, whereas another clone (`010 111 011`) contained two illegal interactions, resulting from incorrect placement of 1 out of 129 knights on all unambiguous 43 boards. This result indicates a 65- to 205-fold selection preference for survival of correct solution strands over incorrect solution strands.[3] This is a real improvement over previous results [**1**], where only three out of the seven clones represented legal solutions.

Acknowledgement

We thank Steve Lawrie for technical assistance and Erik Winfree for helpful discussions. This work was supported in part by AFOSR, NSF/DARPA, and DFG (to DF).

[3]We screened 47 clones. In four ambiguous clones, one of the positions gave no band in the readout gel. Two of these four clones represented legal solutions regardless of the value at the missing position. We made the following estimation of enhancement of solutions over non-solutions:

$$\frac{E_{S_c}}{E_{S_i}} = \frac{[\mathit{legal\ solution}]_{obs} \times 418}{[\mathit{illegal\ solution}]_{obs} \times 94}$$

with $44 \leq [\mathit{legal\ solution}]_{obs} \leq 46$ and $1 \leq [\mathit{illegal\ solution}]_{obs} \leq 3$ and a total of 47 screened clones. E_{S_c} (respectively E_{S_i}) is the expected number of copies of a correct (incorrect) strand that survive the destructive algorithm. The number of all logical legal solutions is 94, and the number of all logical illegal solutions is 418.

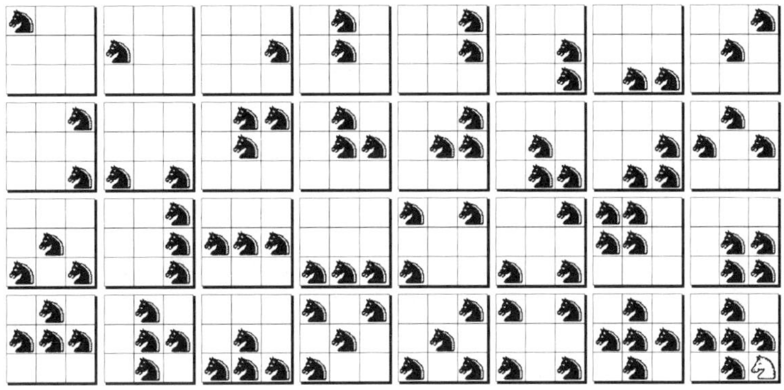

FIGURE 4

References

[1] A. R. Cukras, D. Faulhammer, R. J. Lipton, and L. F. Landweber. *Chess games: a model for RNA based computation.* In L. Kari, editor, DNA Based Computers IV, Proceedings of a DIMACS Workshop, to appear.

[2] D. Faulhammer, A. R. Cukras, R. J. Lipton, and L. F. Landweber. *Molecular computation: RNA solutions to chess problems.* Submitted for publication.

[3] O. Schatz, J. Mous and S. F. J. Le Grice. *HIV-1 RT-associated ribonuclease H displays both endonuclease and $3' \rightarrow 5'$ exonuclease activity.* EMBO J. 9, 1171-1176, 1990.

[4] L. F. Landweber and M. Kreitman. *Producing single-stranded DNA in polymerase chain reaction for direct genomic sequencing.* Methods Enzymol. 218, 17-26, 1993.

[5] J. Sambrook, E. F. Fritsch, and T. Maniatis. *Molecular Cloning: A Laboratory Manual, 2nd edition.* Cold Spring Harbor Laboratory Press, 1989.

[6] A. Meyerhans, J.-P. Vartanian, and S. Wain-Hobson. *DNA recombination during PCR.* Nucleic Acids Res. 18, 1687-1691, 1990.

[7] C. Cazenave and O. C. Uhlenbeck. *RNA template-directed RNA synthesis by T7 RNA polymerase.* Proc. Natl. Acad. Sci. USA 91, 6972-6976, 1994.

[8] R. Jansen and F. Ledley. *Disruption of phase during PCR amplification and cloning of heterozygous target sequences.* Nucleic Acids Res. 18, 5153-5156, 1990.

(D. Faulhammer) Department of Ecology and Evolutionary Biology, Princeton University, Princeton, NJ 08544,

E-mail address: `fadirk@princeton.edu`

(A. Cukras) Department of Ecology and Evolutionary Biology, Princeton University, Princeton, NJ 08544,

E-mail address: `acukras@welch.jhu.edu`

(R. Lipton) Department of Computer Science, Princeton University, Princeton, NJ 08544,

E-mail address: `rjl@cs.princeton.edu`

(L. Landweber) Department of Ecology and Evolutionary Biology, Princeton University, Princeton, NJ 08544,

E-mail address: `lfl@princeton.edu`

DIMACS Series in Discrete Mathematics
and Theoretical Computer Science
Volume **54**, 2000

Solution to 3-SAT by Breadth First Search

Hiroshi Yoshida and Akira Suyama†

ABSTRACT. We have actually solved a very simple instance of 4-variable-10-clause 3-SAT problem with DNA molecules using experimental protocols amenable to automation. The algorithm is based on breadth first search, as can be seen in Ogihara & Ray's [**OR97**], whereby we have succeeded in drastic decrease of the number of DNA molecules that must be present in a test tube. Computer simulations demonstrated that the necessary number of DNA molecules for solving n-variable 3-SAT problems was around $2^{0.58n}$. Hence, compared with the conventional brute force method, our algorithm and implementation can be extended to a larger-scale 3-SAT problem with ease.

1. Introduction

The conventional method of solving combinatorial problems with DNA consists of two phases as proposed by Adleman [**Adl94**]. The first phase is to generate a pool of all possible solutions with DNA molecules, and the next phase is to select DNA molecules representing the answer from the pool according to the solution conditions. There, however, exists a drawback in this brute force method that the larger the size of problems becomes, the larger exponentially the amount of DNA that must be present in the initial pool becomes. For example, the 3-SAT problem with n variables, which is a typical NP-complete problem, requires that there must exist at least 2^n of DNA molecules in the initial pool. This is why DNA computation has been criticized for having no practical feasibility.

In order to overcome the drawback of the conventional brute force method, we followed Ogihara & Ray [**OR97**] and Morimoto *et al.* [**MAS97**], and introduced the breadth first search into DNA computation. Our algorithm does not first generate a large pool of all possible solutions with DNA, but first generates a few short DNA oligomers representing partial assignments: solutions that partially satisfy the solution conditions. DNA oligomers are then extended step by step with pruning unnecessary oligomers as soon as they fail to satisfy the conditions. The *pruning* plays an important role in substantially decreasing the necessary amount of DNA. We have actually succeeded in decreasing the number of DNA molecules to at least around $2^{0.58n}$ for n variable 3-SAT problems. For n = 86, the reduction

2000 *Mathematics Subject Classification.* Primary 68–06; Secondary 68M99, 68W10.

This research was supported in part by a grant from the Japan Society for the Promotion of Science "Research for the Future" Program (JSPS-RFTF96100101).

† To whom correspondence should be addressed.

© 2000 American Mathematical Society

is $2^{0.58*86} \approx 10^{15}$, thereby making the solution of larger scale 3-SAT problems practically feasible.

2. Algorithm

Our algorithm is based on the breadth first search like Ogihara & Ray's [**OR97**] and initially creates a pool of only four partial assignments to the first two Boolean variables x_1 and x_2. Then, these four kinds of partial assignments will be extended variable by variable according to the solution conditions with immediately pruning partial assignments unsatisfying the conditions.

Now we describe the algorithm in more detail. Let F be a formula of 3CNF-SAT with n Boolean variables. Not losing generality, we can assume that none of the clauses of F has both the positive form and the negative form of the same variable and that F does not have two or more clauses consisting of the completely same three literals.

A: Preprocessing
The variables are sorted so that the number of clauses which contain either x_i or its negation is non-increasing, and then for every clause of F, its three literals are reordered so that their indices are increasing.

B: Initializing
Let S be a set of four partial assignments as follows:

$$S = \{(x_1, x_2) \mid x_1 = 0 \vee 1, x_2 = 0 \vee 1\} = \{(0,0),(0,1),(1,0),(1,1)\}$$

C: Main loop
The following procedures are iterated from $k = 3$ to $k = n$.

- Let F_T (F_F) be the two-literal conjunctive normal form expression constructed by concatenating all clauses of F whose third literal is $\neg x_k$ (x_k, respectively), then by eliminating all literals $\neg x_k$ (x_k, respectively).
- Let both of S_T and S_F be empty collections.
- For every partial assignment π in S, add it to S_T if it satisfies F_T.
- For every partial assignment π in S, add it to S_F if it satisfies F_F. Note that if F_T (F_F) contains no clauses, every element π in S belongs to S_T (S_F).
- For every partial assignment π in S_T (S_F), $x_k = 1$ ($x_k = 0$, respectively) is appended to it.
- Let S be a joined collection of S_T and S_F.

D: If S is not empty, F is satisfiable, otherwise unsatisfiable.

The breadth first search algorithm for 3CNF-SAT reported by Ogihara & Ray [**OR97**] requires three collections S_T, S_F and N . In contrast, the present algorithm requires only two collections S_T and S_F, succeeding in decreasing the complexity of actual implementation of the algorithm.

3. Implementation

3.1. DNA Computer Program of 3-SAT. The following four basic DNA commands are used in a DNA computer program implementing the breadth first search algorithm for 3-SAT problems.

- Get($T, +s$), Get($T, -s$):
  get DNA molecules with a subsequence s (without s) in a tube T.

```
/* Problem 3-SAT */
(u_1 ∨ v_1 ∨ w_1) ∧ (u_2 ∨ v_2 ∨ w_2) ∧ ··· ∧ (u_i ∨ v_i ∨ w_i) ∧ ··· ∧ (u_m ∨ v_m ∨ w_m)
u_i ∈ {x_j, ¬x_j}, v_i ∈ {x_k, ¬x_k}, w_i ∈ {x_l, ¬x_l} (1 ≤ j < k < l ≤ n)

/* Program dna3sat */
function dna3sat(u_1, v_1, w_1, ··· , u_m, v_m, w_m)
begin
  T_2 = {X_1^T X_2^T, X_1^F X_2^T, X_1^T X_2^F, X_1^F X_2^F};
  for k = 3 to n do
    Amplify(T_{k-1}, T_w^T, T_w^F);
    for j = 1 to m do
      if w_j = x_k then
        T_w^F = getuvsat(T_w^F, u_j, v_j);
      end
      if w_j = ¬x_k then
        T_w^T = getuvsat(T_w^T, u_j, v_j);
      end
    end
    T^T = Append(T_w^T, X_k^T, \overline{X_{k-1}^{T/F} X_k^T}); T^F = Append(T_w^F, X_k^F, \overline{X_{k-1}^{T/F} X_k^F});
    T_k = Merge(T^T, T^F);
  end
  return Detect(T_n);
end

function getuvsat(T, u, v)
begin
  T_u^T = Get(T, +X_u^T); T_u'^F = Get(T, -X_u^T);
  T_u^F = Get(T_u'^F, +X_u^F); /* or T_u^F = T_u'^F; */
  T_v^T = Get(T_u^F, +X_v^T);
  T^T = Merge(T_u^T, T_v^T);
  return T^T ;
end
```

FIGURE 1. DNA Computer Program of 3-SAT

- Append(T, s, e):
  append a subsequence s at the end of DNA molecules with a splint e in a tube T.
- Merge$(T_1, T_2, \dots, T_n)$:
  merge DNA molecules in tubes $T_1, T_2, \dots, T_n$ into a single tube.
- Amplify$(T, T_1, T_2, \dots, T_n)$:
  amplify DNA molecules in a tube T and divide them into tubes $T_1, T_2, \dots, T_n$.

The program for solving an instance of 3-SAT problems with n variables and m clauses is shown in Fig. 1. The function getuvsat(T, u, v) examines the satisfiability of each clause of two-literal conjunctive normal form expressions F_T and F_F. Tubes T_k $(k = 2, \dots, n)$, T_w^T and T_w^F contain DNA molecules representing partial assignments in the set S, S_T and S_F, respectively. X_i^T and X_i^F $(i = 1, \dots, n)$

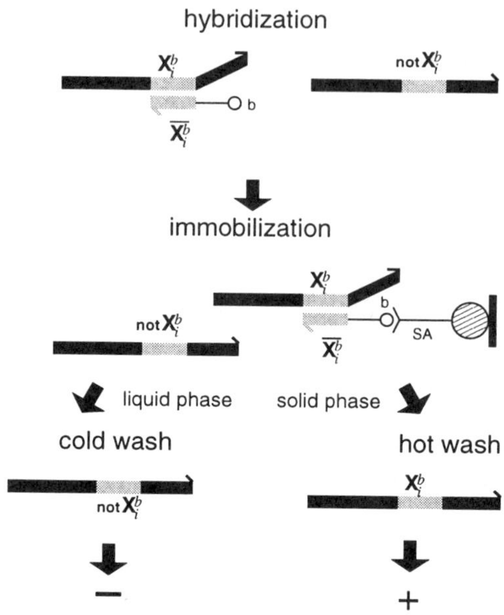

FIGURE 2. Get($T, +X_i^b$) and Get($T, -X_i^b$) ($b \in \{T, F\}$)

designate the DNA base sequence representing ith Boolean variable $x_i = 1$ and $x_i = 0$, respectively. The computation time is $O(n) + O(m)$ because Get, Append, Merge and Amplify commands are executed $3m$, $2n$, $n + m$ and n times, respectively, in the program. Therefore, the NP-complete problem can be solved in the time proportional to the problem size.

3.2. Protocol and Efficiency of Basic DNA Commands.

3.2.1. *Get($T, +X_i^b$) and Get($T, -X_i^b$).* Get($T, +X_i^b$) and Get($T, -X_i^b$) are the commands which select the partial assignments with the i-th variable: $x_i = b$ and $x_i \neq b$ ($b \in \{T, F\}$), respectively. Namely, these commands select the DNA molecules containing X_i^b or not containing X_i^b. The following is the detailed protocol of Get command, which is summarized in Fig. 2. In the protocol, $\overline{X_i^b}$ designates the reverse complement of DNA sequence X_i^b. The reaction volumes were 100 $\mu\ell$ when unspecified.

1. Add b$\overline{X_i^b}$, which is the 5'-end-biotinylated sequence of $\overline{X_i^b}$, to the input sample tube T. The actual amount of b$\overline{X_i^b}$ should be more than the amount of DNA oligomers in the input tube.
2. Heat the tube to 94 ^{o}C and gradually cool it to 25 ^{o}C at 10 ^{o}C /min. b$\overline{X_i^b}$ oligomers anneal to DNA molecules containing X_i^b.
3. Add a sufficient amount (0.1 mg for 40 pmol of b$\overline{X_i^b}$) of Dynabeads M-280 Streptavidin (DYNAL A. S., Norway) and shake the tube moderately for 15 min at room temperature. DNA molecules containing X_i^b are captured on the beads because b$\overline{X_i^b}$ oligomers bind to the beads through a strong and specific interaction between biotin and streptavidin.
4. Separate the beads using a magnet Dynal MPC.

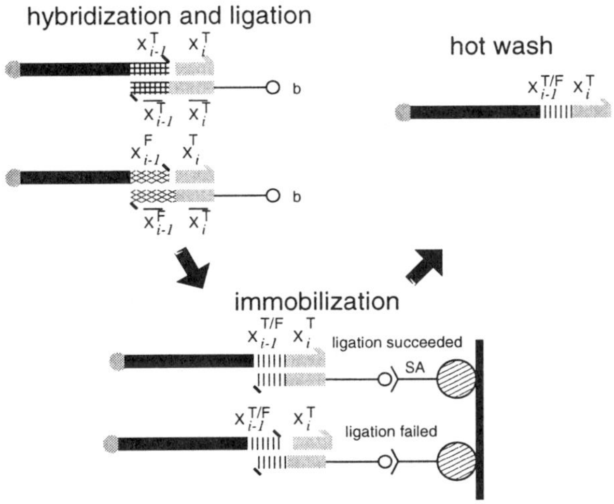

FIGURE 3. Append(T, X_i^b, e) $(e \in \{\overline{X_{i-1}^T X_i^b}, \overline{X_{i-1}^F X_i^b}\}, b \in \{T, F\})$

5. Transfer the supernatant to a tube T^- as the output of Get$(T, -X_i^b)$.
6. Wash the beads with 75 μl of TE buffer. After separate the beads as Step 4, transfer the supernatant to the tube T^-.
7. Repeat Step 6.
8. Add an equivalent volume of 2×B&W buffer (10 mM Tris-HCl (pH 8.0), 1 mM EDTA, 2 M NaCl) to the tube T^- so that its final NaCl concentration is 1 M.
9. Wash the beads with 1×B&W buffer so that the buffers are completely changed, and discard the supernatant.
10. Resuspend the beads in 1×B&W buffer. After incubating the tube at 70 ^oC for 3 min, transfer the supernatant to a tube T^+ as the output of Get$(T, +X_i^b)$.

The efficiency and the error rate of Get command were investigated by the experiments using 50 pmol each of DNA molecules $X_1^T X_2^T$ and X_1^F, and 120 pmol of $b\overline{X_2^T}$ in the input sample tube T. $X_1^T X_2^T$ and X_1^F were labeled with a fluorescent molecule at their 5'-end for detection on a P/ACE5510 capillary electrophoresis system equipped with the laser-induced fluorescence detector (Beckman, USA). They were clearly separated on electropherograms according to their length, and then quantified with a standard marker oligomer added to them just before loading. The efficiency and the false-positive error rate of Get$(T, +X_2^T)$ were 88 ± 3 and 1 ± 1%, respectively. Those of Get$(T, -X_2^T)$ were 99 ± 1 and 12 ± 3%. These results demonstrate that Get$(T, +X_i^b)$ is accurate while Get$(T, -X_i^b)$ is error-prone. Therefore, Get$(T, -X_i^T)$ should always be followed by Get$(T, +X_i^F)$ to reduce false-positive errors as is implemented in the program shown in Fig. 1 if the errors make the following computation nonsence.

3.2.2. *Append(T, X_i^b, e)*. Append(T, X_i^b, e) $(e \in \{\overline{X_{i-1}^T X_i^b}, \overline{X_{i-1}^F X_i^b}\})$ is the command which adds DNA X_i^b to the 3'-end of DNA oligomer representing every partial assignment π in a tube T with splint DNA oligomer(s) e. The append reaction at a high temperature with *Taq* (Thermus aquaticus) DNA ligase facilitates accurate stepwise extension of partial assignments in a tube. Figure 3 summarizes the protocol of Append command, and the following is its details.

1. Create the mixture of 100 pmol each of 5'-end-biotinylated b$\overline{X_{i-1}^T X_i^b}$ and b$\overline{X_{i-1}^F X_i^b}$, 250 pmol of p$X_i^b$, which is the 5'-end-phosphorylated sequence of X_i^b, and 1.5 $\mu\ell$ of 10× *Taq* ligation buffer (200 mM Tris-HCl (pH 7.6), 250 mM CH_3COOK, 100 mM $(CH_3COO)_2Mg$, 100 mM DTT, 10 mM NAD, 1% Triton X-100); and then adjust the volume up to 15 $\mu\ell$ with H_2O.
2. Add DNA oligomers in the input sample tube T to the mixture, and incubate it at 94 ^{o}C for 2 min.
3. Add 8 $\mu\ell$ (320 U) of *Taq* DNA ligase to the mixture.
4. Gradually cool it to 55 ^{o}C , and incubate at 55 ^{o}C for 15 min.
   Note that if the mixture is incubated for 60 min, the efficiency of ligation is improved, but we gave priority to saving time in the experiment.
5. Add 1.8 mg of Dynabeads M-280 Streptavidin, and shake the tube moderately for 15 min at room temperature.
6. Separate the beads at 65 ^{o}C , and discard the supernatant. Note that at 65 ^{o}C the ligation-successful products, which are still captured on the beads, can be separated from the ligation-failed products released in the supernatant.
7. Add 1×ligation buffer, and repeat the step above.
8. Wash the beads with 0.5×B&W buffer, and discard the supernatant.
9. Resuspend the beads in 100 $\mu\ell$ of 0.5× B&W buffer. After incubating the beads at 85 ^{o}C for 3 min, transfer the supernatant to the output tube.

The efficiency and the error rate of Append command were examined by using 50 pmol each of DNA oligomers $X_1^T X_2^T$ and $X_1^T X_2^F$ in a tube T. These oligomers were labeled with a fluorescent molecule at their 5'-end for detection. To their 3'-end, 250 pmol of pX_3^F were appended with 100 pmol each of splint DNA oligomers $\overline{X_2^T X_3^F}$ and $\overline{X_2^F X_3^F}$. The ligation reaction and the separation of ligation-successful products were followed by capillary electrophoresis, and their efficiency and error rate were determined. The efficiency of ligation reaction itself was 43 ± 4%, and the total efficiency of Append command including separation of ligation-successful products was 35 ± 5%. The ligation-failed products in the output tube were below the sensitivity limit of measurement, demonstrating that Append command is accurate.

3.2.3. *Amplify*$(T, T_1, T_2, \ldots, T_n)$. Amplify$(T, T_1, T_2, \ldots, T_n)$ is the command which amplifies DNA molecules in a tube T and divide them into tubes $T_1, T_2, \ldots, T_n$. If the amount of DNA in a tube T is sufficient for subsequent computation, no PCR amplification of DNA is needed. Otherwise, DNA molecules in a tube T should be amplified by PCR according to the following protocol, which was found to increase the amount of DNA about 40 thousand times.

1. Create the mixture of DNA oligomers in a tube T, representing partial assignments to the Boolean variables from x_1 to x_i, 100 pmol each of flX_1^T and flX_1^F, which are forward PCR primers labeled with a fluorescent molecule at the 5'-end, 100 pmol each of 5'-end-biotinylated reverse PCR primers b$\overline{X_i^T}$ and b$\overline{X_i^F}$, 8 $\mu\ell$ of dNTPs mixture (2.5 mM each), and 10 $\mu\ell$ of 10×Pyrobest buffer; and then adjust the volume up to 100 $\mu\ell$ with H_2O.

2. Add 0.5 $\mu\ell$ (2.5 U) of Pyrobest™ DNA Polymerase (TaKaRa, Japan) to the mixture after incubating it at 94 °C for 1 min, and then immediately run the following PCR program: 25 cycles of 94 °C for 0.5 min, 50 °C for 1 min and 72 °C for 1 min.
3. Add 90 $\mu\ell$ of 2×B&W buffer to the PCR mixture.
4. Add 1.4 mg of Dynabeads M-280 Streptavidin, and shake the tube moderately for 15 min at room temperature.
5. While trapping the beads with a magnet Dynal MPC, discard the supernatant.
6. After addition of 1×ligation buffer, incubate the tube at 70 °C for 3 min.
7. Separate the beads and discard the supernatant as Step 5.
8. Wash the beads with 0.5×B&W buffer and discard the supernatant as Step 5.
9. Resuspend the beads in 100 $\mu\ell$ of 0.5×B&W buffer. After incubating the beads at 90 °C for 3 min, transfer the supernatant to the output tube.

4. Design of orthogonal DNA sequences

The following three conditions were adopted in the design of DNA sequences X_i^b representing Boolean variables $x_i = b$ ($b \in \{T, F\}$). We call a set of DNA sequences that satisfy the three conditions "an orthogonal set of sequences"

1: Uniform melting temperature T_m
2: No mishybridization
3: No formation of stable secondary structure by itself

The first condition facilitates uniform efficiency of the basic commands. The last two conditions assure rapid hybridization of the correct DNA oligomers complementary to each other.

In the present study, the following greedy algorithm was used to create an orthogonal set. It can be assured that there are not more than $2 * (6 - 1) = 10$ continuous bases in common between a given sequence and the junction of given two sequences.

- Let S_{pool} be a collection of 15-base DNA sequences with 8/15 GC-content.
- Let S be an empty set.
- Add one sequence picked out of S_{pool} at random to S.
- Pick out one sequence π as a candidate for X_i^b from $S_{pool} - S$. Add π to S if $S \cup \{\pi\}$ satisfies both conditions:
  there are no 6 or more continuous bases in common between X_i^b and $\overline{X_j^b}$ ($i \leq j$), and between X_i^b and X_j^b ($i < j$).
- Repeat the previous step by desire until the sufficient number of DNA sequences are obtained.

The following set of sequence was obtained for solving a 4-variable 10-clause instance of the 3-SAT problem with DNA molecules in the present study. As we shall see later, these sequences caused some problem with verification of partial assignments by graduated PCR though finally we were able to verify all partial assignments in the output tubes.

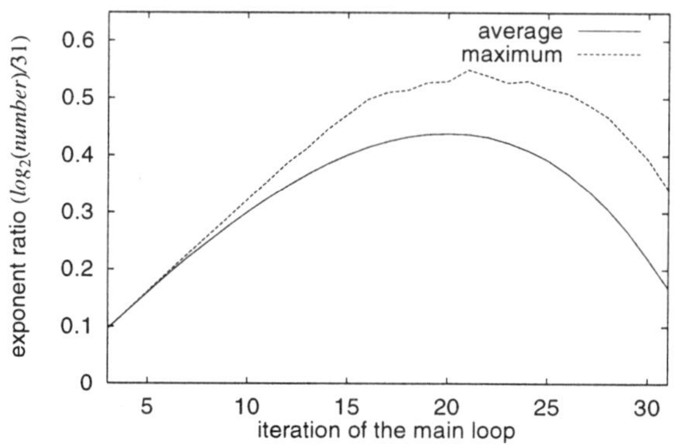

FIGURE 4. Variation of the number of partial assignments with the iteration number of the main loop. The average and the maximum of two thousand instances of 3CNF-SAT problems with 31 variables and 133 clauses are plotted.

X_1^T : 5'-TCCCCTCTACCAATC-3'
X_1^F : 5'-CCCCCCCCAAAAAAA-3'
X_2^T : 5'-CCCCACCCCTAAAAA-3'
X_2^F : 5'-CCCTCCCTCCATAAA-3'
X_3^T : 5'-CCCCCTTCCCTTAAA-3'
X_3^F : 5'-CCCCATCCCACAAAA-3'
X_4^T : 5'-CCCTCACCCACTAAA-3'
X_4^F : 5'-CCCAACCCAACCAAA-3'

5. Results

5.1. Evaluation of our algorithm. We have examined the performance of our algorithm by computer simulations. The clause/variable ratio of the hardest instances of 3CNF-SAT is around 4.3 [**SML96**]. For 20-variable 3-SAT problems, the hardest instances concentrated at the clause/variable ratio of 4.55. We thus generated at random many instances of 3CNF-SAT problems with $n = [20, 31]$ variables and $m = [4.30, 4.35, \ldots, 4.65, 4.70] \times n$ clauses, and then investigated how the number of partial assignments changes while the algorithm runs. Two thousand instances of the problem were examined for each pair of n and m.

Figure 4 shows how the number of partial assignments changes with the iteration number of the main loop when solving 31-variable and 133-clause instances of 3-SAT problems. The average and the maximum of the partial assignment number increased as the main loop was iterated, and reached the largest around an iteration number of 20. The largest value represents the necessary number of partial assignments for solving the problem. Figure 5 shows how the necessary number of partial assignments depends on the size of 3-SAT problems. The maximum of the necessary partial assignment number was found to depend on the number of

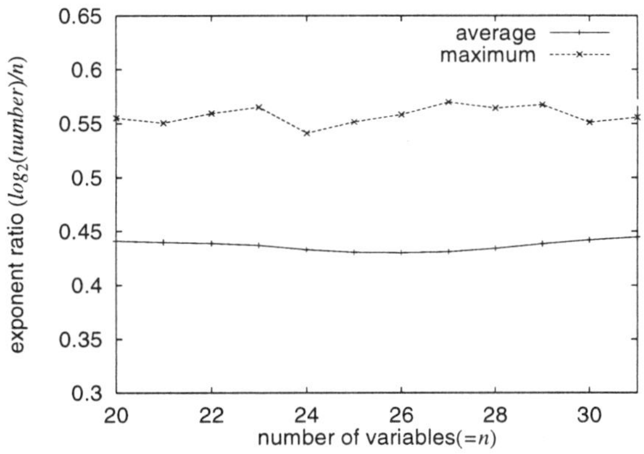

FIGURE 5. The average and the maximum of the necessary partial assignment number for 3CNF-SAT problems with $n = [20, 31]$ variables and $m = [4.30, 4.70] \times\ n$ clauses.

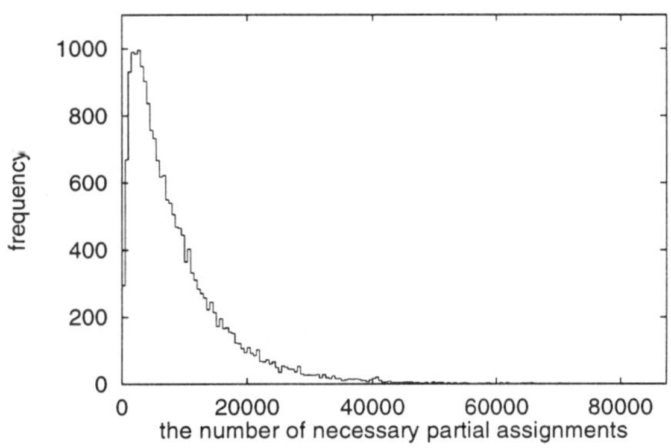

FIGURE 6. Distribution of the necessary partial assignment number made from twenty thousand 3CNF-SAT instances with 31 variables and 133 clauses.

Boolean variables of 3-SAT problems as $2^{0.58n}$, which is much smaller than the necessary number ($= 2^n$) of assignments in the initial pool required for the brute force algorithm. Therefore, the 3-SAT problem can be solved with much smaller amount of DNA compared with the conventional brute force method.

If this relation $2^{0.58n}$ holds true in 3-SAT instances with more variables, our algorithm will make solution of large-scale 3-SAT problems feasible. For $n = 100$, the amount of DNA used in the calculation can be substantially reduced from millions of tons to the practically available amount, such as several milligrams. In addition, Fig. 6 demonstrates that the worst case ($2^{0.58n}$) was rare and that the majority was centered at $2^{0.36n}$. Thus, most instances of 3-SAT problems can be solved with a smaller amount of DNA.

5.2. Lab Experiments. The efficiency and the error rate of the basic commands suggested that an instance of 3CNF-SAT problems was able to be solved

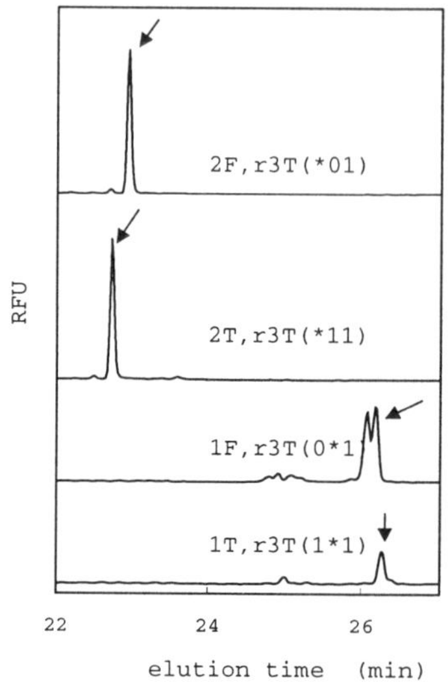

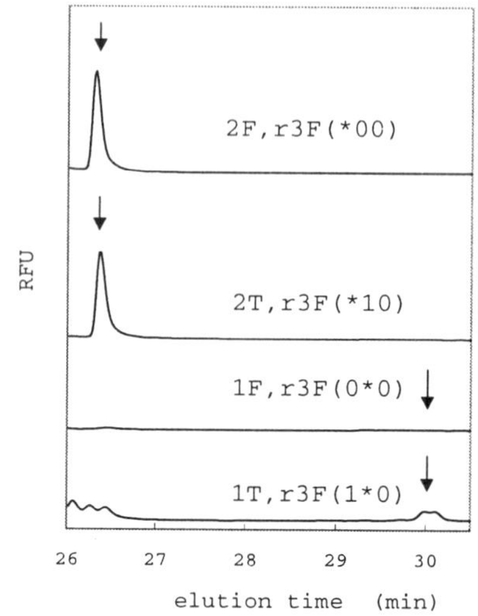

FIGURE 7. (Left) Electropherograms of graduated PCR products in the tube T^T after the first iteration ($k = 3$). '2F, r3T', for example, denotes graduated PCR amplification with the primer pair of X_2^F and $\overline{X_3^T}$ which examine whether DNA molecules representing '*01' exist. The fluorescence intensity was normalized with the peak intensity of the standard marker DNA oligomer added just before sample loading. The arrows designate where PCR products should or should not appear.

FIGURE 8. (Right) Electropherograms of graduated PCR products in the tube T^F after the first iteration ($k = 3$).

with DNA by executing the DNA computer program shown in Fig. 1. We have thus solved a simple instance of 3CNF-SAT problems with DNA. The instance has 4 variables and 10 clauses as follows:

$$(x_1 \vee x_2 \vee x_3) \wedge (x_1 \vee \neg x_2 \vee x_3) \wedge (\neg x_1 \vee x_2 \vee \neg x_3) \wedge (\neg x_1 \vee \neg x_2 \vee \neg x_3) \wedge (x_1 \vee \neg x_3 \vee \neg x_4) \wedge$$
$$(\neg x_1 \vee x_2 \vee \neg x_4) \wedge (\neg x_1 \vee x_3 \vee \neg x_4) \wedge (x_2 \vee x_3 \vee x_4) \wedge (x_2 \vee \neg x_3 \vee x_4) \wedge (\neg x_2 \vee \neg x_3 \vee x_4).$$

The answer to this problem is 'yes'. The partial assignments to the first three Boolean variables after the first iteration ($k = 3$) are

$$\{(x_1, x_2, x_3)\} = \{(0, 0, 1), (0, 1, 1), (1, 0, 0), (1, 1, 0)\}.$$

After the final iteration ($k = 4$), there exists only a single assignment satisfying the CNF expression:

$$\{(x_1, x_2, x_3, x_4)\} = \{(1, 1, 0, 0)\}.$$

The experiments were started with 25 pmol each of DNA oligomers, $X_1^T X_2^T$, $X_1^F X_2^T$, $X_1^T X_2^F$ and $X_1^F X_2^F$. After finishing the first iteration ($k = 3$), we verified the partial assignments in tubes T^T and T^F by graduated PCR. The partial

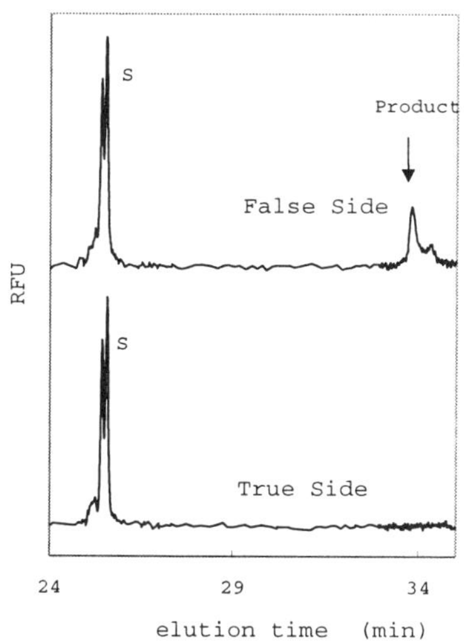

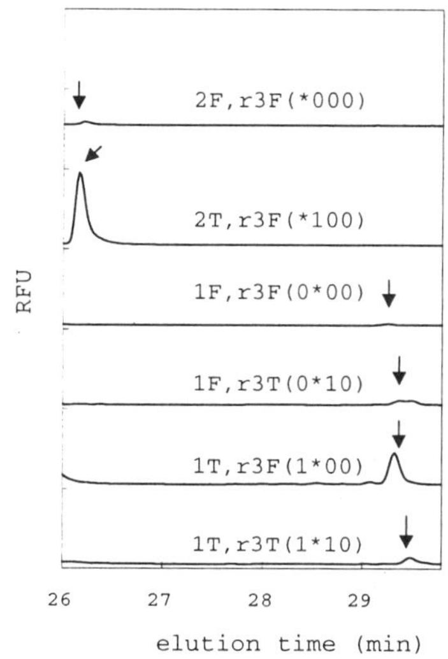

FIGURE 9. (Left) Electropherograms of DNA molecules in the tube T^T (bottom) and T^F (top) after finishing the main loop ($k = 4$). S is a fluorescently-labeled standard marker DNA oligomer used to calibrate the peak intensity. The arrows designate where PCR products should or should not appear.

FIGURE 10. (Right) Electropherograms of graduated PCR products of DNA molecules in the tube T^F after the last iteration ($k = 4$). The fluorescence intensity was normalized with the peak intensity of the standard marker DNA oligomer.

assignments that can exist in the tube T^T are $\{(x_1, x_2, x_3)\} = \{(0,0,1),(0,1,1)\}$. Figure 7 confirms the existence of these correct assignments in the tube. The small peak arrowed on the bottom electropherogram in Fig. 7 implies that there may exist incorrect assignments, either of $(1,1,1)$ or $(1,0,1)$. However, its amount must be sufficiently smaller than that of the correct assignments because the primer pair of X_1^T and $\overline{X_3^T}$ was found to amplify template DNA at a concentration fifty times lower than the other primer pairs do. On the other hand, Fig. 8 confirms that there were only the correct partial assignments, $\{(x_1, x_2, x_3)\} = \{(1,0,0),(1,1,0)\}$, in the tube T^F. The peak of the PCR product amplified with a primer pair of X_1^T and $\overline{X_3^F}$ is smaller than that with X_2^T and $\overline{X_3^F}$ or with X_2^F and $\overline{X_3^F}$. Experiments using synthetic DNA templates demonstrated that it was a tendency of the primer pair of X_1^T and $\overline{X_3^F}$. So the small peak arrowed on the bottom electropherogram is definitely significant. No DNA molecules representing the incorrect partial assignments, $\{(x_1, x_2, x_3)\} = \{(0,1,0),(0,0,0)\}$, existed in the tube T^F as is proven by the second bottom electropherogram in Fig. 8.

The electropherograms of DNA molecules in tubes T^T and T^F after finishing the main loop are presented in Fig. 9. There were 0.14 pmol of DNA molecules of 60 bases long in the tube T^F, while no significant amount of DNA molecule of

60 bases long was observed in the tube T^T. So, we further analyzed the partial assignments in the tube T^F by graduated PCR. The result of the analysis presented in Fig. 10 demonstrates that there was only a single assignment, $\{(x_1, x_2, x_3, x_4)\} = \{(1, 1, 0, 0)\}$, corresponding to the correct answer.

Therefore, we have actually succeeded in solving a 4-variable 10-clause instance of the 3-SAT problem with DNA by executing the DNA computer program based on the breadth first search algorithm. In conclusion, our algorithm and implementation have been proven to do well.

6. Discussion

Our algorithm has been shown to be capable of decreasing drastically the amount of DNA molecules needed for DNA computation of the 3-SAT problem, and the experiments indicate that all procedures do well. These results encourage us to solve larger-scale instances of the 3-SAT problem by repeating many times the main loop of the program in Fig. 1 with the help of automation by robotics.

In the present experiments, we verified whether the computation with DNA had done well by graduated PCR. Unfortunately, the DNA sequences of Boolean variable used caused some problem with verification of partial assignments by graduated PCR though finally we were able to verify all partial assignments in the output tubes. As a matter of fact, non-specific products sometimes existed in PCR products, and the rate of PCR amplification of correct products largely depended on the type of a primer pair. These are partially because many of the DNA sequences designed by us have common DNA bases at the 3'-end, which caused mishybridization followed by strand extension in PCR.

Now, in order to overcome this flaw, we are redesigning DNA sequences using the following strategy. The length of sequences is increased up to 22 mer, which is more suitable for efficient and reliable graduated PCR. A set of approximately a hundred orthogonal 22-mer sequences have been created by the following two filters. The new set of orthogonal DNA sequences will allow us more reliable execution and verification of the DNA computation of 3-SAT problems.

The first filter :

1: Between X_i^b and $\overline{X_j^b}$ $(i \leq j)$, there are no 8 or more continuous bases in common.

2: Between X_i^b and X_j^b $(i < j)$, there are no 8 or more continuous bases in common.

3: The Tm(melting temperature) of the sequence predicted by the nearest neighbor method [**San98**] is 63 ± 1 °C in *Taq* ligation buffer.

4: Between the 3'-end of X_i^b and the 5'-end of X_j^b $(i \neq j)$, there are no 4 or more continuous bases in common.

5: Between the 5'-end of X_i^b and the entire X_j^b $(i \neq j)$, there are no 5 or more continuous bases in common.

- Let S be a empty collection.
- Generate a random number from 0 to $4^{22} - 1$, and translate it into a DNA sequence. Add the sequence to S if it satisfies the condition (3). Repeat this step until S is not empty.
- Generate the next random number, and let π be the sequence translated from it. Add π to S if π satisfies (3) and $S \cup \{\pi\}$ satisfies all of (1),(2),(3),(4), and (5).

- Repeat the previous step as desired.

The good point of using random numbers is that the 3'- and 5'- ends of produced sequences vary moderately. Note that it can be assured there are not more than 8 continuous bases in common between the 5'-end of a given sequence (the 3'-end of its reverse complement) and the junction of any two sequences if the set satisfies all conditions, thereby non-specific amplifications on PCR can be suppressed.

The second filter: From the sequences passing through the first filter, remove the sequences that form stable secondary structure by itself or make a mishybridization. The secondary structure prediction program used in the second filter is based on the algorithm developed by improving Zuker's dynamic programming algorithm for secondary structure prediction [**Suy84**] . The program can predict not only stable secondary structure of one single-stranded chain but also that of two single-stranded chains of any sequences.

7. Future Work

We are planning to solve a large instance of 3CNF-SAT problems such as 50 variable problems. For large-scale DNA computation based on the present method, the automatic execution of the basic DNA commands is indispensable. So we are now developing a DNA computer machine executing the basic commands automatically by improving a MAGTRATION^{TM} DNA extraction system (Precision System Science, Japan).

Another future work is application of our algorithm to other NP-complete problems because the reduction of the amount of DNA needed for computation is also essential to scale up the problem size of other NP-complete problems. We are also interested in 'polynomial isomorphism' [**BH77**], from which it can be revealed how other NP-complete problems are related to the hardest instances of 3-SAT problems [**SML96**].

Acknowledgments

The authors wish to express their gratitude to Mrs. Yoko Sakai for her valuable comments and assistance.

References

[Adl94] L. Adleman. Molecular computation of solutions to combinatorial problems. *Science*, 266(11):1021–1024, 1994.

[BH77] L. Berman and J. Hartmanis. On Isomorphisms and density of NP and other complete sets. *SIAM J. Comput.*, 6(2):305–322, 1977.

[MAS97] N. Morimoto, M. Arita, and A. Suymama. Solid phase DNA solution to the hamiltonian path problem. In *Preliminary Proceedings of 3rd DIMACS Workshop on DNA Based Computers*, pages 83–92, 1997.

[OR97] M. Ogihara and A. Ray. DNA-based parallel computation by "counting". In *Preliminary Proceedings of 3rd DIMACS Workshop on DNA Based Computers*, pages 265–274, 1997.

[San98] J. SantaLucia, Jr. A unified view of polymer, dumbbell, and oligonucleotide DNA nearest-neighbor thermodynamics. *Proc. Natl. Acad. Sci. USA*, 95:1460–1465, 1998.

[SML96] B. Selman, D. Mitchell, and H. Levesque. Generating hard satisfiability problem. *Artificial Intelligence*, 81:17–29, 1996.

[Suy84] A. Suyama. RNA secondary structure and its relation to biological functions. In *Proceedings of the 10th Taniguchi International Symposium*, pages 162–172, 1984.

(H. Yoshida) Department of Physics, University of Tokyo, 3-8-1 Komaba, Meguro-ku, Tokyo 153-8902, Japan

E-mail address: hiroshi@talent.c.u-tokyo.ac.jp

(A. Suyama) Institute of Physics, Graduate School of Arts and Sciences, University of Tokyo, 3-8-1 Komaba, Meguro-ku, Tokyo 153-8902, Japan

E-mail address: suyama@dna.c.u-tokyo.ac.jp

DIMACS Series in Discrete Mathematics
and Theoretical Computer Science
Volume **54**, 2000

In Vitro Selection for a OneMax DNA Evolutionary Computation

David Harlan Wood†, Junghuei Chen†, Eugene Antipov‡, Bertrand Lemieux, and Walter Cedeño

ABSTRACT. Aspects of Evolutionary Computation, DNA computing, and *in vitro* evolution are combined in proposed laboratory procedures. Preliminary experimental results are shown. The traditional test problem for Evolutionary Computation known as the OneMax problem is addressed. The preliminary experimental results indicate successful laboratory "separation by fitness" of DNA encoded candidate solutions.

1. Introduction

Evolution is a concept of obtaining adaptation through the interplay of selection and diversity. Analogies from evolution have been used in both computing and molecular biology. These two areas are called respectively "evolutionary computation" and "*in vitro* evolution." From the beginning of DNA based computing to the present there have been calls [**11, 22, 27**] to carry out evolutionary computations using genetic materials *in vitro*. So far, there have been three such experiment designs proposed, including two in a recent DIMACS Workshop [**4, 7**]. The very first design was presented about two years ago [**10**], but has not yet been carried out in the laboratory.

We design a laboratory implementation of an evolutionary computation. The crucial operation of physically separating DNA strands by their "fitness" is realized using 2-d denaturing gradient gel electrophoresis (2d DGGE). The interest is that candidate solutions are encoded in strands of DNA. Thus, trillions of candidate solutions have their fitness evaluated in parallel. Let us now comment on the possible variety of selection techniques usable after candidates have been physically separated by 2d DGGE. It is not clear if any of these directly correspond to any of the usual selection schemes from classical Evolution Computation. However, we are able to recover separated candidate solutions that are expected to be more fit, on average, than candidate solutions we discard. Our expectation is encouraged by preliminary experimental results included in this paper indicating that even a

---

1991 *Mathematics Subject Classification.* Primary 68Q40, 68U99, 68Q15.

†Partially supported by DARPA/NSF Grant No. 9725021.

‡Partially supported an REU Supplement to DARPA/NSF Grant No. 9725021, and as a Howard Hughes Medical Institute Scholar.

single round of selection significantly enhances the average fitness of a population of candidate solutions.

This paper is based in part on a paper [**34**] for the *Genetic and Evolutionary Computation Conference*. In the present paper we elaborate on the DNA laboratory aspects of our research. In an additional paper we plan to compare our techniques with those used for Evolutionary Computation on conventional computers. The laboratory designs and results discussed in this paper are still preliminary but incorporate (1) details of experimental design, (2) laboratory demonstration of selection and readout, (3) computations confirming alignment specificity, and (4) computer simulation of sample candidate DNA strands in 2d denaturing gradient gel electrophoresis.

2. Evolutionary Computation Outline

In this paper we identify aspects of evolutionary computations and *in vitro* evolution that we combine to address a OneMax problem, a traditional test case. Specifically, we choose Evolutionary Computation manipulating bitstrings using operations of pointwise mutation and crossover. These two operations can be performed by modifying and extending techniques from *in vitro* evolution.

"Evolutionary Computation" comes in many, many different varieties [**3**], but most are variations and/or elaborations on the following very loose outline. Evolutionary Computations typically cycle electronic computers through the following steps, maintaining a population of candidate solutions of a problem.

Evolutionary Computation
Begin with a diverse initial population, perhaps chosen randomly.
Repeat the following steps until convergence.

1. Evaluate fitness of candidates.
2. Select the more fit candidates to breed and the less fit to be replaced.
3. Induce variation by breeding.

2.1. Evolutionary Computation Has Unclear Aspects. We are careful to make the following point. We do not take any stance on the virtues of any particular method of evolutionary computation. Instead, we aspire to *provide the means for assessing* some evolutionary computations using population sizes larger than is practical with conventional computers.

Evolutionary computation makes few assumptions and is ostensibly applicable to very broad classes of problems. Naturally, this makes it difficult to establish many provable guidelines. Just to list a few aspects which are sometimes unclear,

- How does one model fitness?
- How disruptive is crossover?
- What are the pros and cons of transposition, inversions, and introns?
- How long does it take evolutionary computations to converge?
- How does one recognize convergence?

3. Implementing Evolutionary Computation Using DNA

Several means of DNA computation have been addressed. The first was, of course, by Adleman [**1, 2**]. Recent overviews can be found in [**15**] and [**21**]. See also the DNA Computing bibliography of J. H. M. Dassen [**9**].

We anticipate computing time using DNA to be proportional to the number of generations required, but independent of population size. This motivates incorporating both pointwise mutation and crossover, and for that matter any evolutionary analogies that might reduce the number of required generations.

Modifications and extensions of current technology seem to suffice to implement crossover and pointwise mutation. However, selecting DNA strands for "breeding" is moderately challenging because one must *physically separate DNA strands according to their "fitness."*

3.1. DNA Attributes Suit Evolutionary Computation. Of all evolution inspired approaches, certain versions of Evolutionary Computation seem particularly suited to using DNA strands. This is because DNA strands can encode populations of bitstrings suitable for both crossover and pointwise mutation.

DNA computing techniques are desirable for Evolutionary Computation.

- The first main advantage is these techniques might process, in parallel, populations billions of times larger than is usual for conventional computers. this would be beneficial if larger populations might generate high-fitness individuals in fewer generations.
- The second main advantage is the massive information storage available using DNA. For example, a gram of DNA primers contains about 10^{21} bases. The information content is approximately 2×10^{21} bits, greatly exceeding the 200 petabyte storage of all digital magnetic tape produced in 1995 [**33**].
- A third advantage is that methods for implementing crossover using DNA are possible as variations on "Sexual PCR" pioneered by Stemmer [**28**].
- An additional issue is that biolaboratory operations on DNA inherently involve errors. To some extent, errors may be regarded as noisy fitness evaluation. Such errors are sometimes tolerable when executing Evolutionary Computation [**14**], unlike when executing deterministic algorithms.
- Oversimplified estimates indicate [**34**] DNA computing techniques might compare favorably to supercomputers in some cases. These favorable cases include executing Evolutionary Computation having very simple fitness evaluations and very large populations of candidate solutions.

3.2. DNA Evolutionary Computation Compared to *In Vitro* Evolution. Evolutionary Computation are reminiscent of methods in molecular biology referred to as "*in vitro* evolution." Naturally, these methods use fitness criteria constrained to properties of biomolecular interest. Indeed, the necessity of finding methods to physically separate biological materials by "fitness" has determined which problems are addressed by *in vitro* evolution. *In vitro* methods suit some very important but relatively small classes of problems seeking ribozymes [**31, 23, 16, 29**], binding sites [**30, 29**], enhanced enzymes [**26**], etc.

In contrast to *in vitro* evolution, Evolutionary Computation uses *unconstrained* bitstrings as fitness criteria. Consider computations using bitstrings of length 100. Such computations can, in principal, evolve a population of fixed size in such a way as to create any one of $2^{100} \approx 10^{30}$ possible outcomes. Thus, Evolutionary Computation methods using DNA can address larger, less specialized, classes of problems than does *in vitro* evolution.

Evolutionary Computation using DNA would be similar to conventional computers in that (virtually) all 10^{30} possible inputs and outputs would be equally suitable. In contrast, *in vitro* evolution is tailored to the very rare DNA sequences encoding biological functionality. For example, bitstrings might be realized as DNA strands having 100 As and Gs. But as far as we know very few, if any, such DNA bitstrings code for biologically active functions. Or to put it another way, virtually all sequences are equally "meaningful" for the purposes of Evolutionary Computation. In contrast, *in vitro* evolution focuses on variations of the very rare DNA sequences of biological or biochemical interest.

4. The OneMax Problem

Here we present a problem that is relatively simple to address using DNA implementations of Evolutionary Computation. The essence of our implementation is that more fit candidates strands of DNA can be physically separated from less fit candidates according to how well they match (hybridize with) "target" DNA strands.

The OneMax problem is a traditional test problem for Evolutionary Computation. It involves binary bitstrings of fixed length. An initial population (usually randomly generated) is given. The objective is to evolve some bitstrings to match a prespecified "target" (generally taken to be all 1s). The study of OneMax problems has lead to a number of insights. For example, see [**13**] and the references cited there.

4.1. Outline of DNA Implementation. We now outline a DNA implementation of the OneMax problem. The corresponding information is shown in Figure 1. In the remainder of this section, details are given. Throughout this section, information is grouped in the following categories: (1) candidate pool, (2) fitness evaluation, (3) selection and (4) breeding.

Preliminary laboratory results are given later, in Section 5.

The implementation is given by the following outline. The same information, with a few added details, is shown in Figure 1.

DNA Evolutionary Computation for OneMax Problem
Begin with a diverse initial population of candidates.
Repeat the following steps until convergence.

1. Evaluate fitness by hybridizing to target strands and physically separate on a 2-d gel.
2. Select and purify relatively more fit candidates to breed.
3. Amplify fit candidates with pointwise mutation and reserve a portion.
4. Breed candidates, using crossover.
5. Combine reserved and bred candidates, obtaining a new generation.

4.2. Design of Candidate Solutions and Target DNA Strands. Figure 2 shows our design. A target strand and a perfect candidate strand are shown in the figure. Imperfect candidate strands have a mixture of 40 Ts and Cs instead of 40 consecutive Ts. At the $5'$ end of all candidate strands there is a universal section known as the CLAMP section of the target strands. All $3'$ ends of the candidate strands are extended by a universal TAIL sequence. The candidate strands are longer to facilitate eventual separation of target and candidate strands using denaturing gel electrophoresis.

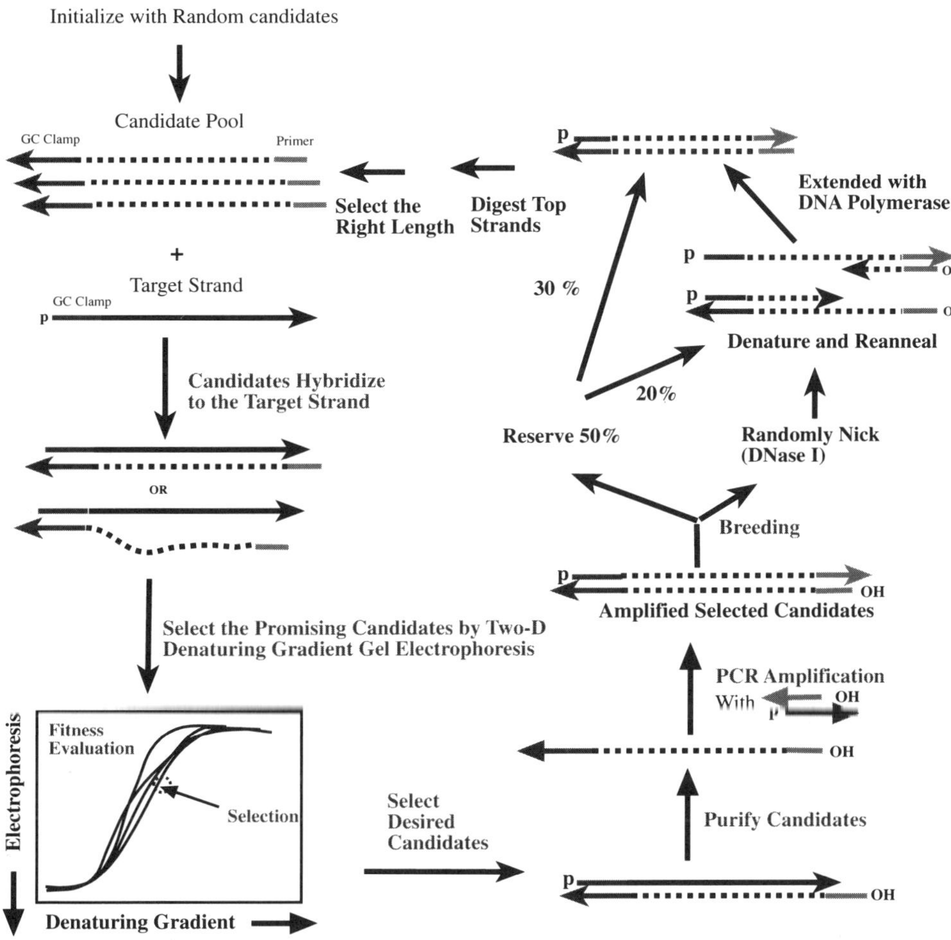

FIGURE 1. Outline of DNA implementation of Evolutionary Computation for OneMax problem. The candidate pool appears in the upper left. Selection using 2-d DGGE appears at the lower left. Purification and amplification of the relatively more fit candidate strands appears at the lower right. Breeding using crossover appears at the upper right.

Since candidate strands have known primer sites at both ends, they can be amplified by PCR. One primer we will call the TAIL-primer, which is the reverse complement of the TAIL. The other primer is simply the CLAMP sequence.

The CLAMP and TAIL sequences have been designed to encourage correct alignment of the PCR primers and to avoid sticking where they are not supposed to stick. Figure 3 concerns how misaligned primer may stick (hybridize) to *candidate strands.* Figure 3 shows melting temperatures T_m (a measure of the strength of hybridization) at all possible alignments of the CLAMP and the TAIL-primer. These short strands are tested against a full-length candidate strand. This figure and the next was generated using the BIND software [**17**] from MIT, for which we have a license for limited use.

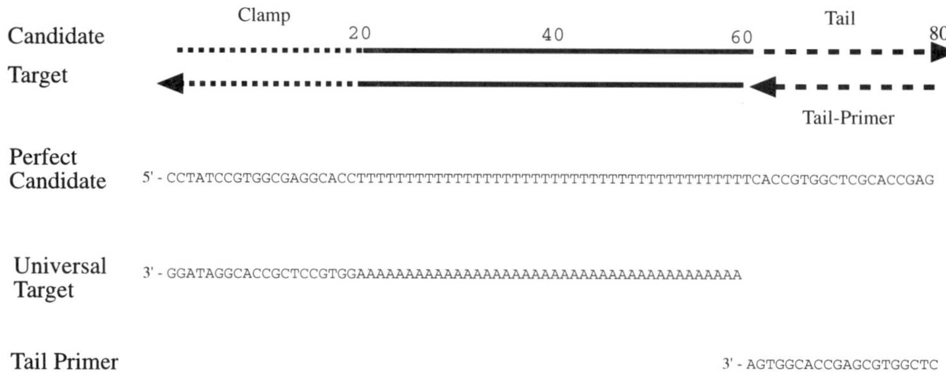

FIGURE 2. Design of target and candidate strands. Perfect candidates have all Ts from position 21 to position 60. Imperfect candidates would have a mixture of Ts and Cs in place of the 40 Ts in the perfect candidate.

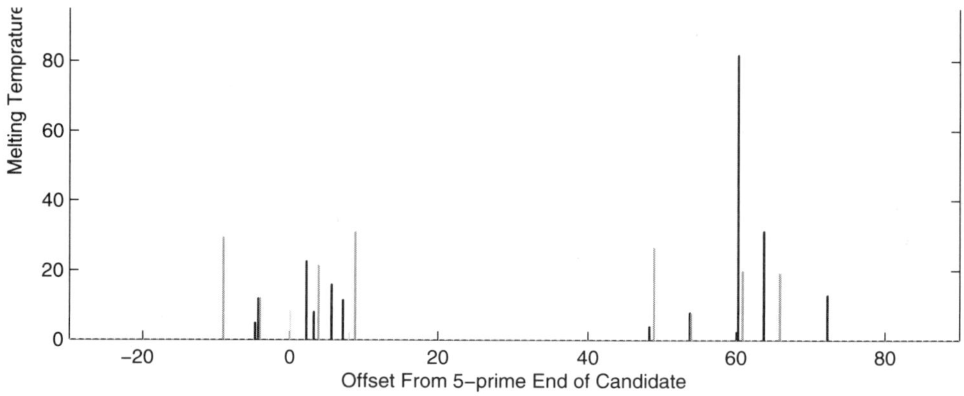

FIGURE 3. At an offset of 60, the correctly aligned TAIL-primer sticks tightly to the candidate strand, providing a melting temperature of 82 degrees. Misalignments of the CLAMP and the TAIL-primer, can only stick weakly with melting temperatures of 31 degrees or less.

Figure 4 concerns how misaligned primer may stick (hybridize) to *complements of candidate strands*. Figure 3 shows melting temperatures T_m at all possible alignments of the CLAMP and the TAIL-primer. These short strands are tested against the full-length reverse complement of a candidate strand.

4.3. Fitness Evaluation by DGGE Physical Separation of DNA. Our fitness evaluation is carried out in the laboratory using so-called 2-d denaturing gradient gel electrophoresis (DGGE) [**18**]. Let us first review the nature of DGGE. Figure 5 shows DGGE from our laboratory having perfect candidates combined with target strands. The target strands hybridize (stick) to the perfect candidate strands. Each candidate is extended by a universal TAIL at its $3'$ end. The mixture of hybridized strands is placed uniformly along the top of the gel. The hybridized

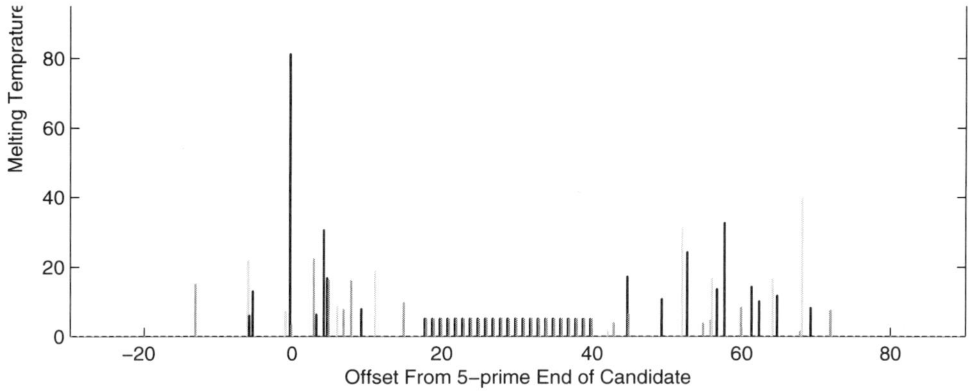

FIGURE 4. At an offset of 0, the correctly aligned CLAMP sticks tightly to the reverse complement candidate strand, having a melting temperature of 81 degrees. Misalignments of the CLAMP and the TAIL-primer, can only stick weakly with melting temperatures of less than 40 degrees.

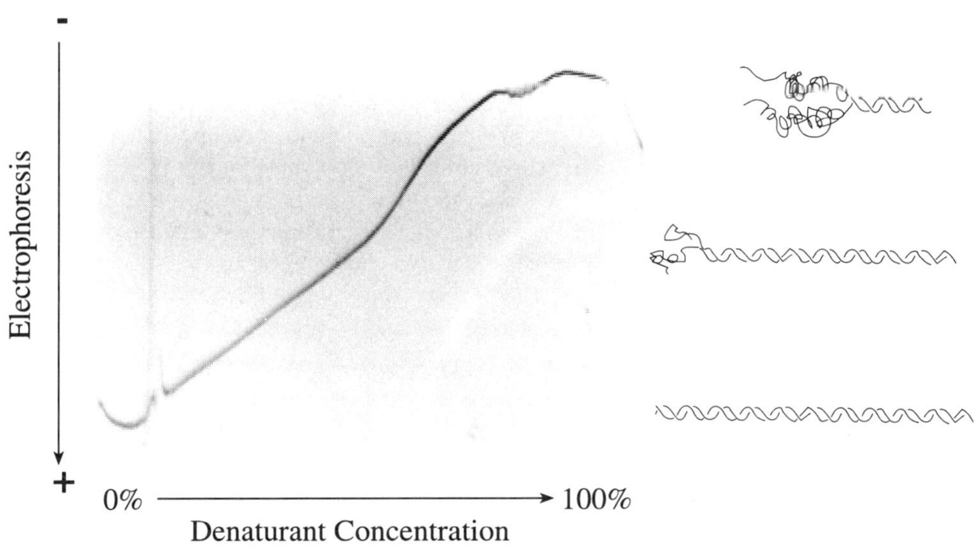

FIGURE 5. DGGE using perfect candidates. DNA strands move downward from a reservoir at the top of the figure. The speed of vertical strand migration is retarded as strands come apart (denature) as shown schematically on the right edge of the figure.

strands travel vertically downward in the gel as a result of an applied electric field. However, their speed of migration is determined by their initial placement from left to right; that is, by how strongly they are denatured (pulled apart). On the left, where no denaturant is encountered, the strands move relatively quickly downward. In the center, they move more slowly because they encounter intermediate denaturing. At the extreme right, the stands are able to move only very slowly because the strands are almost completely pulled apart except in the more resistant CLAMP region.

An important fact for DNA computing is that *DGGE can detect a single base mismatch.* Indeed, this was an early application of DGGE in molecular biology [**18**].

4.4. Selection of Relatively More Fit Candidate Solutions. Selection is done by literally cutting out a portion of the 2-d gel and extracting the DNA strands from it. The most fit candidates are presumably lowest on any vertical line.

This allows a wide latitude for selection criteria. Experience in Evolutionary Computation computing demonstrates the desirability of maintaining genetic diversity to prevent the loss of genetic materials which may be needed in later stages of evolution.

4.5. Amplification With Mutation via PCR. The purified candidate strands are amplified by PCR, which also induces pointwise mutation at at a rate of about 10^{-4} nucleotide per cycle when *Taq* DNA polymerase is used [**12**]. Special PCR protocols can be used to induce pointwise mutation at intermediate rates 10^{-2} to 10^{-4} [**5, 6, 8, 19, 20, 32, 35**].

Two primers are used. One primer we will call the TAIL-primer, which is the reverse complement of the TAIL. The other primer is a copy of the CLAMP. These two primers are chosen to produce repeated doubling of copies of the candidate strands (and their complements). In contrast, target strands will not double in number because they can hybridize with *only one* of the two primers, namely the CLAMP primer. Extension of this primer can form a complementary copy of the target. We see that candidate strands and their complements amplify at an exponential rate, doubling in number with each PCR cycle. They would increase a million-fold in 20 PCR cycles. But any target strands present can only make complementary target strands, and only at a linear rate. They would increase twenty-fold in 20 PCR cycles. It is important to note that these complements of target strands are both *rare and short* compared to candidate strands.

4.6. Purification of Mutated Candidates. The PCR product can be purified by length using conventional denaturing gel electrophoresis. This gets rid of any target strands and complements of target strands because these will be shorter than the candidates and their complements. Any excess primers are also eliminated. The purified candidates and their complements and be further amplified with mutation using additional primers and PCR. A relatively large quantity is generated to allow for the very low yield expected from the subsequent crossover step.

A portion of the double stranded product is temporarily reserved; the remainder is used for crossover.

4.7. Breeding Using Single Point Crossover. The double stranded product to be used for breeding is partially digested with *DNase I* to mostly nick (cut only one strand) at random locations about once per strand. The nicked strands are combined with a similar amount of reserved unnicked strands. The mixture is denatured (strands are melted apart) and allowed to reanneal forming new combinations. Many, many possible configurations] could be formed. But among these, some will have portions of complements of candidate strands annealed at many positions to a $5'$ end of a candidate strand. The TAIL of the candidate strands is designed to enforce alignment of such hybridizations. A typical desirable hybridization is illustrated in the upper right corner of Figure 1. By adding *DNA Polymerase I*, the partial candidate strand is extended to a full length candidate. In this process it combines its genetic information with that encoded in the intact strand. *DNA Polymerase I* is used because it will remove any annealed fragments in its way as it proceeds in the 5′ to 3′ direction. This ensures the maximum possible extension.

The net result is single point crossover. The offspring candidate strand produced by the polymerase has a block of genetic information from one parent followed by another block from a different parent. In particular, our single point crossover extends results due to Stemmer [**28, 25**].

Single point crossover is a common approach used in Evolutionary Computation, but multiple point crossover can be achieved by iterating the above procedure.

4.8. Purification to Select Candidate Strands. The reaction products from crossover are combined with the reserved material. Purification by length (using denaturing gel electrophoresis) completes the breeding operation.

4.9. The New Generation. The new generation is now complete and ready to be processed.

4.10. Readout of Candidate Pool. We want to monitor the generation-to-generation progress of our *in vitro* evolution. We assume we are given a pool of candidate strands. (Any contaminating complementary strands, or contaminating target strands, will not affect the readout results.)

4.10.1. *Estimation of Average Fitness.* Figure 6 shows the method of generating radiolabeled primer extensions for readout. Our method is a variation on the Sanger method of DNA sequencing.

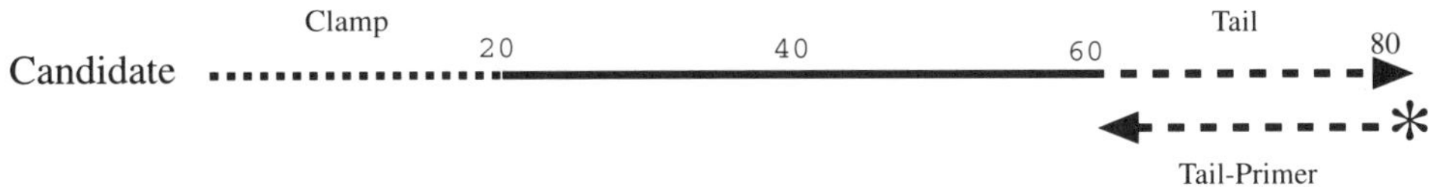

FIGURE 6. Readout of a candidate strand by means of primer extension. A radiolabeled TAIL-primer is annealed to the candidate strand and extended by *Taq* polymerase. With a certain probability, primer extension may halt when a C is encountered.

A large number of primers complementary to the TAIL section of candidates are radiolabeled and added to the pool of candidates. By first raising then lowering the temperature, each candidate preferentially anneals to a primer. The primer is extended in the 3′ direction by including *Taq* polymerase. However, we use only the nucleotides dATP together with a mixture of dGTP and ddGTP. On one hand, primer extension does not stop upon encountering a T in a candidate. This is because the nucleotide A, complementary to T, is contributed by dATP and incorporated at that point. On the other hand, *primer extension can stop* upon encountering a C in a candidate. This is because there is a mixture of dGTP and ddGTP. Since G is complementary to C, either dGTP or ddGTP can contribute a complement to C. But whenever ddGTP is used, the strand can not then be extended any further.

The result is: for each candidate strand, we obtain a primer extended by some number of bases. Each temperature cycle extends a radiolabeled primer on each candidate strand. The resulting extended primers can be separated by their various lengths and read out by using denaturing gel electrophoresis. This indicates the average density and location of the Cs in the candidate strands. Thus, a estimate of the average fitness of a candidate pool can be obtained.

5. Experiment Demonstrating Selection and Readout

5.1. Candidate Solutions Used for Readout. In our experiments the 40 variable positions between the CLAMP and the TAIL of each imperfect candidates are independently chosen to be Ts with probability .8 or Cs with probability .2. Figure 7 shows how the average total number of Ts in each candidate varies in the randomly generated population of candidates.

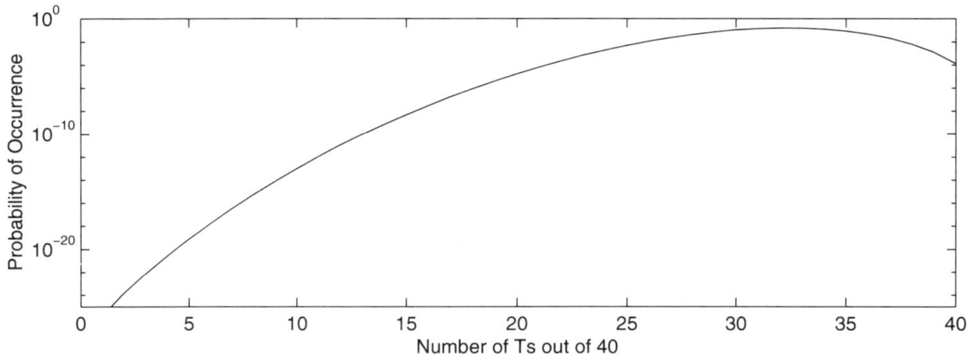

FIGURE 7. Relative frequency as a function of the number of Ts. The case shown is a randomly generated population of candidates having 40 bases which consist of 80% Ts and 20% Cs.

5.2. Experimental Fitness Evaluation by DGGE Physical Separation of DNA. Recalling the nature of 2d DGGE discussed in section 4.3, we heuristically reason about a mixture of targets and imperfect candidates. We expect that everywhere across the gel the candidate strands that best match (hybridize to) the targets will migrate downward relatively faster. In fact, imperfect matches exhibit vertical spreading in our experiments. See Figure 8. We assume that on any vertical line the most fit candidates are lowest on the 2d DGGE gel. However, the nature of variation from left to right is not clear. In our preliminary experiments, we have selected from several places at the lower edge of the 2d DGGE in Figure 8 and combined the extracted DNA. Further experiments will be needed to optimize a selection strategy.

Using the on-line POLAND software, we obtain estimates of the mobility of our particular candidate design, having (selected) mismatches. Figure 8 shows DGGE predictions from the POLAND software [**24**].

5.3. Experimental Readout of the Result of First Round of Selection. Figure 9 contains data from a preliminary experiment demonstrating readout before and after selection. Deliberate overexposure of the lower half of the gel allows detail

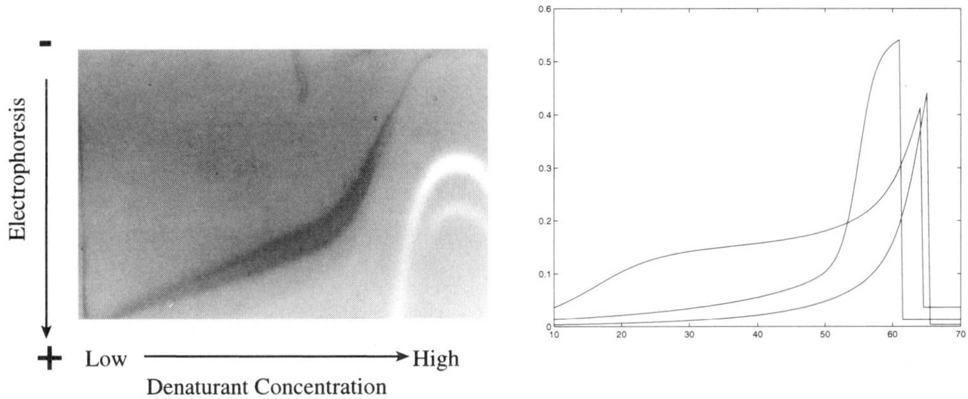

FIGURE 8. DGGE vertically separates imperfect candidates. The speed of vertical strand migration is retarded as strands come apart (denature). This is due to two factors: increasing denaturant concentration and decreasing quality of target-candidate matching (hybridization). On the right, DGGE predictions for selected imperfect candidates using the POLAND software. At 50°, the curves correspond, from the bottom up, to perfect candidate, single mismatch at position 31, and single mismatch at position 50.

in the upper half to be observed. Lane 2 of the gel shows primer extensions for the initial population of randomized candidate strands. Lane 3 shows primer extensions for the population resulting from the first selection.

When a C is encountered during primer extension, there is a probability, call it p, of termination. That is to to say,

$$\text{Probability of terminating} = p \;\text{ Probability of encountering a C.}$$

Thus when the probability of encountering a C is a uniform constant,

$$\text{Probability of terminal length } k = p \;\; [\, 1 - p \;\text{ Probability of encountering a C}\,]^{k-1}.$$

Recall that for the initial population every position between 20 and 60 has .8 probability of being a T and .2 probability of being a C. Using this fact together with the previous equation, we have

$$\text{Probability of terminal length } k = p \;\; [\, 1 - 0.2\, p\,]^{k-1}. \tag{5.1}$$

In summary, long extensions for the initial population are exponentially decreasing in likelihood. This corresponds exactly to the question "What is the frequency of observing runs of all tails using a fixed but biased coin?"

For the initial population the graphical readout in Figure 9, aside from the overexposed left half, shows the expected behavior of exponentially decreasing numbers of long extensions.

After the first selection, the right half of the graphical readout in Figure 9 shows that Cs are now much less common, allowing longer extensions. That is to say, no longer is the probability of encountering a C given by a uniform 0.2. Clearly, Cs are now much less likely, and they are no longer uniformly distributed.

The pool of candidates has been dramatically altered by the first round of selection. Now one observes many more long primer extensions. Therefore, the strands

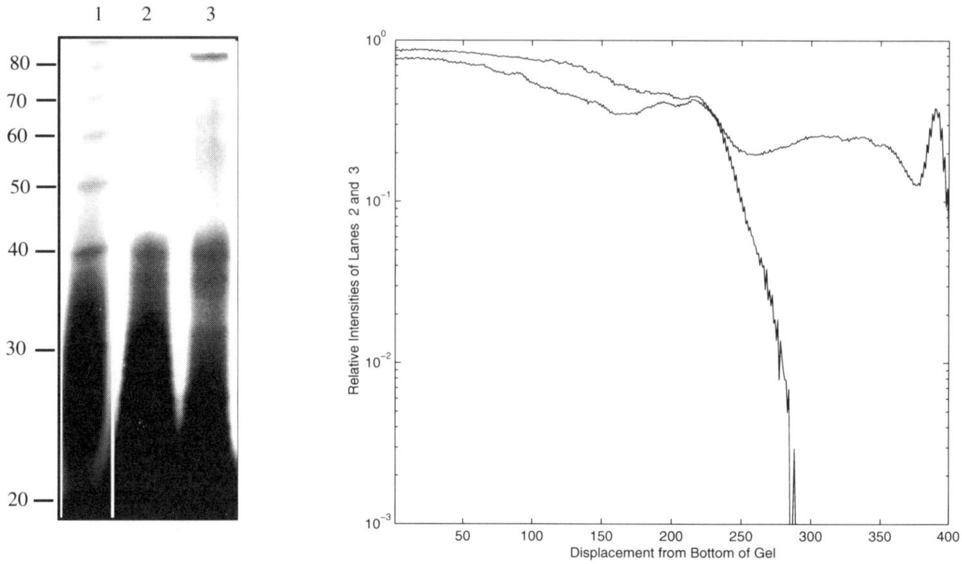

FIGURE 9. The gel on the left: Lane 1 contains a reference set of strands differing in length by 10 bases. Lane 2 is a readout generated by the pool of initially random candidates. Lane 3 is a readout generated after the first round of selection. The graph on the right: A logarithmic scan of Lanes 2 and 3. The uniform distribution of Ts in the initial population leads to exponential decay with increased length. After the first selection, Ts are much more rare which allows full length extensions.

being read out represent much more nearly homogeneous sequences. This represents significant progress toward our goal of evolving some completely homogeneous strands, which would represent solutions to the OneMax problem. Relatively many primer extensions run to the maximum possible length in Figure 9. That is to say, it appears that solutions of the OneMax problem have been selected. Regrettably, not all the full length extensions need represent perfect solutions because of unwanted substitutions by the polymerase. However, these substitutions can be made relatively rare. In any case, though, substitution effects are equally present in Lanes 2 and 3. It is the comparison of these two lanes that demonstrates the significant improvement resulting from the first round of selection. However, it must be said that the experiment shown in Figure 9 is still only preliminary and is in need of further refinement and calibration.

6. Discussion

Although much work remains to be done, it seems clear that even a single round of selection using 2d DGGE is capable of significantly enhancing the "fitness" within a population of candidate solutions of the OneMax problem. Since candidate solutions are encoded in strands of DNA, we are dealing with populations billions of times larger than is the usual practice with conventional computers. We reemphasize that candidate solutions in the entire population have their fitness evaluated at the same time. Separation by fitness is merely a single laboratory step.

Of course, laboratory steps are slow compared to computer cycle times, but DNA can represent populations so large that their computational power can compare favorably with supercomputers [**34**].

References

[1] Leonard M. Adleman. Molecular computation of solutions to combinatorial problems. *Science*, 266:1021–1024, November 1994.

[2] Leonard M. Adleman. Computing with DNA. *Scientific American*, 279(2):54–61, August 1998.

[3] Thomas Bäck, David B. Fogel, and Zbigniew Michalewicz, editors. *Handbook of Evolutionary Computation*. Institute of Physics Publishing, Philadelphia, 1997.

[4] Thomas Bäck, Joost N. Kok, and Grzegorz Rozenberg. Evolutionary computation as a paradigm for DNA-based computing. In Laura Landweber, Erik Winfree, Richard Lipton, and Stephan Freeland, editors, *Preliminary Proceedings DIMACS Workshop on Evolution as Computation*, pages 67–88, DIMACS, Piscataway NJ, January 1999. Available on request from DIMACS. Paper found at URL: http://www.wi.LeidenUniv.nl/~joost.

[5] R. Craig Cadwell and Gerald F. Joyce. Randomization of genes by PCR mutagenesis. *PCR Methods and Applications*, 2:28–33, 1992.

[6] R. Craig Cadwell and Gerald F. Joyce. Mutagenic PCR. In Carl W. Dieffenbach and Gabriela S. Dveksler, editors, *RCP Primer: A Laboratory Manual.* Cold Spring Harbor Laboratory Press, Plainview, New York, 1995.

[7] Junghuei Chen, Eugene Antipov, Bertrand Lemieux, Walter Cedeño, and David Harlan Wood. DNA computing implementing genetic algorithms. In Laura Landweber, Erik Winfree, Richard Lipton, and Stephan Freeland, editors, *Preliminary Proceedings DIMACS Workshop on Evolution as Computation*, pages 39–49, DIMACS, Piscataway NJ, January 1999. Available on request from DIMACS. Paper found at URL: http://www.cis.udel.edu/~wood/papers/DIMACS_99.ps.

[8] R. Cheynier, S. Gratton, J. P. Vartanian, A. Meyerhans, and S. WainHobson. G→A hypermutation does not result from polymerase chain reaction. *AIDS Research and Human Retroviruses*, 13(12):985–986, August 10, 1997.

[9] J. H. M. Dassen. A bibliography of molecular computation and splicing systems. HTML source: http://www.wi.LeidenUniv.nl/ jdassen/dna.html, BibTeX source: http://www.wi.LeidenUniv.nl/ jdassen/dna.bib. This bibliography is also hooked into http://liinwww.ira.uka.de/bibliography/, The Collection of Computer Science Bibliographies.

[10] R. Deaton, R. C. Murphy, J. A. Rose, Max H. Garzon, Donald R. Franceschetti, and S. E. Stevens Jr. A DNA based implementation of an evolutionary search for good encodings for DNA computation. In *IEEE International Conference on Evolutionary Computation*, pages 267–271, Indianapolis, Illinois, April 13–16, 1997.

[11] Alan Dove. From bits to bases: Computing with DNA. *Nature Biotechnology*, 16(9):830–832, September 1998.

[12] K. A. Eckert and T. A. Kunkel. High fidelity DNA synthesis by the *Thermus aquaticus* DNA polymerase chain reaction. *Nucleic Acids Research*, 18:3739–3744, 1990.

[13] Philippe Gigure and David E. Goldberg. Population sizing for optimum sampling with genetic algorithms: A case study of the Onemax problem. In John R. Koza, Wolfgang Banzhaf, Kumar Chellapilla, Kalyanmoy Deb, Marco Dorigo, David B. Fogel, Max H. Garzon, David E. Goldberg, Hitoshi Iba, and Rick Riolo, editors, *Genetic Programming 1998: Proceedings of the Third Annual Conference*, pages 496–503, University of Wisconsin, Madison, Wisconsin, USA, 22-25 July 1998. Morgan Kaufmann.

[14] David E. Goldberg, Kalyanmoy Deb, and James H. Clark. Genetic algorithms, noise, and the sizing of populations. *Complex Systems*, 6(4):333–362, August 1992.

[15] Tino Gramß, Stephan Bornholdt, Michael Gramß, Melanie Mitchell, and Thomas Pellizzari. *Non-Standard Computation.* Wiley-VCH, Weinheim, 1998.

[16] Rachel Green, Andrew D. Ellington, David P. Bartel, and Jack W. Szostak. *In vitro* genetic analysis: Selection and amplification of rare functional nucleic acids. *Methods*, 2:75–86, 1991.

[17] Alexander J. Hartemink and David K. Gifford. Thermodynamic simulation of deoxyoligonucleotide hybridization for DNA computation. In Harvey Rubin and David Harlan Wood, editors, *DNA Based Computers III: DIMACS Workshop, June 23-25, 1997*, volume 48 of

DIMACS series in discrete mathematics and theoretical computer science, pages 25–37, Providence, 1999. American Mathematical Society.

[18] L. S. Lerman, K. Silverstein, and E. Grinfeld. Searching for gene defects by denaturing gradient gel electrophoresis. *Trends in Biochemical Sciences*, 172(3):89–93, 1992.

[19] M. M. Ling and B. H. Robinson. Approaches to DNA mutagenesis: An overview. *Analytical Biochemistry*, 254(2):157–178, December 15, 1997.

[20] J. L. LinGoerke, D. J. Robbins, and J. D. Burczak. PCR-based random mutagenesis using manganese and reduced dNTP concentration. *Biotechniques*, 23(3):407–, September 1997.

[21] Carlo C. Maley. DNA computation: Theory, practice, and prospects. *Evolutionary Computation*, 6(3):201–229, 1998.

[22] Robert Pool. Forget silicon, try DNA. *New Scientist*, 151(2038):26–31, July 13, 1996.

[23] D. L. Robertson and F. G. Joyce. Selection *in vitro* of an RNA enzyme that specifically cleaves single-stranded DNA. *Nature*, 344(6265):467–468, March 29, 1990.

[24] Gerhard Steger. Thermal denaturation of double-stranded nucleic acids: Prediction of temperatures critical for gradient gel electrophoresis and polymerase chain reaction. *Nucleic Acids Research*, 22(14):2760–2768, July 25, 1994. Online server is found at the URL: http://www.biophys.uni-duesseldorf.de/POLAND/poland.html.

[25] Willem P. C. Stemmer. DNA shuffling by random fragmentation and reassembly: *In vitro* recombination for molecular evolution. *Proceedings of the National Academy of Science, U.S.A.*, 91:389–391, 1994.

[26] Willem P. C. Stemmer. Rapid evolution of a protein by DNA shuffling. *Nature*, 370:389–391, 1994.

[27] Willem P. C. Stemmer. The evolution of molecular computation. *Science*, 270:1510–1510, December 1, 1995.

[28] Willem P. C. Stemmer. Sexual PCR and Assembly PCR. In Robert M. Meyers, editor, *The Encyclopedia of Molecular Biology and Molecular Medicine*, volume 5, pages 447–457. VCH, New York, 1996.

[29] Jack W. Szostak. *In vitro* genetics. *Trends in Biochemical Sciences*, 172(3):89–93, 1992.

[30] H. J. Thiesen and C. Bach. Target detection assay (TDA) — A versatile procedure to determine DNA-binding sites as demonstrated on SP1 protein. *Nucleic Acids Research*, 18(11):3203–3209, 1990.

[31] C. Tuerk and L. Gold. Systematic evolution of ligands by exponential enrichment — RNA ligands to bacteriophage-T4 DNA-polymerase. *Science*, 249(4968):505–510, August 3, 1990.

[32] J. P. Vartanian, M. Henry, and S. WainHobson. Hypermutagenic PCR involving all four transitions and a sizeable proportion of transversions. *Nucleic Acids Research*, 24(14):2627–2631, July 15, 1996.

[33] Roy Williams. Data powers of ten. Web page at http://www.ccsf.caltech.edu/~roy/dataquan.

[34] David Harlan Wood, Junghuei Chen, Eugene Antipov, Walter Cedeño, and Bertrand Lemieux. A DNA implementation of the Max 1s problem. In Wolfgang Banzhaf, A. E. Eiben, Max H. Garzon, Vasant Honavar, Mark Jakiela, and Robert E. Smith, editors, *GECCO-99: Proceedings of the Genetic and Evolutionary Computation Conference, July 13-17, 1999, Orlando, Florida USA.*, pages 1835–1842, San Francisco, 1999. Morgan Kaufman.

[35] M. Zaccolo, D. M. Williams, D. M. Brown, and E. Gherardi. An approach to random mutagenesis of DNA using mixtures of triphosphate derivatives of nucleoside analogues. *Journal of Molecular Biology*, 255(4):589–603, February 2, 1996.

(D. Wood) Department of Computer Science, University of Delaware, Newark, DE
E-mail address: wood@cis.udel.edu
URL: http://www.cis.udel.edu/~wood

(J. Chen) Dept. of Chemistry & Biochemistry, University of Delaware, Newark, DE
E-mail address: junghuei@udel.edu
URL: http://www.udel.edu/chem/chen/chen.html

(E. Antipov) Dept. of Chemical Engineering, University of Delaware, Newark, DE
E-mail address: ultra@udel.edu
URL: http://copland.udel.edu/~ultra/

(B. Lemieux) Dept. of Plant & Soil Sciences, University of Delaware, Newark, DE
E-mail address: blemieux@udel.edu
URL: http://bertrand.ags.udel.edu

(W. Cedeño) Hewlett-Packard Company, 2850 Centerville Road, Wilmington, DE
E-mail address: wcedeno@lf.hp.com
URL: http://www.personal.psu.edu/users/w/x/wxc28

DIMACS Series in Discrete Mathematics
and Theoretical Computer Science
Volume **54**, 2000

Liposome Mediated Biomolecular Computation

Brian Bloom and Carter Bancroft

ABSTRACT. Biological vesicles are composed of a phospholipid membrane bilayer, which has folded on itself to form a spherical structure with an aqueous cavity. We propose that synthetic vesicles, termed liposomes, composed of specific species of phospholipids might provide a useful architecture for the execution of some existing algorithms for DNA based computation. In this model, the liposomes would serve simultaneously as water-filled chambers containing specific DNA reactants; and as valves to permit, via controlled fusion, interaction of only the DNA molecules and possibly enzymes appropriate for a particular stage of computation.

1. Introduction

One of the greatest potentials of DNA based computation is to carry out massively parallel operations. In order to exploit fully this potential, these parallel operations should be performed on different subsets of molecules, with the resultant strings all eventually combined into one solution tube. To address the technological problem of differentially affecting desired vs. undesired interactions of DNA reactants, Reif et al have recently proposed a Micro Flow Bio-Molecular Computation (MF-BMC) architecture. In the MF-BMF model, a decrease in human intervention is brought about through automation of miniature chambers, pumps, valves and sensors [**6**]. Kurtz et al and Hagiya have previously proposed that biological membranes might prove useful in forming compartments for organizing DNA based computation [**8**], [**9**]. We present a model based on this idea, termed Liposome-Mediated Biomolecular Computation (LMBMC). The fundamental concept underlying LMBMC is that liposomes could serve simultaneously as chambers and valves for directing DNA based computation. Our preliminary theoretical analysis of this model suggests that, for certain types of computation, this approach might provide a useful alternative to MF-BMC.

2. Biochemistry Of Biological Membranes

Cells employ biological membranes to separate their internal contents from the surrounding environments [**10**], [**20**]. Intracellular organelles use membranes to further compartmentalize specific molecules and processes. Lipids, mainly phospholipids and sphingolipids, are the primary contributors to biological membrane

2000 *Mathematics Subject Classification.* Primary 68–04; Secondary 68U99.

Supported by DARPA/NSF Grant CCR-9724012.

© 2000 American Mathematical Society

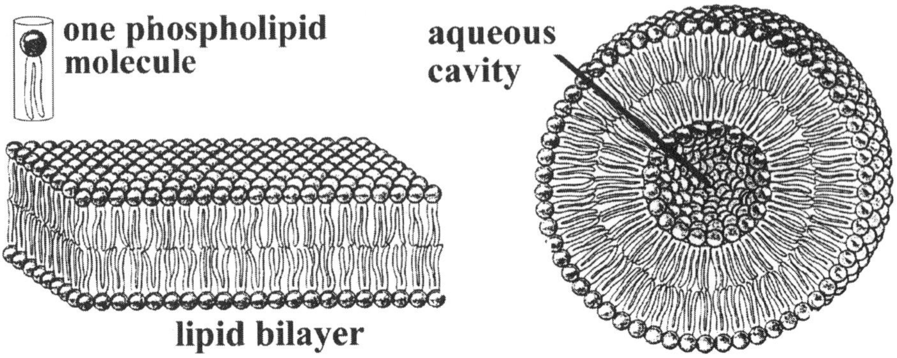

FIGURE 1. Lipids associate to form a bilayer (left). The bilayer can fold on itself to form a spherical structure with an aqueous cavity (right). From Lehninger et al, 1993, [**10**].

From *Principals of Biochemistry*, Second Edition, by A. L. Lehninger, D. L. Nelson, and Michael M. Cox, © 1983 by Worth Publishers, Inc. Used with the permission of Worth Publishers.

structure. Membrane lipids have a polar or ionic (water-loving) head group and two or more hydrophobic (water-hating) acyl chains. As Figure 1 illustrates, when the cross sectional area of the head group (dark circles) is equal to the area of the tails (light chains), each phospholipid molecule will adopt a cylindrical shape. In aqueous solutions, the tails will avoid high-energy interactions with water by aggregating together forming a two-dimensional bilayer. If the bilayer is large enough, it will fold on itself and form a three-dimensional, spherical vesicle enclosing an aqueous cavity.

In addition to lipids, biological membranes are composed of peripheral and integral proteins (often attached to lipids or sugar moieties) and cholesterol. The presence and amount of these constituents determine the function or behavior of the membrane. Since membranes are held together by weak, as opposed to covalent interactions, all components are free to diffuse laterally in the two-dimensional plane. The stable, flexible nature of biological membranes allows cellular vesicles to undergo either fission, splitting into two compartments, or fusion, which combines two aqueous cavities into one large compartment. Finally, the hydrophobic region within the bilayer restricts the passage of large polar solutes. Thus, membranes are semi-impermeable.

3. Advantages Of LMBMC

Computation using this type of experimental system, would have the following advantages:

3.1. Cost. The components (both lipids and proteins) of synthetic vesicles, or liposomes are inexpensive and available in large, pure quantities. Furthermore, manipulation of membranes does not require equipment other than that found in a typical molecular biology laboratory.

3.2. Synthesis. The formation of artificial liposomes has been researched extensively. For example, well-characterized methods exist which produce one large unilammellar vesicle, which is then divided into hundreds of identical small unilammellar vesicles or liposomes. These liposomes can then be separated by gel filtration, without disrupting membrane integrity, to provide a homogenous preparation. Synthesis can be performed in one reaction solution in a matter of hours [**17**].

3.3. Diversity. Liposomes can be synthesized with various membrane phospholipids and with or without integrated proteins and entrapped solutes. In addition, like all organic molecules, membrane phospholipids can be covalently attached to antibodies, magnetic components or fluorophores. By combining all of these factors, one can generate a finite set of liposomes, each with its own topologies, behaviors and characteristics [**18**].

3.4. Shelf Life. When a solution of liposomes is produced, it can be stored for weeks, with little spontaneous fusion or rupture.

3.5. Achievement Of High Initial Concentration. When preparing liposomes, solutes up to 1×10^6 Da (DNA strands up to 3000-4000 bases) can be enclosed at the same concentration that they appear in the starting solution [**18**]. Therefore, one can produce a high initial concentration of internal solutes. This is important, because after many traditional DNA computation operations, the effective concentration of reactants can become homeopathically low [**9**]. Conversely, when two liposomes of equal size fuse, the concentration of entrapped solutes *increases*. Following fusion, the surface area increases by a factor of 2, while the volume increases by a factor of $2\sqrt{2}$. As a consequence, the new surface area cannot hold the initial contents and some water is excluded, increasing the internal concentration [**18**].

3.6. Specificity. Specificity is the key to any biological or DNA computing operation. By varying the membrane lipid and/or protein content of liposomes, one can direct the behavior (fusion or other manipulations) of a distinct subset of the liposome population.

3.7. Parallelism. There are a number of physical limitations to the amount of connectivity between Reif's MF-BMC chambers and channels [**6**], because of their solid architecture. By contrast, if liposomes are synthesized correctly, a very large number of parallel fusion "connections" (merging of liposome contents) can form in solution. This is because the liposomes are free in solution, so that every liposome can potentially interact with all others. On the other hand, there are also limitations on the rates of parallel processes with liposomes, since the rate of fusion in diffusion-limited.

3.8. Thermostability. Many unicellular organisms have evolved to withstand a wide variety of harsh environments, such as extreme temperatures. *Thermophiles* can employ enzymes with high temperature optimal activities, long-chain fatty acids and special proteins in their membranes to thrive in 85 to 95°C environments [**11**]. It is thus possible to synthesize liposomes with thermostable membranes, permitting high temperature DNA computation operations such as melting and PCR, to occur without destroying the liposomes.

3.9. Automation. As described in the following section, most of the proposed LMBMC operations can be readily automated.

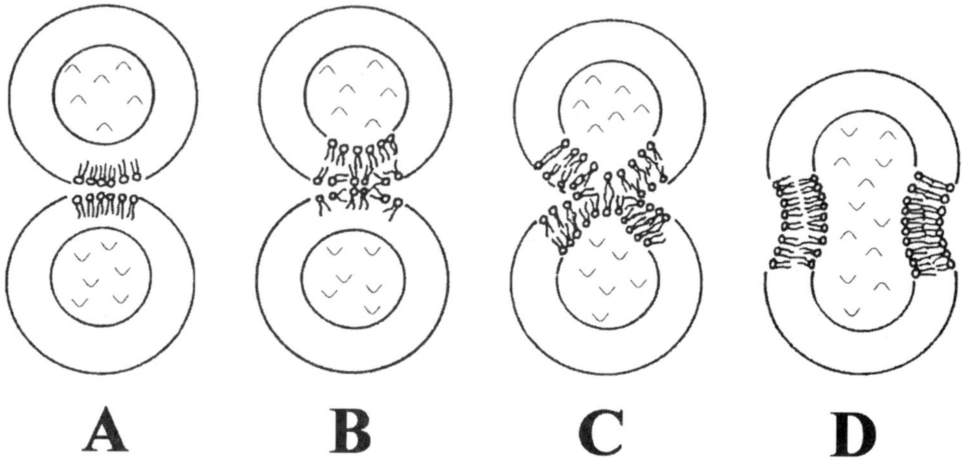

FIGURE 2. Steps involved in liposome fusion. A, aggregation; B, contact; C, lipid mixing; D, fusion of contents. Adapted from Duzgunes, 1990, [**5**].

From Duzgunes, N., "Molecular Control of Vesicle Aggregation and Membrane Fusion: Hypothesis on Bile Vesicles", *Hepatology* **12** (3), 70, 1990. Used with the permission of W. B. Saunders Company and the author.

4. The Operations

A set of liposomes is analogous to a set of microscopic test tubes. Thus, most novel operations will involve the micromanipulation of T, as opposed to specific sequences, c. (As defined in [**2**])

4.1. Modified or Novel Operations For LMBMC.

4.1.1. *Initialization.* This LMBMC operation involves the synthesis of liposomes containing appropriate integrated proteins and entrapped solutes, including DNA strands and possibly enzymes. The initialized vessel would contain *all* of the different synthesized liposomes.

4.1.2. *Liposome Merge/Union.* This is the most important operation in LMBMC. It is traditionally defined as the combination of two or more sets of strings. This operation is very common in solution-based DNA computation, and is also error prone, since DNA strands can be lost in the process. Merging of strands in LMBMC is accomplished by one of three different mechanisms; lipid controlled fusion, protein mediated fusion or viral insertion.

4.1.2.1. *Lipid-Controlled Fusion.* The most attractive mechanism for the *Merge/Union* operation is *lipid-controlled fusion.*

As depicted in Figure 2, a fusion event involves the aggregation of liposomes (A), interbilayer contact or lipid mixing (B and C) and fusion of aqueous contents (D). It should be noted that the specificity and rate-limiting step of fusion is dependent on the rate of liposomal aggregation. For example, liposomes composed of phosphatidylserine (PS) fuse with a half-life of days, while phosphatidylcholine (PC) liposomes are never allowed to aggregate or fuse, regardless of temperature or ionic conditions [**5**].

In *lipid controlled fusion*, simple physical/chemical changes in, for example, [Ca^{2+}], [Mg^{2+}], pH or temperature, can selectively induce fusion of liposomes that contain specific phospholipid components [**4**]. For example, if an algorithm required the fusion of liposomes A and B (in one step) and subsequently C, one would initialize A and B with pure phosphatidic acid (PA), and D with a 1:1 ratio of PA:PC. It is known that PA liposomes selectively fuse at a specific [Ca^{2+}], while PA:PC liposomes can fuse with pure PA liposomes at higher [Ca^{2+}] [**5**]. The addition of Ca^{2+} is a reversible manipulation (since it can be selectively precipitated), as are pH and temperature changes.

As Figure 3 illustrates, when a tree-like algorithm is required, phospholipid fusion rules, can be employed to devise a scheme for selective fusion at each *level* of the tree. Lipid-controlled fusion involves easy, hands-off manipulations amenable to automation. On the other hand, it appears the number of levels (depth) in a *fusion-tree algorithm* is limited, because after many levels, the phospholipid content of the growing liposome would become too complex, rendering it susceptible to side reactions and undesired fusion events.

4.1.2.2. *Protein-Assisted Fusion.* Biological cellular vesicles employ integrated proteins to assist in selective fusion. For example, studies performed both in vitro and in vivo have shown that membrane-integrated SNARE proteins are capable of docking or aggregating two specific liposomes together in *preparation* for fusion [**13**], [**15**]. Furthermore, aqueous external factors necessary for the fusion event (NSF and SNAPS) could be applied at the specific level where fusion is desired. There are numerous species-specific pairs of SNARES (types v and t SNARES), which can increase the combination of distinct fusion events. Thus, *protein-assisted fusion* may be combined with *lipid controlled fusion* to increase the possible complexity of a fusion-tree algorithm.

4.1.2.3. *Viral Insertion.* In the third LMBMC *Merge/Union* operation, viral insertion can be used to add a single string to a set of liposomal strings. Certain viruses are capable of encapsulating double stranded or single stranded DNA. In solution, the virus interacts with its receptor, inserting a single copy of its DNA into an accepting liposome. Accepting or "host" liposomes would be synthesized to contain integral membrane protein virus receptors [**10**]. This approach would be useful if only a small number of DNA molecules are sufficient for a computation operation. Viral insertion and protein-assisted fusion are illustrated in Figure 4.

4.1.3. *Filter.* This is another novel LMBMC operation. By adding certain factors to solution, one can selectively (depending on phospholipid components) filter small solutes out of the liposomes, while retaining large ones. For example, Sphingomyelinase induces leakage of solutes in sphingomyelin-containing liposomes [**1**]. The peptide sticholysin II creates a pore of approximately 1 nm diameter, which allows solutes up to 1 kDa to exit sphingomyelin and PC-containing liposomes [**3**]. Finally, antibacterial cecropin-A-melittin hybrid peptides can slowly induce leakage of 20 kDa solutes in PS or PC liposomes [**12**]. The *Filter* operation could possibly be used to sieve small strands from the liposomes, that have not been integrated into longer solution strands (for example, in a constructive primer extension NP-Complete algorithm). As more selective and specific leakage-inducing agents are discovered, they would be integrated into LMBMC algorithms.

4.1.4. *Liposome Separate.* If one expects the solution molecule of a computation to reside in the largest liposome (following many fusion events), the *Liposome*

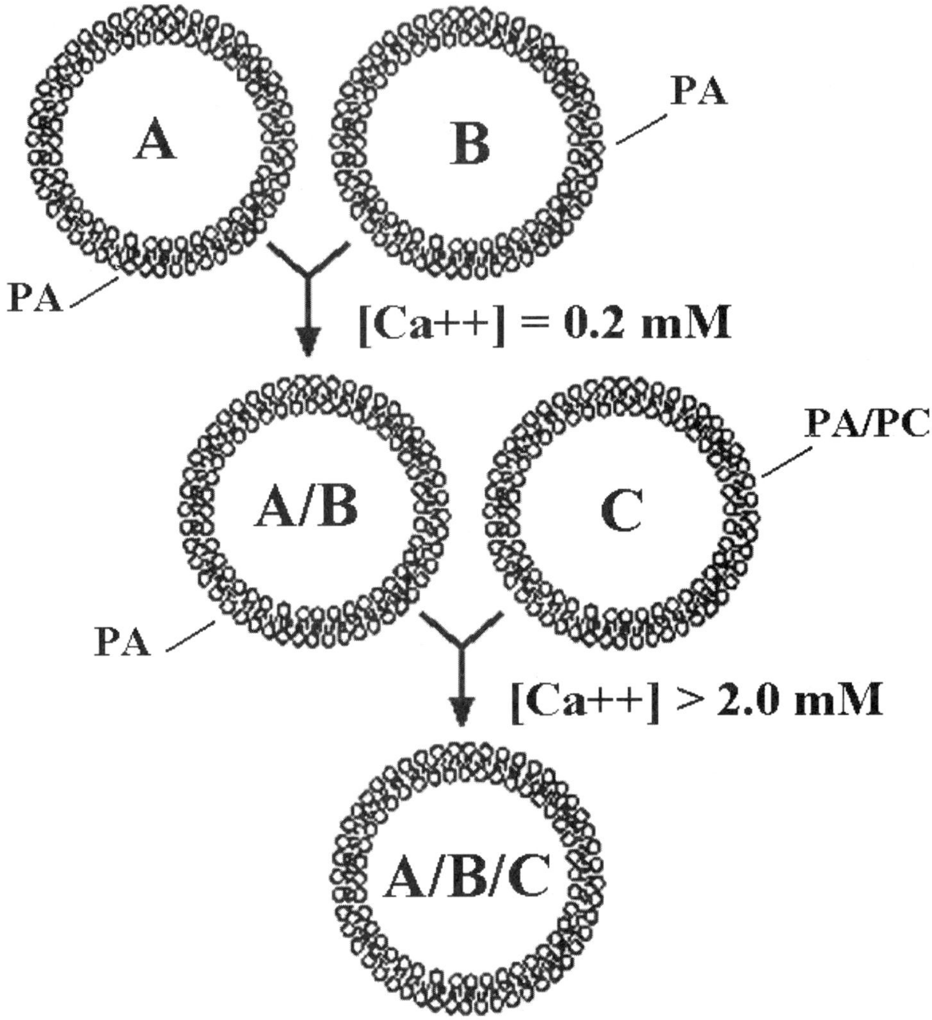

FIGURE 3. Simple fusion tree algorithm. Liposomes A and B are composed of pure phosphatidic acid, PA, while liposome C is composed of a 1:1 ratio of PA and phosphatidylcholine, PC. A/B represents the fusion product of liposomes A and B. A/B/C represents the fusion product of liposomes A/B and C.

Separate operation, achieved via gel filtration, could be used to separate and isolate liposomes according to size.

4.1.5. *Fusion Detect.* Since there are many assays available to confirm the fusion of two liposomes, a *Fusion Detect* operation can be defined. This would be employed to signal the completion of an algorithm or the incorrect fusion of two

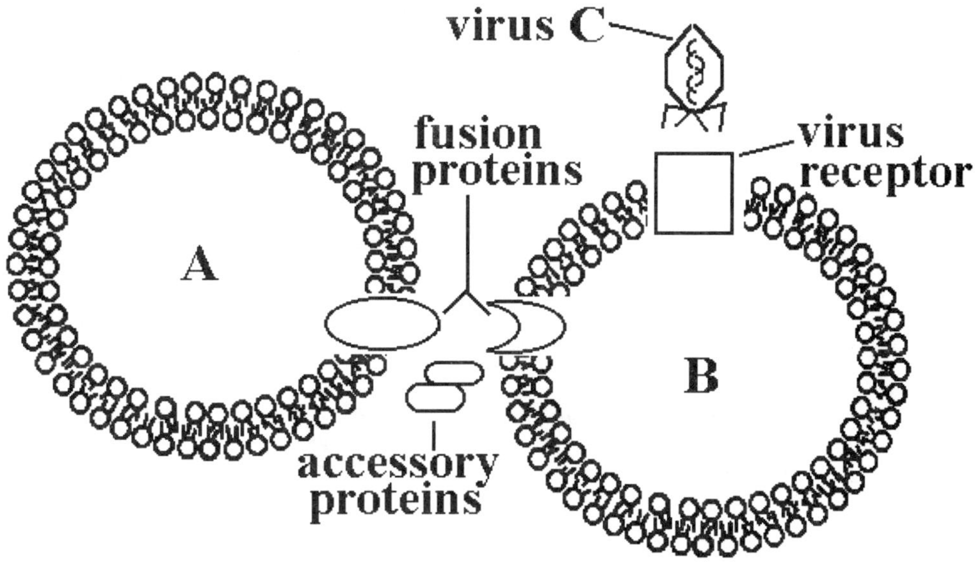

FIGURE 4. DNA contained in liposomes A and B can fuse via protein-assisted fusion and DNA from virus C can merge with liposome B contents via viral insertion.

liposomes. For example, if liposomes fuse producing a visual or fluorescent signal, the final fusion event of a *fusion-tree algorithm* can be readily detected.

4.1.6. *Lyse.* Detergents can be added to solubilize all membrane lipids and proteins, releasing the DNA strands for purification or detection. This is defined is the *Lyse* operation.

4.2. Potential Application Of Traditional Sequence-Specific DNA Computation Operations To LMBMC. Most operations described in the DNA based computation literature involve the identification or manipulation of a specific DNA sequence. These operations, *Append, Ligate, Digest* and *Cut,* are all catalyzed by enzymes. In solution based DNA computation, enzymes (such as ligase, polymerase or exo/endonuclease) are simply added to test tubes. However, in LMBMC, sequence-specific manipulation would require fusion events that combine DNA/enzyme with new DNA. Alternatively, specialized *catalytic liposomes* could be fused with DNA-containing liposomes for specific operations.

4.3. Operations That Are Difficult To Achieve In LMBMC. The most commonly used operation that would be difficult with LMBMC is *Split,* since artificial liposome fusion is an irreversible event. Cellular processes do exist that induce budding or splitting of membrane contents into two compartments, but the factors involved, such as receptors, protein coats and signals, are probably too complex for use with LMBMC.

Unlike solution or surface-based computation, termination of the enzymatic operations described above cannot be achieved by physical means, such as washing or gel separation. Furthermore, traditional operations such as *Length, Detect* and *Test,* which operate exclusively on DNA strands, can only be executed after the

Lyse operation. Finally, the *Amplify* (PCR) operation can be conducted only with thermostable liposomes. If these liposomes are used, it should be noted that information regarding selective fusion is not available. Therefore, amplification should also be performed after the final *Lyse* operation. Successive *Melt* and *Append* operations are acceptable, as the temperatures required should not disrupt liposome integrity.

5. Potential Applications Of LMBMC To Existing Algorithms

It seems likely that the LMBMC architecture described above will prove useful in a number of existing algorithms for DNA-based computation. Three such examples are described below.

5.1. State Transition Model (Hagiya et al). This seems the most attractive algorithm for application of LMBMC [**16**]. This group has experimentally demonstrated that Boolean expressions and NP-Complete problems can be successfully solved with their unique *successive localized polymerization* operation. Although this operation involves successive primer extension reactions, Hagiya et al have developed a new constant temperature protocol, which would allow the application of LMBMC, without extreme thermophile membrane components [**7**]. Solutions could be generated *intramolecularly*, in a massively parallel fashion, as the program, transitions and current state can reside on a single molecule. One limitation of the state transition model is efficient dataflow. Hagiya et al note that the size and complexity of their approach would be increased dramatically if the program and state resided on one molecule, while the inputs are encoded on others [**8**]. If the program were accomplished it could produce an output state, which would be passed on to other programs as input. It seems likely that, with LMBMC, simple *fusion-trees* would readily increase the dataflow and complexity of the state-transition algorithms developed by this group.

5.2. Self-Assembly DNA Computation Model (Winfree). LMBMC appears to provide the ideal architecture to mediate large-scale application of this computation approach. It has been shown that NP-Complete solutions can be synthesized by the self-assembly of DNA tiles [**19**]. When N is large, the complexity of the solution increases to the point where the time required for solution generation is immense. However, if partial solutions were generated in enclosed liposomes, the entire population could be induced to slowly fuse into one large unilammellar vesicle, where all partial solutions could interact like completed sub-sections of a jigsaw puzzle. This would not be possible with a traditional solution-based Merge operation, because large, non-ligated solution arrays would be disturbed. Furthermore, as the liposomes fuse, the reaction volume would increase, allowing room for larger structures to form, while the concentration of DNA would also increase, compensating for the increased complexity.

5.3. Boolean Circuit Algorithm (Ogihara and Ray). This approach [**14**] already employs a tree-like strategy. Unfortunately, each level requires a *Length* operation, which cannot be performed in LMBMC, until the completion of all liposomal fusion events. If this step were replaced with another operation (such as endonuclease *Digest* or exonuclease *Destroy* of wrong solutions), LMBMC might supply the ideal medium for their computation.

6. Conclusion

We have described here an LMBMC model for DNA based computation, based upon manipulations of DNA reactants enclosed in liposomes possessing specific chemical properties. We have considered briefly the possible application of this model to several known DNA based algorithms. However, further analysis, both theoretical and experimental, will be required to ascertain the specific liposome structures and manipulations required for the execution of each algorithm. It also seems likely that the model presented here can be extended to the development and application of new algorithms for DNA based computation, that are based specifically upon the particular strengths of LMBMC. It is also notable that conditions for specific phospholipid fusion (e.g., precise Ca^{2+} concentrations required for fusion of specific lipids) have not previously been examined in detail, probably because there was no reason to do so. It thus seems possible that further experimental studies will reveal more specific information about the precise lipid combinations and environmental conditions that will lead to controlled fusion of liposomes. It may even prove possible to synthesize phospholipid head groups not found in nature, that would permit specific "fusion rules" not provided by natural phospholipids. Finally, the experimental execution of a simple example of LMBMC would provide a strong "proof-of-principle" for this model.

References

[1] Basenez G, Ruiz-Arguello MB, Alonso A, Goni FM, Karisson G, Edwards K, Morphological Changes Induced by Phospholipase-C and by Sphingomyelinase on Large Unilamellar Vesicles: A Cryo-Transmission Electron Microscopy Study of Liposome Fusion, Biophysical Journal, 72: 2630-2637, 1997

[2] Beigel R, Fu B, Solving Intractible Problems with DNA Computing, http://www.eecs.lehigh.edu/~beigel/papers/bf-dnasurvey-complexity98.PS.gz, 1998

[3] De Le Rios V, Mancheno JM, Lanio ME, Onaderra M, Gavilanes JG, Mechanism of the Leakage Induced on Lipid Model Membranes by the Hemolytic Protein Sticholysin II From the Sea Anemone Stichodactyla helianthus, Eur. J. Biochem, 252: 284-289, 1998

[4] Duzgunes N, Wilschut J, Fraley R, Papahadjopoulos D, Studies on the Mechanism of Membrane Fusion. Role of Head-Group Composition in Calcium- and Magnesium-Induced Fusion of Mixed Phospholipid Vesicles, Biochim. Biophys. Acta., 642 (1): 182-195, 1981

[5] Duzgunes N, Molecular Control of Vesicle Aggregation and Membrane Fusion: Hypothesis on Bile Vesicles, Hepatology, 12 (3),: 67S-74S, 1990

[6] Gehani A, Reif J, Micro Flow Bio-Molecular Computation, In Proceedings of the Fourth International Meeting on DNA Based Computing, 1998

[7] Hagiya M, Arita M, Kiga D, Sakamoto K, Yokoyama S, Towards Parallel Evaluation and Learning of Boolean m-Formulas with Molecules, In Proceedings of the Third DIMACS Meeting on DNA Based Computers, 1997

[8] Hagiya M, Towards Autonomous Molecular Computers, Genetic Programming 1998: Proceedings of the Third Annual Conference, 691-699, 1998

[9] Kurtz S, Mahaney S, Royer J, Simon J, Active Transport in Biological Computing, In Second DIMACS Workshop on DNA Based Computers, 1996

[10] Lehninger AL, Nelson DL, Cox MM, Principles of Biochemistry, Second Edition, Worth Publishers, 1993

[11] Luzzati, V, Gambacorta, A, Derosa, M, Gulik, A, Polar Lipids Of Thermophilic Prokaryotic Organisms: Chemical And Physical Structure, Annu. Rev. Biophys. Biomol. Struct., 16: 25-47, 1987

[12] Mancheno JM, Onaderra M, Martinez del Pozo A, Diaz-Achirica P, Andreu D, Rivas L, Gavilanes JG, Release of Lipid Vesicle Contents by Antibacterial Cecropin A - Melittin Hybrid Peptide, Biochemistry, 35: 9892-9899, 1996

[13] O'Connor V, Augustine GJ, Betz H, Synaptic Vesicle Exocytosis: Molecules and Models, Cell, 76: 785-787, 1994
[14] Ogihara M, Ray A, Simulating Boolean Circuits on a DNA Computer, Technical Report TR 631, University of Rochester, Computer Science Department, 1996
[15] Rothman JE, Wieland FT, Protein Sorting by Transport Vesicles, Science, 272: 227-234, 1996
[16] Sakamoto K, Kiga D, Komiya K, Gouzu H, Yokoyama S, Ikeda S, Sugiyama H, Hagiya M, State Transitions by Molecules, in Proceedings of the Fourth International Meeting on DNA Based Computers, 1998
[17] Sharata HH, Katz KH, Liposomes, International Journal of Dermatology, 35 (11): 761-767, 1996
[18] Szoka Jr. F, Papahadjopoulos D, Comparative Properties and Methods of Preparation of Lipid Vesicles (Liposomes), Ann. Rev. Biophys. Bioeng., 9: 467-508, 1980
[19] Winfree E, Algorithmic Self-Assembly of DNA, PhD Thesis, California Institute of Technology, Pasadena, California, 1998
[20] Yeagle PL, The Membranes of Cells, Academic Press, Inc., 1993

(B. Bloom) Mount Sinai School of Medicine, Dept. of Physiology and Biophysics, New York, NY, 10128

E-mail address: `bloomb01@doc.mssm.edu`

(C. Bancroft) Mount Sinai School of Medicine, Dept. of Physiology and Biophysics, New York, NY, 10128

E-mail address: `cbancro@smtplink.mssm.edu`

DIMACS Series in Discrete Mathematics
and Theoretical Computer Science
Volume **54**, 2000

Error Correction in DNA Computing: Misclassification and Strand Loss

Kevin Chen and Erik Winfree

ABSTRACT. We present a method of transforming an extract-based DNA computation that is error-prone into one that is relatively error-free. These improvements in error rates are achieved without the supposition of any improvements in the reliability of the underlying laboratory techniques. We assume that only two types of errors are possible: a DNA strand may be incorrectly processed or it may be lost entirely. We show how to deal with each of these errors individually and then analyze the tradeoff when both must be optimized simultaneously.

1. Introduction

Despite the significant theoretical achievements made in the field of DNA computing, no physical DNA computer has yet been used to successfully solve any real problem. For researchers trying to make DNA computation a physical reality, a central problem is that of error correction since the obvious implementations of DNA algorithms using current lab techniques all produce errors at rates unacceptable for useful computation. In particular, two major sources of errors are misclassification errors caused by faulty extracts and strand loss.

However, the unreliability of the underlying biotechnology does not mean that DNA computing must remain a purely theoretical exercise. Even without improvements in lab techniques, research has shown that error rates can be significantly reduced solely by algorithmic methods [**14, 15**]. In other words, an algorithm for a DNA computer can be mapped to an equivalent one that is comparatively error-free, admittedly at the cost of more time and space. This paper presents such a transformation for extract-based DNA algorithms and an analysis of the tradeoff between the reduction of errors it produces and the extra space and time it requires.

2. Model of Computation

Although many models of DNA computation have been proposed, the canonical model remains the *extract model* of Adleman [**1**], as generalized by Lipton [**2**]. In this model we start with a test tube of DNA strands encoding every possible solution to the problem and we make sure that the sequences are chosen such that the strands will be well-behaved in the laboratory [**16, 17, 18, 19**]. The goal of the

2000 *Mathematics Subject Classification.* Primary 68Q05.

© 2000 American Mathematical Society

computation then is to separate the true solutions (the "good strands") from the non-solutions (the "bad strands").

The algorithm uses only three primitives: *extract*, *combine* and *detect*. An extract step takes as input a test tube T and a subsequence S, and produces as output a *yes-tube* containing the strands in T containing the subsequence S and a *no-tube* containing the other strands. A combine step takes two tubes and produces a single tube containing the contents of both. A detect step is used to check if a test tube contains any DNA strands and, if so, to determine the sequence of one (randomly chosen) strand. We will consider only computations where *detect* is used only at the end of the computation to check for the existence of a solution. Note that since the number of combine steps done is bounded by the number of extracts performed, it is reasonable to use the number of extract steps as our measure of time complexity. If we assume the algorithm divides all strands into a final yes-tube and a final no-tube, then ideally, the yes-tube would contain all the good strands and the no-tube all the bad strands. In a real computation, however, the yes-tube may contain both types of strands, so we must choose a small number of strands from the yes-tube, sequence them and verify their correctness, in effect repeating the detect step several times. By the definition of NP, this verification can be carried out efficiently. There is also a fourth primitive called the *duplicate* step, in which PCR is used to make a single duplicate copy of each strand of DNA in the system. This primitive is not used in the conventional extract-based model but will be introduced later as a way to correct errors in the computation.

Formally, an algorithm in the extract model can be represented as a directed acyclic graph with a source and two sinks. Each node represents a test tube; the source node represents the original tube containing all possible solutions and the two sinks represent the final yes- and no-tubes. All non-sink nodes are labeled with some symbol σ and have two outgoing edges labeled σ and $\overline{\sigma}$. $S(\sigma)$ is the sequence which we use to perform an extract on the node. If more than one edge enter a node, we implicitly do a combine step. We shall assume that at the end of the computation, each strand is in one of the two sink nodes.

Figure 1 shows how we can solve 3-SAT with an extract-based DNA computer; specifically, it illustrates the step that extracts putative solutions that satisfy the clause ($x = 0$ or $y = 1$ or $z = 1$). Starting at the node at the top-left, all strands not containing "x=1" are extracted and put in the yes-tube at the bottom of the figure. Next, from the remaining strands, all those containing "y=1" are extracted and put in the yes-tube, and similarly for the last extract. At the end of this computation, the yes-tube contains all strands satisfying the clause. For 3-SAT computations we can immediately discard all other strands, but in general both the yes- and no-tube strands may be used later on, so we assume that these remaining strands are implicitly kept in a no-tube until the end of the computation. For the next clause, we can do a similar computation but starting with the yes-tube of the previous clause. Clearly, this algorithm requires 3 extract steps and 1 combine step for each clause plus a constant amount of time for the detect step. This gives a running time complexity linear in the size of the formula compared to the exponential time required by the best known algorithm for the classical Turing-machine model. However, an exponential number of DNA strands must be used.

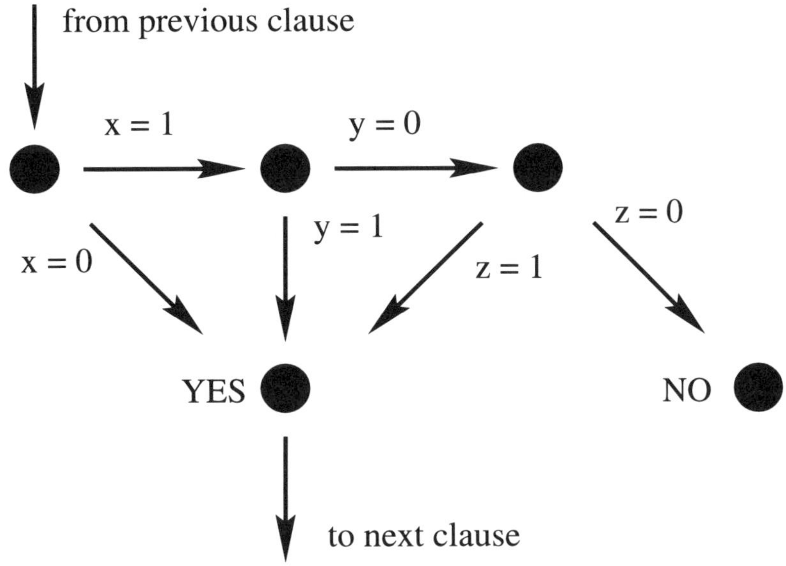

FIGURE 1. Solving 3-SAT with an extract-based DNA computer. Subroutine for "... & ($x = 0$ or $y = 1$ or $z = 1$) & ..."

3. Error model

From the previous section, it is clear that the error rate of the computation is dominated by the error rates of the extract and combine steps. Furthermore, since an algorithm can have at most one combine step per extract, errors that occur in a combine step (loss of DNA strands) will be attributed to the preceding extract. Therefore, from now on, we will assume that errors in processing are caused only by extract steps and we will treat an algorithm as simply a sequence of E extracts. In general the number of extracts performed on each strand may be different. However, if E is the maximum number of extracts performed on any strand, then we can think of E as an upper bound on the number of extracts performed on any strand. We can then transform any algorithm A in which each strand goes through $\leq E$ extracts to an equivalent algorithm A' in which each strand goes through exactly E extracts simply by adding redundant steps. A' is then a layered graph, with E layers. The error rates derived for A' are an upper bound on the error rates for A so from now on we will assume that all strands undergo the same number of extracts.

In this paper, we will consider only two sources of errors in a DNA computation. First, as stated above, each extract step may simply not work correctly, with the result that some strands containing the sequence are classified as not containing it while some strands not containing the sequence are classified as containing it. These errors are called false negatives and false positives respectively. In practice, an implementation of the extract step may have false negative and false positive rates that are several orders of magnitude different. Also, the error rate may be different depending on which DNA sequence is used for the extract since, for example, some sequences bind better to separation beads than others. However, for the sake of simplifying our calculations, we will assume in this paper that both of these error rates are in fact equal. Let the probability of a strand being correctly processed by a single extract step be p, and the probability that the strand is correctly processed by the entire algorithm be $P_{correct}$. For an indication of the overall error rate of a computation where no error correction is used, consider that molecular biologists are generally comfortable with handling around 2^{50} DNA strands at a time. This allows us to reasonably handle an instance of 3-SAT with 50 variables and 200 clauses since this would require 2^{50} strands. To do this we need 3 extracts per clause for a total of 600 extracts, so even with $p = 0.99$, $P_{correct}$ is only $(0.99)^{600} = 0.002$.

The second type of error we consider is caused by a strand being lost during the course of the computation. DNA strands, being physical objects and fragile ones at that, may be lost by getting stuck in the apparatus, by cleavage or cross-linking reactions or by any number of other possible means. Let q be the probability of a strand surviving one step of processing, and let $k = \log_q \frac{1}{2}$. The number of strands will be halved after every k steps; that is, $q^k = .5$. In an extreme case one may have to remove a considerable fraction of the strands after each step for debugging and $k \approx 1$, but we expect that in general $k > 5$ can be achieved. We denote the probability of at least one copy of (perhaps initially many copies of) a strand surviving to the end of the computation as $P_{survival}$.

4. Related Work

Although previous research has been done on improving $P_{correct}$ and $P_{survival}$ individually, to our knowledge no one has taken the next step of studying them together. Intuitively, we need to do more computational steps to improve the rate of correctly processing the strands, but doing so will only increase the chances for a strand to be lost in the computation. If we wish to study the overall rate of error, it is clearly important to study the tradeoff between the two error rates.

In his paper on DHPP [1] Adleman [**1**] briefly mentioned two methods for dealing with errors. For false positives, he suggested that the extract could simply be repeated on the yes-tube, thus weeding out any bad strands that were wrongly misclassified. However, if the no-tube must be used later, repeated separation is useful only if the false negative rate is satisfactory already. So far, this has been shown only for separation techniques that incur substantial strand loss [**20, 21, 22, 23**]. For false negatives – which in his case amounted to strand loss – Adleman suggested using PCR to amplify good (and possibly bad) strands. This is essentially the approach we develop quantitatively in this paper.

Karp et al [**14**] first introduced the notion of a compound extract as a way to deal with processing errors from the extract step. The idea is that one can repeat a faulty extract operation many times to simulate a reliable one. We present essentially the same algorithm but with a more concrete analysis of the construction. Karp claimed that if δ is the desired error rate of the compound extract and $\epsilon = 1-p$ the error rate of the simple extract, then we can achieve the desired error rate by using an extra $O(\log_\epsilon^2 \delta)$ steps, $O(\log_\epsilon \delta)$ of which can be performed in parallel. Our analysis gives a complete proof of the upper bound and shows that the constant factors are in fact not large. Later, Roweis et al [**24, 15**] presented essentially the same algorithm as Karp but with a different analysis.

[1] Directed Hamiltonian Path Problem. Adleman's algorithm is sometimes referred to as an algorithm for the *Travelling Salesman Problem* (TSP) although strictly speaking, TSP involves edge-weights and DHPP does not.

Although no algorithm for improving $P_{survival}$ as defined here has been studied, Boneh et al [**25**] proposed and analyzed the use of PCR to reduce errors in decreasing volume computations, a class of DNA algorithms that includes the standard DHPP and 3-SAT algorithms but unfortunately not algorithms for formula-SAT, circuit-SAT or breaking DES [2]. In decreasing volume computations, once a strand is known to not contain a solution, it is discarded, so that the total volume of DNA in the system is decreasing, hopefully at a constant rate. In this case, if bad strands are removed at a faster rate than good strands are lost, then PCR can be used to amplify the good (and bad) strands, making it unlikely for all copies of a good strand to be lost. We adapt Boneh's algorithm to the general problem, where all strands are retained, and we analyze how it affects the probability of a strand surviving the computation. Note that our analysis is not the same as Boneh's because in his case, a bad strand has a much higher probability of being discarded than a good strand, while in our case, both types of strands have equal probabilities of becoming lost.

Although, Karp, Roweis and Boneh are our primary models, it is perhaps worth mentioning a few other error-correcting methods. Amos et al [**26**] proposed an alternative implementation of the extract step which used a restriction enzyme to digest away unwanted strands instead of extracting out the good ones. Ouyang et al [**10**] has successfully implemented this model for a 6-variable SAT problem, and Faulhammer et al [**13**] performed a 9-variable computation using RNA strands and RNAse H as a "universal RNA restriction enzyme." From an algorithmic standpoint the bead-separation and digestion implementations of *extract* are equivalent if *duplicate* is included as a primitive step. An extract step that separates strands into a yes-tube and a no-tube can simulate a digest step by simply discarding the no-tube. Similarly, assuming we are using an encoding (such as Lipton's) where each strand contains either $S(\sigma)$ or $S(\overline{\sigma})$ but not both, we can simulate an extract on sequence S by using a duplicate step to make two copies of the tube, and then digesting the $S(\sigma)$ in one to get the no-tube and digesting $S(\overline{\sigma})$ in the other to get the yes-tube. Although the restriction enzyme digestion approach was proposed to take advantage of the high specificity and efficiency of restriction enzymes, the fact that duplication steps are necessary and that strands may be lost in incidental lab procedures complicates the error analysis. The results in this paper should provide a framework for rigorous analysis of errors in digestion-based DNA computing.

Finally, it is possible for DNA strands to interact with each other in undesired ways and even alter each other's sequences during the duplicate step [**13**]. This paper cannot address these problems; however, the problems may be minimized by experimental protocol modifications and by good choices of coding sequences [**16, 17, 18, 19**]. Good sequences and encodings may also improve the error rates for extracts since some sequences are more likely to be properly extracted than others depending on their biochemical properties [**25**].

---

[2]The problem is to break the Data Encryption Standard (DES) using a *chosen plain-text attack*, which means that an adversary can obtain a (plain-text, cipher-text) pair where the plain-text is chosen by the adversary.

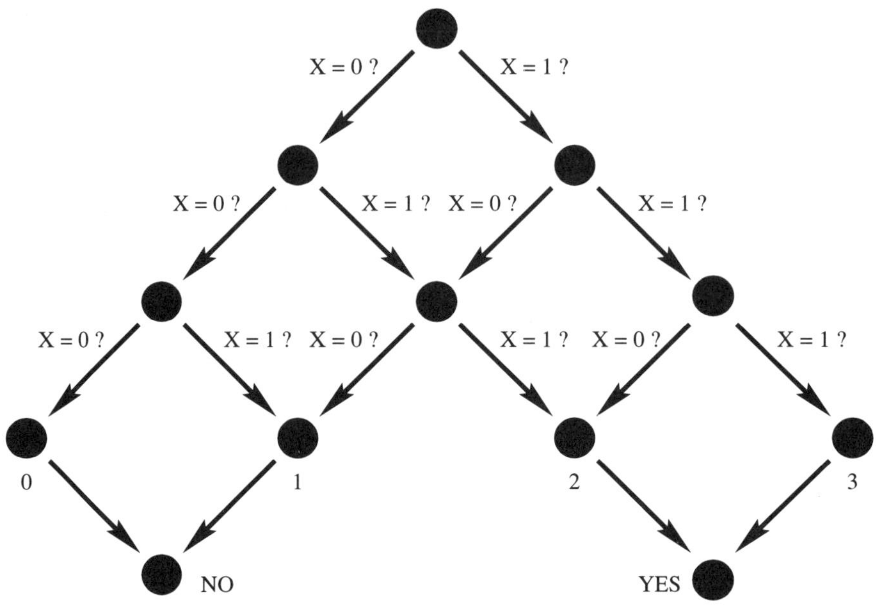

FIGURE 2. The compound extract subroutine, for $n = 3$.

5. Probability of a strand being correctly processed

A compound extract is a *subroutine* designed to be used in place of a simple extract, which we have been referring to up to now simply as an *extract.* A *compound extract* works by repeating the same simple extract many times as outlined in Figure 2, which shows a compound extract on "x=1". Each node represents a test tube and each pair of arrows leaving a node represents a simple extract on "x=1". The question marks have been added to suggest that the simple extract step is not 100% accurate. A compound extract is organized in n phases: in the above example there are three phases each corresponding to a level in the lattice. After each phase, all the strands from the original tube are distributed among the tubes in that level: the jth tube from the left starting at 0 contains strands encoding "x=1" which were extracted correctly exactly j times, as well as strands encoding "x=0" which were extracted incorrectly exactly j times. After the nth phase, we take the right-most tubes and combine them into a single yes-tube, and take the left-most tubes and combine them into a single no-tube. If n is even, and hence after the last phase of extracts there are $n + 1$ tubes, then the central tube may be combined arbitrarily with the yes- or no-tube. The greatest number of tubes needed at any time is n and in principle we can perform all the extracts in a given phase in parallel[3]. Therefore, we measure the time complexity for the compound extract by the number of phases, n.

[3]In fact, we can reduce the number of simple extracts in the compound by a factor of 2 without degrading performance. Observe that the fate of strands leaving the central $\frac{n}{2} \times \frac{n}{2}$ diamond is at that point determined, so all extracts outside the diamond may be eliminated, and all strands may be routed directly to the yes-tube and the no-tube. However, the discussion is simplified by considering the strictly layer phases as described above.

Strands processed according to the Figure 2 perform a random walk. The strands containing the sequence "x=1" have a bias toward the right and those with the sequence "x=0" have a bias (which we assume for this derivation is equal to the rightward bias) towards the left. Therefore the probability of being correctly processed by this method is given by the tail of a binomial distribution. The analysis that follows looks complicated but is in fact straightforward. For a full treatment, see Feller [**27**] Chapter 6.2-3 for the binomial distribution and Chapter 2.9 for Stirling's formula.

We want to find an upper bound on the probability of error due to a compound extract. Define P_e to be this probability and define $P_c = 1 - P_e$. Let

$$b(h; n, p) = \binom{n}{h} p^h (1-p)^{n-h}$$

be the probability that n Bernoulli trials with probabilities p for success and $(1-p)$ for failure result in h successes and $n-h$ failures. The probability of error of the compound extract is given by the sum of the terms of the Binomial distribution from 0 to the central term, $\frac{n-1}{2}$ or $\frac{n}{2}$ (in which case we over estimate either false negatives or false positives). We assume that p is greater than 0.5. Note that the error bound for even $n = 2m$, $P_e \leq \sum_{h=0}^{m} b(h; 2m, p)$, is strictly greater than the error for odd $n = 2m+1$, $P_e = \sum_{h=0}^{m} b(h; 2m+1, p)$, by observing that the strands in the latter tubes are a subset of those that reach the former. Consequently, for mathematical convenience we develop the bound for even n only. To get an upper bound on this quantity, we find an upper bound on the largest, central term, and then bound the rest of the tail with a geometric series.

To get an upper bound on $b(\frac{n}{2}; n, p)$, we use the conventional upper and lower bounds for $n!$ given by Stirling's formula:

$$\sqrt{2\pi n}\; n^n e^{-n} e^{1/(12n+1)} < n! < \sqrt{2\pi n}\; n^n e^{-n} e^{1/(12n)}.$$

Since we assume that n is even,

$$b(\frac{n}{2}; n, p) = \frac{n!}{[(n/2)!]^2} p^{n/2} (1-p)^{n/2}.$$

Plugging the bounds for Stirling's formula in this equation and simplifying (the derivation is tedious but entirely straightforward and hence omitted), we get

$$\begin{aligned} b(\frac{n}{2}; n, p) &\leq \frac{1}{\sqrt{2\pi n}} 2^{n+1} e^{1/(12n) - 2/(6n+1)} p^{n/2} (1-p)^{n/2} \\ &< 2^n p^{n/2} (1-p)^{n/2} \\ &= [4p(1-p)]^{n/2}. \end{aligned} \tag{1}$$

Now to find a good geometric series bound on the tail, consider the ratio between consecutive terms in the tail. If we can bound this ratio with some β, then the probability of an error occuring is

$$\begin{aligned} P_e &= b(\frac{n}{2}; n, p) + b(\frac{n}{2} - 1; n, p) + ... + b(0; n, p) \\ &\leq b(\frac{n}{2}; n, p)[1 + \beta + \beta^2 + ... + \beta^{n/2}] \\ &= b(\frac{n}{2}; n, p)\{1 - \beta^{n/2+1}\}/\{1 - \beta\}. \end{aligned} \tag{2}$$

Typical values and estimates for P_e

	$p = 0.90$		$p = 0.95$		$p = 0.99$	
	$\approx$	$\leq$	$\approx$	$\leq$	$\approx$	$\leq$
$n = 6$	1.6×10^{-2}	1.4×10^{-1}	2.2×10^{-3}	2.1×10^{-2}	2.0×10^{-5}	1.9×10^{-4}
$n = 10$	1.6×10^{-3}	1.8×10^{-2}	6.4×10^{-5}	7.4×10^{-4}	2.4×10^{-8}	2.9×10^{-7}
$n = 20$	7.2×10^{-6}	1.1×10^{-4}	1.1×10^{-8}	1.8×10^{-7}	1.7×10^{-15}	2.8×10^{-14}

TABLE 1

For $h \leq \frac{n}{2}$ and $p > .5$, the ratio between consecutive terms is

$$(3) \qquad \frac{b(h-1;n,p)}{b(h;n,p)} = \frac{h(1-p)}{(n-h+1)p} \leq \frac{(n/2)(1-p)}{(n/2+1)0.5} < 2(1-p).$$

Thus we get our upper bound on the ratio of consecutive terms, $\beta = 2(1-p)$. From (1), (2) and (3) we can get an upper bound on the probability of error as follows:

$$P_e \leq b(\frac{n}{2};n,p)\frac{1-\beta^{n/2+1}}{1-\beta}$$
$$< [4p(1-p)]^{n/2}/(2p-1).$$

And from this we see immediately that for $p > 2/3$,

$$(4) \qquad P_e < 3[4p(1-p)]^{n/2} \qquad \text{and} \qquad P_c \geq 1 - 3[4p(1-p)]^{n/2}.$$

So we have shown that using the compound extract method with n phases gives us a compound extract with an error rate decreasing exponentially in n, at the cost of requiring n-fold more time and n-fold greater parallelism. Table 1 shows values for P_e given typical values of n and p, as estimated numerically from the tail distribution and from the analytical bound (4).

Our analysis thus far shows the error rate achievable for a given number of phases, n, of the compound extract. However, for a given target error rate, δ, we would like to find the minimum n that gives us this error rate. Introducing $\epsilon = 1-p$, we can write (4) as

$$P_e \leq 3[4p\epsilon]^{n/2}.$$

Using $n > 2\log_{4p\epsilon}\delta/3$ ensures that $P_e \leq \delta$. When p is close to unity, our choice for n is roughly $2\log_{4\epsilon}\delta/3 = O(\log_\epsilon \delta)$. This is the result Karp presented; our derivation shows, in addition, that the constant factors inherent in the big-O notation are small.

6. Probability of survival

The algorithm is simply to apply one cycle of PCR to all the tubes each time the number of DNA strands falls by half. We assume that exactly one duplicate copy of each strand of DNA is produced. Since we are assuming that strand loss occurs at a constant rate, we need to apply PCR once every k steps for some constant k. If k is not an integer, then the PCR should be applied enough times to keep the volume of DNA constant on the average over the course of the computation. For example, if $k = 4.2$, PCR should be applied after the 4th, 8th, 12th, 16th and 21st steps. For our analysis, however, we will simply take the floor function of k – equivalent to decreasing the stepwise survival probability q. The true error rate will be better than our estimate. Therefore from now on, we will assume that k is an integer.

We will follow a single DNA strand through the computation and use a branching process to analyze P_s, the probability that at least one copy of that strand survives the computation. For more on branching processes, see Feller Chapter 12.3-4. Briefly, a branching process describes particles which are able to reproduce (for example, atoms in a nuclear reaction or, more to the point, DNA strands in a DNA computation). For our algorithm we take a generation to be the number of extract steps required to make the number of strands drop in half, which means that at the end of each generation we do a PCR step. An algorithm with E layers has $\lfloor E/k \rfloor$ full generations.

Let Z_m be the number of strands in the mth generation. Then starting with a single strand as generation zero, let $P_{m,i}$ be the probability that $Z_m = i$. $P_{m,i}$ characterizes the loss of strands and the action of PCR. For example, if PCR perfectly duplicates each strand exactly once, $P_{1,0} = 0.5, P_{1,2} = 0.5$ and $P_{1,i} = 0$ for $i \neq 0, 2$. That is, each strand has probability 0.5 of being lost during any one generation (in which case it produces 0 strands in the next generation, i.e. $P_{1,0} = 0.5$) and it has probability 0.5 of surviving (in which case it is duplicated and produces 2 strands in the next generation, i.e. $P_{1,2} = 0.5$).

We now introduce the notion of a generating function, which is simply a way of expressing a discrete probability distribution in algebraic form. The generating function, $G_m(s)$, of a probabilty distribution is defined to be

$$G_m(s) = \sum_{i \geq 0} P_{m,i} s^i$$

where the dummy variable, s, is simply a place holder without any meaning of its own.

Suppose there were t strands on the first generation. Then the number of strands in the mth generation can be expressed as the sum of t random variables, $X_1, \cdots, X_t$, each representing the number of descendents produced by a particular strand in the first generation. By assumption, X_i all have the same probability distribution and therefore the same generating function. This generating function is just $G_{m-1}(s)$ because X_i are the result of similar branching processes that started at generation one instead of generation zero. Now the probability distribution for the mth generation is given by

$$Prob(Z_m = j) = \sum_{t \geq 0} Prob(Z_1 = t) Prob(X_1 + ... + X_t = j | Z_1 = t). \tag{5}$$

For a fixed t, the distribution of $Z_m = X_1 + \cdots + X_t$ is the convolution of the distributions for $X_1, \ldots, X_t$. Since all X_i have the same distribution $G_{m-1}(s)$, the generating function of $Z_m = X_1 + \cdots + X_t$ is $[G_{m-1}(s)]^t$. We can see this by observing that when we multiply two generating functions, $A(s) = \sum a_i s^i$ and $B(s) = \sum b_i s^i$, term-by-term and collect terms with equal powers of s, the coefficient of s^r is $a_0 b_r + a_1 b_{r-1} + ... + a_r b_0$, which is exactly the rth term in the convolution of $A(s)$ and $B(s)$. Therefore $G_m(s)$ is given by

$$G_m(s) = \sum_{j\geq 0} Prob(Z_m = j)s^j$$
$$= \sum_{j\geq 0}\sum_{t\geq 0} Prob(Z_1 = t)Prob(X_1 + ... + X_t = j|Z_1 = t)s^j \tag{6}$$
$$= \sum_{t\geq 0} P_{1,t} \sum_{j\geq 0} Prob(X_1 + ... + X_t = j|Z_1 = t)s^j$$
$$= \sum_{t\geq 0} P_{1,t}[G_{m-1}(s)]^t.$$

But the right hand side of this equation is just the generating function for G_1 substituting $G_{m-1}(s)$ for s. Therefore

$$G_m(s) = G_1(G_{m-1}(s)). \tag{7}$$

Recall that $G_m(0) = P_{m,0} = Prob(Z_m = 0)$ is the probability of the strand being extinct in the mth generation. By (7) we have

$$P_{m,0} = G_1(P_{m-1,0}).$$

Given a particular distribution $G_1(s)$, this polynomial recurrance can be solved to find how quickly (or slowly) $P_{m,0}$ approaches 1.

Here we find bounds for the example discussed above, where PCR is assumed to perfectly duplicate each surviving strand: $P_{1,0} = P_{1,2} = 0.5$, and all other values are zero. Therefore,

$$G_1(s) = \sum_{i\geq 0} P_{1,i}s^i = 0.5 + 0.5s^2$$

and the recurrence is

$$P_{m,0} = G_1(P_{m-1,0}) = 0.5 + 0.5P_{m-1,0}^2.$$

If we take Q_m to be the probability of a strand surviving (i.e. $Q_m = 1 - P_{m,0}$) we get

$$Q_m = 0.5 - 0.5(1 - Q_{m-1})^2 = Q_{m-1} - 0.5Q_{m-1}^2. \tag{8}$$

Now all that remains is to find a lower bound on the recurrence (8). We will show by induction that $Q_m > 1/m$ for $m \geq 4$. For the base case, we compute the first few values of Q_m and $1/m$ by hand: $Q_1 = .5$ and $Q_4 > 1/4$.

For the inductive step, suppose $Q_{m-1} > 1/(m-1)$. Then

$$Q_{m-1} - 0.5Q_{m-1}^2 > 1/(m-1) - 0.5[1/(m-1)]^2$$

if the function $f(Q) = Q - 0.5Q^2$ is strictly increasing. Taking the derivative of $f(x)$, we have $f'(Q) = 1 - Q$ which is positive if $Q < 1$. Since the Q_m's are all probabilities, they are all strictly less than 1, so f is strictly increasing on this interval. This gives us

$$\begin{aligned} Q_m &= Q_{m-1} - 0.5Q_{m-1}^2 \\ &> 1/(m-1) - 0.5[1/(m-1)]^2 \\ &> (m^2 - 2m + 1)/[m(m-1)^2] \qquad \text{for } m > 2 \\ &= 1/m \end{aligned} \tag{9}$$

which proves that $Q_m > 1/m$ for $m \geq 4$. Similar arguments show that $\frac{1}{m+1} \leq Q_m \leq \frac{2}{m+2}$ for all m; thus our bound is reasonably tight. For convenience, we use $Q_m > \frac{1}{m-1}$ for $m \geq 7$. By our assumption that one generation consists of k extract steps (the final generation may have fewer), we have $m = \lceil nE/k \rceil \leq nE/k + 1$ for the entire computation, so the probability of a strand surviving to the end of the computation with at least one copy remaining is (for $m \geq 7$)

$$P_s > \frac{1}{m-1} > \frac{k}{nE}. \tag{10}$$

This suggests that, if we were to start with *many* copies of each sequence, the probability that *at least one* of the original strands will have descendents in the final tubes will be reasonably high. We show this in the next section.

7. Combining survival and correctness probabilities

To complete the computation, we select a small number of strands from the final yes-tube at random, sequence them and verify them for correctness. Our goal is to guarantee with high probability that there is at least one good strand in the final yes-tube and that the ratio of good strands to bad strands in the yes-tube is high. For concreteness, we show that we can guarantee a ratio of at least 1:1, in which case we should only need to sequence a small number of strands to find a solution. Our terms of merit, therefore, are $P_{correct}$, the probability that a strand that survives the entire computation was processed correctly by each compound extract, $P_{survival}$, the probability that at least one copy of a given sequence survives the entire computation, and $\langle N_{bad} \rangle$, the expected number of bad strands in the final yes-tube.

Unfortunately, the expected number of bad strands in the yes-tube is highly dependent on the particular algorithm. For example, for arbitrary Boolean formulas, it is possible that a single extract error can make a bad strand end up in the final yes-tube. On the other hand, for a practical problem like breaking DES, many extract errors typically would have to be made before a bad strand will arrive in the final yes-tube. To resolve this issue, we will follow Boneh et al **[25]** in requiring that each DNA algorithm A come with information $|M_i|$, where M_i is the set of input sequences that could end up in the yes-tube if it were to suffer i steps of incorrect processing (in the worst-case). By definition, $|M_0|$ is the total number of good sequences and we will assume the worst case that $|M_0| = 1$. Note that a single strand can occur in many M_i.

Our criterion can now be stated as follows: given an algorithm A (and hence E and $|M_i|$) and error rates p and q (and hence k) for correctness and survival, we would like to choose n, the number of phases in the compound extract, and R, the number of redundant copies of each sequence in the original input tube, such that:

1. Each strand is probably correctly processed: $P_{correct} > 0.95$.
2. Each original sequence probably has at least one representative in the final tubes: $P_{survival} > 0.95$.
3. There is probably no more than one bad strand in the yes-tube: $Prob(N_{bad} \leq 1) > 0.95$.

We can now find conditions on n and R that guarantee that each clause be satisfied. First, we note that

$$P_{correct} = (P_c)^E = (1 - P_e)^E > 0.95 \Rightarrow P_e < 1 - 0.95^{1/E}.$$

Thus, the bound (4) on P_e guarantees that #1 is satisfied if

$$n > 2\log_{(4p\epsilon)}[(1 - 0.95^{1/E})/3].$$

The threshold value of n depends on p and E but not on q, R, or $|M_i|$.

Next, we use $(1 - \frac{1}{x})^x < \frac{1}{e}$ for $x > 0$ to obtain from (10) that,

$$P_{survival} = 1 - (1 - P_s)^R > 1 - e^{-P_s R} > 1 - e^{-Rk/En}.$$

Thus, #2 is satisfied if

$$R \geq \frac{3En}{k}.$$

The threshold value of R depends on q and E but not on p or $|M_i|$, and it is linear in n. Criteria #1 and #2 delineate a region of possible values with a corner at n_{min}, R_{min}.

Finally, note that if $N_{bad} \geq 2$ at least $1/20^{th}$ of the time, then the expected value $\langle N_{bad} \rangle \geq 1/10$. Thus, $\langle N_{bad} \rangle < 1/10$ would imply that #3 is satisfied. Furthermore, we can get an upper bound on $\langle N_{bad} \rangle$ by presuming that *any* i errors routes the strands in M_i into the yes-tube.

$$\begin{aligned} \langle N_{bad} \rangle &\leq b(1; E, P_e)|M_1|R + b(2; E, P_e)|M_2|R + \cdots + b(E; E, P_e)|M_E|R \\ &< \sum_{i \geq 1} \binom{E}{i} P_e^i |M_i| R \doteq \gamma(n, R). \end{aligned} \tag{11}$$

Thus, #3 is satisfied if $\gamma(n, R) < 1/10$; this final constraint imposes an additional boundary in the space of possible n and R. The shape of this boundary depends the values of $|M_i|$, which can take on almost arbitrary values, but nonetheless certain properties can be predicted. Suppose $\gamma(n_{min}, R_{min}) > 1/10$. Note that simply increasing n decreases all P_e^i exponentially, while the concomitant increase in R is only additive. Therefore, the desired values for n and R are easily found.

Consider the worst case for a b-bit satisfiability problem, where all $2^b - 1$ bad strands are in M_1. Equation (11) reduces to

$$\langle N_{bad} \rangle \leq E P_e M_1 R \leq 9E^2 [4\epsilon]^{n/2} 2^b n / k$$

and therefore, $n = -2\log_{4\epsilon}(9E^2 2^b / k / 1000)$ satisfies #3 so long as $n < 100$. In particular, for the 50-variable, 200 clause 3-SAT problem discussed earlier, where $p = 0.99$, no strand loss, and no error correction resulted in $P_{correct} = 0.002$, we can now state that even with $q = 0.90$ (thus $k = 6$), $n = 31$ results in $P_e = 2.6 \times 10^{-24}$ (satisfying #1 with overkill); $R = 9300$ ensures #2 is satisfied; and altogether, $\langle N_{bad} \rangle \leq E P_e |M_1| R = 0.016 < 1/10$, satisfying #3. If each strand is 2000 bases long, then the total population of $2^b R$ strands weighs a substantial but not outrageous 12 grams.

In contrast, consider the plaintext-ciphertext attack on DES considered by Boneh et al [**3**] and Adleman et al [**28**], for which $b = 56$. Surprisingly, for reasonable assumptions the average number of bad strands in the yes-tube is simply $\langle N_{bad} \rangle = R/256$, independent of the error rate [**28**]. For $E = 6719$, $p = 0.99$, and $q = 0.99$ (thus $k = 69$), we can satisfy #1 using $n = 7$, for which $P_e = 3.4 \times 10^{-7}$ and $P_{correct} > 0.997$; $R = 2048$ satisfies #2 with $P_{survival} > 0.95$; and altogether, $\langle N_{bad} \rangle \approx R/256 = 8$, so sequencing 26 strands from the yes-tube should ensure that the good strand is sequenced and identified. If each strand is 10,000 bases long, then the total population of $2^b R$ strands weighs 810 grams – substantially better than the 23 Earth masses estimated in [**28**] for $pq = 0.99$.

8. Discussion

The essential point of this investigation is that strand loss and misclassification errors do not present insurmountable obstacles to DNA-based computation. Even in the presence of what would be unthinkable error rates for electronic computers – say 1% per-step error rates – DNA computation can succeed, at the cost of only a modest increase in the number of extract steps and the volume of DNA being handled. 1% error rates are within reach for current biotechnology. Of course, the best solution, if it can be done without affecting the time or cost of each step, would be to improve the fundamental biotechnology of the extract and combine steps directly. Even with such advances, however, the overall error rates required for problems like SAT are unlikely to be achieved without error-correcting algorithms such as the ones discussed here.

Several of the assumptions we used in our derivations, chosen to simplify the mathematics, can be relaxed. One is our assumption that PCR is error-free. It is clear that PCR can make errors by simply copying a strand incorrectly or by failing to make a duplicate copy of a particular strand at all. To account for this, our branching process analysis can be extended to any distribution of values for the P_i's. We can also adapt our algorithm to account for strand loss from PCR by integrating this loss rate into the probability of loss due to extract steps.

The assumption that the combine step is error free can be weakened to the assumption that each combine step has a constant probability of strand loss. Then, just as we can add the strand loss rates of PCR and the extract step, we can also add to their sum the error rate from the combine steps. The assumption that the rate of false-negatives and false-positives are equal is also not necessary and the exact same analysis will hold for this case; Karp et al [**14**] and Adleman [**29**] have shown a simpler variety of compound extract whose false-postive and false-negative rates are both roughly equal to the minimum of the original rates.

However, some non-idealities are not so easy to rectify using our approach. A good example would be *systematic bias* due to sequence-dependent biophysical or biochemical factors – for example, certain DNA sequences may extract with unusually low or high rates, may be amplified particularly reliably or unreliably, or may be lost due to hydrolysis at a faster or slower rate. It is a significant open question whether such systematic bias can be corrected for or avoided, for example, by appropriate strand design and experimental protocols.

In conclusion we have shown that it is theoretically possible to reduce errors due to false negative, false positives, and lost strands to tolerable levels in any extract-based DNA computation with only small extra time and space factors. Although these techniques should be useful immediately for larger problems than have been demonstrated to date, it is clear that more significant applications of DNA computing will require substantial improvements in the underlying biotechnology.

References

[1] Leonard M. Adleman. Molecular computation of solutions to combinatorial problems. *Science*, 266:1021–1024, November 11, 1994.

[2] Richard J. Lipton. DNA solutions of hard computational problems. *Science*, 268:542–544, 1995.

[3] Dan Boneh, Christopher Dunworth, and Richard J. Lipton. Breaking DES using a molecular computer. In Lipton and Baum [**30**], pages 37–65.

Notation used in this paper.

symbol	interpretation
q	$Prob$(strand survives an extract step)
$\epsilon = 1 - p$	$Prob$(false negative or false positive | strand survives the step)
P_s	$Prob$(strand survives entire computation)
$P_{survival}$	$Prob$(at least one of R copies survives the computation)
P_c	$Prob$(strand is correctly processed by a compound extract)
P_e	$= 1 - P_c$
$P_{correct}$	$Prob$(strand is correct after the computation | strand survives)
$k = \log_q \frac{1}{2}$	the number of extract steps per duplicate step
E	the number of layers in the original algorithm
R	the number of copies of each strand initially (the redundancy)
n	the number of phases (layers) in a compound extract
δ	target error rate per step
m	the number of generations of "lose-half-then-duplicate"
Z_m	the number of strands in generation m
$P_{m,i}$	$Prob$(generation m has exactly i strands)
$G_m(s)$	$= \sum_{j \geq 0} P_{m,j} s^j$, the generating function for $P_{m,i}$
$Q_m = 1 - P_{m,0}$	$Prob$(at least one strand survives to generation m)
b	for SAT problems, the number of binary variables in the input
M_i	set of inputs that can reach the yes-tube only after i errors
N_{bad}	the number of bad strands in the final yes-tube

TABLE 2

[4] Paul W. K. Rothemund. A DNA and restriction enzyme implementation of Turing Machines. In Lipton and Baum [**30**], pages 75–119.

[5] Warren D. Smith. DNA computers in vitro and in vivo. In Lipton and Baum [**30**], pages 121–185.

[6] Donald Beaver. A universal molecular computer. In Lipton and Baum [**30**], pages 29–36.

[7] Erik Winfree. On the computational power of DNA annealing and ligation. In Lipton and Baum [**30**], pages 199–221.

[8] Erik Winfree, Xiaoping Yang, and Nadrian C. Seeman. Universal computation via self-assembly of DNA: Some theory and experiments. In Landweber and Baum [**31**].

[9] Erik Winfree, Furong Liu, Lisa A. Wenzler, and Nadrian C. Seeman. Design and self-assembly of two-dimensional DNA crystals. *Nature*, 394:539–544, 1998.

[10] Qi Ouyang, Peter Kaplan, Shumao Liu, and Albert Libchaber. DNA solution of the maximal clique problem. *Science*, 278:446–449, 1997.

[11] Qinghua Liu, Anthony G. Frutos, Liman Wang, Anne E. Condon, Robert M. Corn, and Lloyd M. Smith. DNA computating on surfaces. In Winfree and Gifford [**32**].

[12] Hiroshi Yoshida and Akira Suyama. Solution to 3-SAT by breadth first search. In Winfree and Gifford [**32**].

[13] Dirk Faulhammer, Anthony R. Cukras, Richard J. Lipton, and Laura F. Landweber. When the knight falls: On constructing an RNA computer. In Winfree and Gifford [**32**].

[14] Richard M. Karp, Claire Kenyon, and Orli Waarts. Error-resilient DNA computation. In *Proceedings of the Seventh Annual ACM-SIAM Symposium on Discrete Algorithms*, pages 458–467, Providence, RI, January28–30 1996. AMS/SIAM.

[15] Sam Roweis and Erik Winfree. On the reduction of errors in DNA computation. *Journal of Computational Biology*, 6(1):65–75, 1999.

[16] Eric B. Baum. DNA sequences useful for computation. In Landweber and Baum [**31**], pages 235–241.

[17] Kalim U. Mir. A restricted genetic alphabet for DNA computing. In Landweber and Baum [**31**], pages 243–246.

[18] R. Deaton, R. C. Murphy, M. Garzon, D. R. Franceschetti, and S. E. Stevens, Jr. Good encodings for DNA-based solutions to combinatorial problems. In Landweber and Baum [**31**], pages 247–258.

[19] R. Deaton, M. Garzon, R. C. Murphy, D. R. Franceschetti, and S. E. Stevens, Jr. Reliability and efficiency of a DNA-based computation. *Physical Review Letters*, 80(2):417–420, 1998.

[20] Takashi Ito, Cassandra L. Smith, and Charles R. Cantor. Sequence-specific DNA purification by triplex affinity capture. *Proc. Natl. Acad. Sci. USA*, 89:495–498, January 1992.

[21] H. Orum, P. E. Nielsen, M. Jorgensen, C. Larsson, C. Stanley, and T. Koch. Sequence-specific purification of nucleic acids by PNA-controlled hybrid selection. *BioTechniques*, 19:472–480, 1995.

[22] Julia Khodor and David K. Gifford. The efficiency of sequence-specific separation of DNA mixtures for biological computing. In Rubin and Wood [**33**], pages 39–46.

[23] Junghuei Chen and David Harlan Wood. A new DNA separation technique with a low error rate. In Rubin and Wood [**33**], pages 47–56.

[24] Sam Roweis, Erik Winfree, Richard Burgoyne, Nickolas V. Chelyapov, Myron F. Goodman, Paul W. K. Rothemund, and Leonard M. Adleman. A sticker based architecture for DNA computation. In Landweber and Baum [**31**], pages 1–29.

[25] Dan Boneh, Christopher Dunworth, Richard J. Lipton, and Jiří Sgall. Making DNA computers error resistant. In Landweber and Baum [**31**], pages 163–170.

[26] Martyn Amos, Alan Gibbons, and David Hodgson. Error-resistant implementation of DNA computations. In Landweber and Baum [**31**], pages 151–161.

[27] William Feller. *An Introduction to Probability Theory and Its Applications*, volume 1. John Wiley & Sons, 3 edition, 1968.

[28] Leonard M. Adleman, Paul W. K. Rothemund, Sam Roweis, and Erik Winfree. On applying molecular computation to the data encryption standard. *Journal of Computational Biology*, 6(1):53–63, 1999.

[29] Leonard M. Adleman. On constructing a molecular computer. In Lipton and Baum [**30**], pages 1–21.

[30] Richard J. Lipton and Eric B. Baum, editors. *DNA Based Computers: DIMACS workshop, April 4, 1995*, volume 27, Providence, RI, 1996. American Mathematical Society.

[31] Laura F. Landweber and Eric B. Baum, editors. *DNA Based Computers II: DIMACS Workshop, June 10-12, 1996*, volume 44, Providence, RI, 1998. American Mathematical Society.

[32] Erik Winfree and David K. Gifford, editors. *Proceedings of the* 5th *DIMACS Meeting on DNA Based Computers, held at MIT , June 14-15, 1999*, preliminary, 1999.

[33] Harvey Rubin and David Harlan Wood, editors. *DNA Based Computers III: DIMACS Workshop, June 23-25, 1997*, volume 48 of *DIMACS: Series in Discrete Mathematics and Theoretical Computer Science.*, Providence, RI, 1999. American Mathematical Society.

(K. Chen) Department of Computer Science, Princeton University, Princeton, NJ 08544,

E-mail address: kcchen@phoenix.princeton.edu

(E. Winfree) Department of Molecular Biology, Princeton University, Princeton, NJ 08544, and, Departments of Computer Science and Computation and Neural Systems, California Institute of Technology, Pasadena, CA 91125

E-mail address: winfree@hope.caltech.edu

DIMACS Series in Discrete Mathematics
and Theoretical Computer Science
Volume **54**, 2000

DNA Analog Vector Algebra and Physical Constraints on Large-Scale DNA-Based Neural Network Computation

A. P. Mills Jr., B. Yurke, and P. M. Platzman

ABSTRACT. An analog neural network may be represented by chemical operations performed on strands of DNA. This new type of DNA computing has the possible advantage of being fault tolerant and thus more immune to DNA hybridization errors than a Boolean DNA computer. We describe a particular set of DNA operations to effect the interconversion of electrical and DNA data and to represent the Hopfield associative memory. We consider the constraints imposed by Poisson statistics and diffusion times on the amount of DNA needed and the time required to perform computations using a DNA neural network. For an m-unit Hopfield neural network with $m \times m$ connections, we derive expressions for these quantities as a function of m. A Hopfield network with a memory capacity of 10^{14} bits (comparable to the human brain) would have a cycle time of a few hours and would require about 100 μmoles of DNA per cycle and 8 cycles for convergence.

1. Introduction

Adleman[1] first pointed out in his pioneering paper that Watson-Crick hybridization of pairs of complementary DNA strands makes possible a representation of highly parallel selective operations that could be one key to the utility of molecular computation. However, small departures from the ideal selectivity of DNA hybridization will inevitably lead to significant undesired pairings of strands and thus to difficulties in schemes for implementing large scale Boolean functions using DNA.

One approach to fault-tolerant computation is the use of neural networks, since they do not need the high precision associated with digital computing. Recently Deaton et al.[5] showed that it should be possible to find a large enough set of mutually non-hybridizing DNA strands to in fact allow digital molecular computation of high complexity with tolerable error rates. As an alternative possibility we have suggested[8] that networks in which the usual axons and neurons are replaced by the diffusion and molecular recognition of DNA can make practical use of the massive parallelism associated with the simultaneous reactions of up to 10^{24} molecules[1,7]. In this paper we discuss the physical limitations imposed on a DNA neural network by the reaction kinetics and by the statistics associated with sampling finite volumes. We show that for a network cycle time of order hours, a connectivity

2000 *Mathematics Subject Classification.* Primary 92C40, 92B20, 82C32.

© 2000 Lucent Technologies

approaching the size of the human brain is possible. As an example we present a DNA embodiment of a Hopfield neural network or associative memory[6,10,11].

2. DNA Vectors

The elements of memory in the Hopfield network are represented as m-component vectors $\mathbf{V}=\sum_{i=1}^{m} V_i \mathbf{e}_i$ in a space with basis vectors e_i $(i=1,2,\ldots m)$. The items of experience, a set of vectors $\mathbf{V}^{(a)}$ (with $a = 1, 2, \ldots s$), are stored in memory by summing the outer product matrices of the experience vectors:

(1) $$T_{ij} = \sum_{a=1}^{s} V_i{}^{(a)} V_j{}^{(a)}.$$

The removal of all the diagonal components, so that $T_{ij} = 0$ for $i = j$, is required along with symmetry, $T_{ij} = T_{ji}$, for unconditional stability of the recall algorithm[4]. Recall of a particular experience $V_i^{(b)}$, imperfectly represented as $U_i^{(b)}$, is effected by solving the non-linear equations[6]

(2) $$V_i = S(\sum_{j=1}^{m} T_{ij} V_j),$$

where the function $S(x)$ is a saturating function such as $\lambda \tanh(x)$ acting separately on each component of its vector argument. Typically solutions to Eq (2) are found by iteration starting with $V_j = U_j^{(b)}$ and with the small-signal gain λ being adjusted to facilitate convergence. If the $V_i^{(a)}$ are sufficiently different, i.e. are part of a nearly orthogonal set, the system will settle into a state closely resembling $V_i^{(b)}$. This necessitates having some negative amplitudes in the vector components and thus in the weight matrix T_{ij}. Hopfield found that the number of memories that can be stored simultaneously using the learning algorithm of Eq (1) is roughly 10% of the number of independent basis vectors, i.e. $s \approx \frac{m}{10}$, and that only a few iterations are required for convergence.

In order to implement such a neural network in a DNA language we need a representation of Oliver's analog matrix algebra[9] that includes negative amplitudes. We envision working with a set of single-stranded DNA oligomers of the form

(3) $$E_i = 5'\mathtt{R}_1\mathtt{R}_2 \cdots \mathtt{R}_r\mathtt{N}_1^i\mathtt{N}_2^i\mathtt{N}_3^i \cdots \mathtt{N}_q^i\mathtt{S}_1\mathtt{S}_2 \cdots \mathtt{S}_r 3',$$

where there will be one oligomer E_i associated with each basis vector $\mathbf{e}_i$, $i = 1, 2, \ldots, m$ in our m-dimensional vector space.

In Eq (3) the central q-mer is used to identify the basis vector, while two invariant r-mers $5'\mathtt{R}_1\mathtt{R}_2 \cdots \mathtt{R}_r 3'$ and $5'\mathtt{S}_1\mathtt{S}_2 \cdots \mathtt{S}_r 3'$ with $r \approx 10$ are placed at the ends to assist in carrying out various biochemical operations such as hybridization, ligation and cutting via a restriction enzyme. Since the j-th nucleotide of the sequence of $q \approx 20$ nucleotides $\mathtt{N}_j^i$ is chosen from the four bases A, G, C, T, the number of different sequences is 4^q. If $4^q \gg m$, we can select a subset of q-mers which will give a set of basis strands that are non-palindromic, relatively free of secondary structure and have minimal overlap with the other basis vectors[5]. Any vector $\mathbf{V} = \sum_{i=1}^{m} V_i \mathbf{e}_i$ in our neural net is then represented by a DNA sample containing E_i strands with concentrations $[E_i]$ proportional to the amplitudes V_i. The analog accuracy will be limited by Poisson fluctuations in the numbers of molecules in a finite sample volume. Suppose we have 1 pmole of DNA oligomers $(6 \times 10^{11}$ molecules) drawn randomly from a much larger set of strands representing a typical vector of dimension $m = 10^8$. The sampled strands will represent the vector with roughly 1% errors in the individual amplitudes.

Since concentrations are always positive, we also need an appropriate representation for negative amplitudes. We choose to represent negative amplitudes in the direction of a unit vector $\mathbf{e}_i$ by a concentration of oligomers having a sequence of bases $\overline{E}_i$ complementary to E_i. As a result, when two vectors are added by combining the two DNA samples representing them into one container, any positive and negative contributions to the amplitude for a given unit vector will hybridize and can be removed by column separation or by digestion using a suitable enzyme.

3. DNA Input and Output

The first thing one needs to make a practical DNA computer is a means for interconverting DNA strands and digital data. The standard DNA hybridization chip[12,13] presents a simple solution. The DNA chip is currently available as an array of up to 10^6 pixels as small as $4\mu m$ on a side, each of which is coated with a unique type of single stranded DNA at a density of about 10^4 molecules per $(\mu m)^2$. For our purposes we require a chip with $2m$ pixels carrying the complete set of E_i's and $\overline{E}_i$'s. The conversion of digital data to DNA strands is accomplished by starting with our DNA hybridization chip completely saturated with hybridized complementary strands: each tethered DNA strand is bound to its complementary strand to form a Watson-Crick double helix. We now flash a digital image corresponding to the a-th data vector $\mathbf{V}^{(a)}$ on the chip to cause local heating of the double stranded DNA. A number n_i of single stranded E_i or $\overline{E}_i$ molecules is released that depends on the temperature dependent rate of dissociation of the double stranded DNA at the i-th pixel. By suitable calibration of the image intensity scale, one or more bits of amplitude information could be encoded into the concentrations of each of the E_i's and $\overline{E}_i$'s. The chip would be washed following exposure to the image to collect the DNA representation of the data vector amplitudes $V_i^{(a)}$.

Suppose that the DNA chip consists of a Si wafer covered with a layer of porous silica. On the silica surface are amorphous Si pads with radius r and separated center to center from each other by a distance $3r$. The amorphous Si pads are good absorbers for Ar laser light and may thus be selected individually using a focused laser beam or in parallel using the projected image of a mask. The pads are covered with a thin silica layer upon which are fixed the DNA strands. The thermal conductivity of the porous silica is roughly the same as that of the water environment (6 mW $\mathrm{cm}^{-1}\mathrm{K}^{-1}$) so that the pads are effectively in a uniformly thermally conducting medium. Heating a single pad by δT above the ambient temperature will raise the temperature of the neighbors by roughly $\delta T/9$, a difference sufficient to select the release of DNA from only the desired pad. For parallel addressing, the thermal isolation between pads can be increased at the expense of needing more heating power by making the porous silica layer thickness comparable to r.

4. Addition of Vectors

The steps in the addition of two vectors are as follows.

(I) Combine in one container equal quantities from the two collections of DNA representing the two vectors at twice the standard concentration. Positive and negative contributions to the concentration of oligomers corresponding to any given basis vector will hybridize to yield double-stranded DNA with blunt ends. After the reaction is complete, the surviving single-stranded DNA for each basis vector will be virtually all of the positive or negative type of strands.

(II) Digest the double-stranded DNA, using an appropriate enzyme such as Exonuclease III (E. coli). Following the Exonuclease reaction we will be left with a container of single-stranded oligonucleotides representing the sum of the two vectors plus unwanted single-stranded DNA fragments that are significantly shorter than the strands representing a vector.

(III) Remove the unwanted fragments by a high-performance liquid chromatography (HPLC) purification step. One can take the sum of any number of vectors simultaneously in the same manner. The individual vectors may be multiplied each by a different scalar by adjusting the concentrations.

5. Outer Product of Two Vectors

An outer product matrix $V_i^{(a)}V_j^{(b)}$ is formed by joining the single-stranded DNA corresponding to $V_i^{(a)}$ at their 3' ends to the 5' termini of the $V_i^{(b)}$. The strands join randomly so that the concentration of the joined strands is proportional to the product of the concentrations of the individual strands. If $a = b$, the sum of all such strands becomes a contribution $T_{ij}^{(a)}$ to the memory matrix of Eq (2). The strands corresponding to the diagonal components of $T_{ij}^{(a)}$ must be removed, and what remains is a solution of strands of the form E_iE_j, $\overline{E}_iE_j$, etc. with $i \neq j$. The correlations of the strand concentrations give the DNA representation of the memory matrix $T_{ij}^{(a)}$. The specific steps in one realization of this operation are as follows.

(I) The vector strands $V_i^{(a)}$ are divided onto two portions which we call V_i and W_i. To ensure that only the desired type of connection is made, the 5' phosphate residues are removed from the V_i strands using bacterial alkaline phosphatase, while the 5' termini of the W_i are phosphorylated using bacteriophage T4 polynucleotide kinase.

(II) The W_i strands are further modified by ligating to their 3' termini long strands F that do not hybridize significantly to the set of E_i's.

(III) The modified V_i and W_j strands are ligated using the four types of linker strands

$$L1 = 5'\mathtt{S}_1\mathtt{S}_2\cdots\mathtt{S}_r\overline{\mathtt{S}}_r\cdots\overline{\mathtt{S}}_2\overline{\mathtt{S}}_1 3' \tag{4}$$

$$L2 = 5'\overline{\mathtt{R}}_r\cdots\overline{\mathtt{R}}_2\overline{\mathtt{R}}_1\overline{\mathtt{S}}_r\cdots\overline{\mathtt{S}}_2\overline{\mathtt{S}}_1 3' \tag{5}$$

$$L3 = 5'\overline{\mathtt{R}}_r\cdots\overline{\mathtt{R}}_2\overline{\mathtt{R}}_1\mathtt{R}_1\mathtt{R}_2\cdots\mathtt{R}_r 3' \tag{6}$$

$$L4 = 5'\mathtt{S}_1\mathtt{S}_2\cdots\mathtt{S}_r\mathtt{R}_1\mathtt{R}_2\cdots\mathtt{R}_r 3' \tag{7}$$

to obtain strands of the form E_iE_jF, $\overline{E}_iE_jF$ and so forth. The number of ij strands is proportional to the product of the concentrations of the V_i and W_j strands and hence to the desired outer product. This is approximately true even if the reaction is allowed to run to completion since there are so many different reactions competing for the same strands. The diagonal components of T_{ij} may be removed if desired by allowing the strands representing T_{ij} to hybridize with an equal total concentration of strands representing a unit tensor δ_{ij}, i.e. strands of the form E_iE_i and $\overline{E}_i\overline{E}_i$. The T_{ii} strands of the form $E_i\overline{E}_i$ and $\overline{E}_iE_i$ will have already formed hairpins and may be removed by a first purification stage using electrophoresis or chromatography at room temperature. The desired T_{ij} strands lacking any diagonal components may then be extracted by a second purification stage performed at a temperature such that single duplex pairs of E's melt but double length hybridized segments corresponding to $T_{ii}\delta_{ii}$ do not.

6. Product of a Matrix and a Vector

To implement the Hopfield neural network we need to find the matrix inner product $\sum_{j=1}^{m} T_{ij}X_j$, where T_{ij} is the total memory matrix represented by strands of the form E_iE_jF, $E_i\overline{E}_jF$, etc. and X_i is an imperfect query vector $U_i^{(b)}$ represented by strands of the form E_i and $\overline{E}_i$. We allow the vector strands to hybridize randomly with complementary 3' ends of the matrix strands. The hybrids are cleaved with a restriction enzyme that recognizes the central portion of the DNA sequence $5'\mathtt{S}_1\mathtt{S}_2\cdots\mathtt{S}_r\mathtt{R}_1\mathtt{R}_2\cdots\mathtt{R}_r3'$. The single-stranded regions of the form E_i and $\overline{E}_i$ from the 5' ends are collected and their concentrations represent the vector $\sum_{j=1}^{s} T_{ij}U_i^b$. The specific steps are as follows.

(I) Obtain a sample of the complement to X_j, $\overline{X}_j$.

(II) To both the vector strands and their complements we add via ligation to their 5' ends strands G twice as long as the 3' termination string F on the matrix strands and having their 3' halves complementary to F.

(III) One of the resulting two strand collections, GX is incubated with a sample of the matrix stands, the set of r-mers

$$P_1 = 5'\overline{\mathtt{S}}_r\cdots\overline{\mathtt{S}}_2\overline{\mathtt{S}}_1 3' \tag{8}$$

$$P_2 = 5'\mathtt{R}_1\mathtt{R}_2\cdots\mathtt{R}_r 3' \tag{9}$$

and ligase. The ligase is inactivated, and the double strands are cut using a restriction enzyme that cuts in the center of the sequence

$$5'\overline{\mathtt{R}}_3\overline{\mathtt{R}}_2\overline{\mathtt{R}}_1\mathtt{R}_1\mathtt{R}_2\mathtt{R}_3 3' = 5'\mathtt{S}_{r-2}\mathtt{S}_{r-1}\mathtt{S}_r\overline{\mathtt{S}}_r\overline{\mathtt{S}}_{r-1}\overline{\mathtt{S}}_{r-2}3'. \tag{10}$$

One possibility would be to use the enzyme Bst1107 I which cuts in the center of the sequence 5'GTATAC. Since the recognition site is only present between the E_i and E_j, etc., the short segments when purified will be a collection of strands representing an unchanged sign contribution to the product.

(IV) The other of the two strand collections, $G\overline{X}$, is incubated with a second sample of the matrix stands, T4 polymerase and a suitable buffer solution. Incubation results in the extension of the $\overline{X}$ strands at their 3' ends. The strands are melted and the long segments of the form E_iXF are separated. These strands are then hybridized with the set of linker strands (4)-(7) and cut using the restriction enzyme. The short strands are separated and represent a changed sign contribution to the inner product. They are then added using the addition routine to the similar strands obtained in step (III) above to give a representation of desired matrix inner product. Note that the presence of the long sequence F allows the hybridizations in this scheme to be done at a temperature such that unwanted hybridizations of matrix strands with each other and vector strands with the improper halves of matrix strands will not be stable.

7. Saturating Function

To implement the saturating function $S(x)$ the strands representing the vector x are captured on a sub-stoichiometric complete set of tethered E's and E's, for example using a hybridization chip. After hybridization with the tethered strands, the array is washed to remove the excess strands, which are discarded. The collection of DNA strands representing the saturated function is obtained by melting the hybridized strands from the array and collecting the strands released. If we have chosen $x = \sum_{j=1}^{m} T_{ij} U_i^{(b)}$, the resulting solution is the DNA representation of the first iteration of Eq (2), which is presumably approaching one of the initial memorized vectors $\mathbf{V}^{(\mathrm{b})}$.

We have now described in outline form all the steps necessary for implementing the first iteration of Eq (2) using a DNA representation. If we choose to stop after one iteration, we can now take the solution representing the vector $S(\sum_{j=1}^{m} T_{ij} U_i^{(b)})$ and read it. The reading step involves exposing the DNA of our liquid sample to a single stranded DNA hybridization chip containing the complete set of 2m E_i's and $\overline{E}_i$'s. The amount of the different types of DNA captured at each location of the chip is proportional to their concentrations in solution for a short enough exposure and is determined by the position-dependent fluorescence of a molecule such as ethidium bromide that binds selectively to double-stranded DNA. We have now in principle completed one full computation cycle. It is important to note that in a real application, a simple Hopfield associative memory would not be very useful for a task such as image recognition because it would not have the ability to recognize different views of the same subject. A classifying form of network would be more suitable for such a task[14].

8. Cycle Time

Although the network involves many operations acting in parallel, the times for a complete network cycle would not be particularly fast. The various chemical operations such as separation via gel electrophoresis, ligation, and cutting via a restriction enzyme would presently require times t_{min} of the order of an hour for completion. It seems reasonable that t_{min} might be reduced to less than 10^3 sec after some development. To this minimum time must be added the query time of the neural network algorithm determined by the matrix multiplication step, namely by the rate of forming hybridized (double-stranded) DNA from the single stranded DNA representing the T's and V's of Eqs (1) and (2). It is important to note that the time for rehybridization[2] of a collection of single stranded DNA segments is proportional to the total number of different types of strands and will limit the size of any DNA computation based on hybridization. If we start at time t=0 with unhybridized (single-stranded) E's and $\overline{E}$'s each having a concentration C, the fraction of strands remaining single after a time t is[3]

$$f_s(t) = (1 + t/\tau)^{-1}, \tag{11}$$

where $\tau^{-1} = Ck_2$, and where k_2 is the second-order rate constant. The rate of reassociation of DNA is optimal for sodium ion concentrations of order 1 M (mole/liter) and temperatures about 25 °C below the melting point of the DNA. The rate constant is then[2] $k_2 \approx 7.0 \times 10^6$ liters per mole per second, and is relatively insensitive to the strand length and composition. The total concentration of DNA bases pairs is $C_{tot} = m \times (q+2r) \times C$ and the maximum practical value of C_{tot} is of order 1 M. Here,

$q + 2r$ is the length of the basis oligomers given in Eq (3). If we require $\tau = t_{min}$, then the number of independent vectors is given by $m = C_{tot}k_2t_{min}/(q + 2r)$. For the case $q + 2r$=40 and $t_{min} = 10^3$ sec, $m = 1.7 \times 10^8$. In this case the number of parallel computation steps involved in performing the multiplication of a matrix times a vector is roughly $m^2 = 3 \times 10^{16}$. Our estimated time for carrying out one iteration of Eq (2) would correspond to an effective rate roughly an order of magnitude faster than a teraflop super computer.

9. Statistical Limitations

We are now in a position to calculate the amounts of DNA needed for faithful retrieval of an image. The memory matrix T_{ij} has m^2 components. Suppose that in the process of accumulating memories, each component of the outer product of a single memory vector is represented by an average of t molecules. If we end up storing s vector memories, the total memory matrix will consist of m^2st molecules. With $m = 10^8$ and $st \approx m$ the total memory matrix will consist of about m^3 molecules or about 1 mole of DNA. Since the maximum practical concentration of bases is about 1 M and the molecules of T_{ij} are about 80-mers, this translates into about 80 liters of solution. However, an accurate solution of Eq (2) requires us to withdraw and consume only a small part of our 80 liter library.

First of all, for maximal data storage the images would have to be optimally compressed to achieve data representations in the form of nearly random sequences that are virtually orthogonal to each other. From now on we suppose the set of memories is a collection of s m-dimensional vectors with nominally duo-binary two-valued components having mean values $\langle V_i \rangle = \pm 1$. The third possible value, namely zero, is only taken by the largely missing amplitudes of a partial representation of a memory vector used to initiate a recall operation. Because we can sample only a finite volume of liquid, the non-zero components inevitably have a Poisson distribution of values with a mean number of molecules in a sample n. The root mean square [rms] standard deviation of the sampled component V_i is thus $\sigma = n^{-1/2}$. Here we represent the mean or ideal value of a variable z by angle brackets, $\langle z \rangle$, and its rms standard deviation by appending it following a $\pm$ sign, as in $z = \langle z \rangle \pm \sigma$. Since the vectors represent optimally compressed data, they will be virtually indistinguishable from a random sequence of $+1$'s and -1's. The inner product of two sampled vectors is therefore

$$\sum_{i=1}^{m} V_i^{(a)} V_i^{(b)} = m\delta_{ab} \pm m^{1/2}(1 + \sigma^2). \tag{12}$$

We are assuming the two vectors are represented by independent samples so that the sampling errors σ are uncorrelated, and have used the fact that the rms standard deviation of a random sequence of $+1$'s and 1's is 1. The outer product matrix of the a-th vector with itself is the a-th contribution to the Hopfield memory matrix, and is an $m \times m$ matrix with entries that are all ± 1. We now suppose that each component of $T_{ij}^{(a)}$ is highly under sampled, so that the mean number of components present in any sample is m^2/g, with $g \gg 1$. An imperfect query vector $U^{(c)}$ has $p < m$ entries that are the same as those of $V^{(c)}$, but with sampling errors σ uncorrelated to those of $V^{(c)}$, and with the rest of the entries of $U^{(c)}$ being zero. If $U^{(c)}$ is in stoichiometric excess, the inner product of the b-th memory matrix with the query vector is

$$\sum_{j=1}^{m} T_{ij}^{(b)} U_j^{(c)} = (p/g)\delta_{bc}\langle V_i^{(c)}\rangle \pm [p/g]^{1/2}[1+\sigma^2]^{1/2}. \tag{13}$$

The total Hopfield memory is the sum of s different memory outer product matrices. The inner product of the total Hopfield memory matrix with a stoichiometric query vector is thus

$$X_i^{(1)} = \sum_{j=1}^{m} T_{ij} U_j^{(c)} = (p/g)\langle V_i^{(c)}\rangle \pm (ps/g)^{1/2}, \tag{14}$$

neglecting σ^2 compared to 1. Each component of $X_i^{(1)}$ is buried in the noise if $p < s$, but since there are $m > s$ components, there is still information contained in the whole vector. The inner product of the memory matrix with $X_i^{(1)}$ is

$$X_i^{(2)} = \sum_{j=1}^{m} T_{ij} X_j^{(1)} = (mp/g^2)\langle V_i^{(c)}\rangle \pm (mp/g^2)^{1/2} s. \tag{15}$$

After y iterations, we have

$$X_i^{(y)} = m^{(y-1)} p g^{-y} \{\langle V_i^{(c)}\rangle \pm p^{-1/2} m^{-1/2(y-1)} (gs)^{y/2}\} \tag{16}$$

For $X_i^{(y)}$ to be a representation of the target vector $V_i^{(c)}$ with errors less than $\epsilon \geq \sigma$, we need $(gs/m)^y (m/p) < \epsilon^2$, and therefore we require a number of iterations $y > \ln(m/p\epsilon^2)/\ln(m/gs)$. For y to be finite, we require $m/gs > 1$. The small signal gain λ is related to y, $\lambda \approx 1+y^{-1}$, since the process of extracting the signal from the noise requires linearity for roughly y iterations. In practice, one would apply the saturating function only once in y iterations. As an example, if $m = 10^7$, $s = 10^5$, $p = 100$, $g = 10$ and $\epsilon = 0.2$, then we need about 7 iterations for convergence. For operation of the Hopfield algorithm, we need samples of the total Hopfield memory matrix T_{ij} containing of order $s/g = 10^4$ molecules per matrix element. If $m = 10^7$, we thus need $m^2 s/g = 10^{18}$ molecules or about 1 μmole of DNA to represent T_{ij} in each step of the recall operation. The amount of DNA required increases rapidly with memory size. A practical upper limit might correspond to $m = 10^9$, $s = 10^5$ and $g = 10^3$. In this case the cycle time would be of order hours, one would need about 100 μmoles of DNA in each step, and the memory capacity would be $ms = 10^{14}$ bits, nearly equal to the capacity of the human brain.

10. Conclusions

We have attempted to show that a DNA neural network could have enticingly large capacities and not unduly long cycle times using reasonable scale chemical operations. A demonstration of a simple Hopfield network in DNA would serve as an indicator for the feasibility of making more versatile networks using DNA matrix algebra.

11. Acknowledgments

The authors are grateful to John Hopfield, David Tank, Andrew Turberfield and Erik Winfree for stimulating discussions.

12. References

1. Adleman, L.M., 1994, Molecular computation of solutions to combinatorial problems, *Science* **266**, 1021-1024.
2. Britten, R.J. and Kohne, D.E., 1968, Repeated sequences in DNA, *Science* **161**, 529-540.
3. Cantor, C.R. and Schimmel, P.R., 1980, *Biophysical Chemistry, Part III: the behavior of biological macromolecules* (Freeman) ISBN 0-7167 1191-5, p. 1226-1234.
4. Cohen, M.A. and Grossberg, S.G., 1983, Absolute stability of global pattern formation and parallel memory storage by competitive neural networks, *IEEE Transactions on Systems, Man and Cybernetics* **13**, 815-826.
5. Deaton, R., Garzon, M., Murphy, R.C., Rose, J.A., Franceschetti, D.R. and Stevens Jr., S.E., 1998, Reliability and efficiency of a DNA-based computation, *Phys. Rev. Lett.* **80**, 417-420.
6. Hopfield, J.J., 1982, Neural networks and physical systems with emergent collective computational abilities, *Proc. Nat. Acad. Sci. USA* **79**, 2554-2558.
7. Lipton, R.J., 1995, DNA solution of hard computational problems, *Science* **268**, 542-545.
8. A. P. Mills, Jr., B. Yurke and P. M. Platzman, 1999, to be published in the proceedings of the 4th DIMACS conference on DNA computing, Philadelphia (June, 1998).
9. Oliver, J.S., 1997, Matrix multiplication with DNA, *J. Molec. Evol* **45** , 161-167.
10. Wasserman, P.D., 1989, *Neural Computing, theory and practice* (Van Nostrand Reinhold, New York) ISBN 0-442-20743-3.
11. E. B. Baum, *Science* **268**, 583-585 (1995).
12. S. P. A. Fodor, J. L. Read, M. C. Pirrung, L. Stryer, A. T. Liu and D. Solas, Light-directed, spatially addressable parallel chemical synthesis, *Science* **251**, 767-773 (1991).
13. R. J. Lipshutz et al., Using oligonucleotide probe arrays to access genetic diversity, *Biotechniques* **19**, 442-447 (1995).
14. G. A. Carpenter and S. Grossberg, *Comp. Vision, Graph. and Im. Pro.* **37**, 54 (1987).

(A. Mills) Bell Laboratories, Lucent Technologies, 600 Mountain Avenue, Murray Hill, NJ 07974

E-mail address: apm@physics.lucent.com

(B. Yurke) Bell Laboratories, Lucent Technologies, 600 Mountain Avenue, Murray Hill, NJ 07974

E-mail address: yurke@physics.lucent.com

(P. Platzman) Bell Laboratories, Lucent Technologies, 600 Mountain Avenue, Murray Hill, NJ 07974

E-mail address: pmp@physics.lucent.com

DIMACS Series in Discrete Mathematics
and Theoretical Computer Science
Volume **54**, 2000

On Combinatorial DNA Word Design

Amit Marathe, Anne E. Condon, and Robert M. Corn

ABSTRACT. We consider the problem of designing *DNA codes*, namely sets of equi-length words over the alphabet $\{A, C, G, T\}$ that satisfy certain combinatorial constraints. This problem is motivated by the task of reliably storing and retrieving information in synthetic DNA strands, for use in DNA computing or as molecular bar codes in chemical libraries. The primary constraints that we consider, defined with respect to a parameter d, are as follows: for every pair of words w, x in a code, there are at least d mismatches between w and x if $w \neq x$ and also between the reverse of w and the Watson-Crick complement of x. Extending classical results from coding theory, we present several upper and lower bounds on the maximum size of such DNA codes and give methods for constructing such codes.

An additional constraint that is relevant to the design of DNA codes is that the free energies and enthalpies of the code words, and thus the melting temperatures, be similar. We describe dynamic programming algorithms that can (i) calculate the total number of words of length n whose free energy value (as approximated by a formula of Breslauer et al.) falls in a given range, and (ii) output a random such word. These algorithms are intended for use in heuristic algorithms for constructing DNA codes.

1. Introduction

The design of codes that satisfy combinatorial constraints has long been studied, motivated by the problem of sending information reliably over a noisy channel [**23**]. In this paper, we study code design problems that are motivated by the task of storing and retrieving information in short DNA strands, which we refer to as DNA *code words.* A (single) DNA strand is a sequence of nucleotides; there are four possible nucleotides, denoted A, C, G, and T, at each position of the sequence, which has chemically distinct ends known as the $5'$ and $3'$ ends. Since a DNA strand of length n can be used to represent one of up to 4^n possible values, and since short DNA strands can be quickly and cheaply synthesized, DNA code words can be used to store information at the molecular level, thus providing a basis for biomolecular computation [**1**]. DNA code words are also used as molecular bar codes, or tags, for the purpose of manipulating and identifying individual molecules in complex chemical libraries [**3, 4, 28**].

These applications require success in achieving specific hybridization between a DNA code word and its Watson-Crick complement, while minimizing false positive and false negative signals, as we now explain. The Watson-Crick complement of a

2000 *Mathematics Subject Classification.* Primary 92E99, 94B65, 94B25, 68W01.

© 2000 American Mathematical Society

DNA strand is the strand obtained by replacing each A by a T and vice versa, each C by a G and vice versa, and switching the $5'$ and $3'$ ends. For example, the Watson-Crick complement of $5'-AACATG-3'$ is $3'-TTGTAC-5'$. Specific hybridization is the process whereby a strand and its Watson-Crick complement bond to form a double helix. Specific hybridization can be used (along with other methods) to identify and retrieve target DNA code words from a set of such code words. A false positive results when non-specific hybridization occurs, such as between a DNA strand and the Watson-Crick complement of a distinct DNA strand, in which case there are *mismatches*. For example, there are two mismatches between $5'-AACATG-3'$ and $3'-TAATAC-5'$, in the second and third positions from the $5'$ end of the first strand. Non-specific hybridization may also occur between a DNA strand and the reverse of a distinct strand. A false negative occurs when hybridization between a DNA strand and its complement does not take place as intended.

Several papers have proposed the use of combinatorial constraints on the composition of a set of DNA code words, in order to limit false positives and false negatives in specific applications [**1, 2, 3, 4, 7, 8, 9, 10, 14, 15, 16, 24, 26, 28, 29, 31**]. Our premise is that a theoretical framework for designing sets of DNA code words should be useful for scalable use of DNA code words.

We focus on sets of words satisfying one or more of four constraints, which we next define and motivate. In our study, we represent a DNA code word simply as a string over the alphabet $\{A, C, G, T\}$ and assume that the leftmost (or low order) end of the string corresponds to the $5'$ end of the associated DNA code word. Thus, CCGAT represents $5'-CCGAT-3'$, for example. It is useful to define a *word* to be a string over a finite alphabet, where the alphabets of most interest to us are of size 2 or 4 (such as $\{A, C, G, T\}$). Let $x = x_1x_2 \ldots x_n$ be a word. The *reverse* of x, denoted by x^R, is the word $x_nx_{n-1} \ldots x_1$. If x is over the alphabet $\{A, C, G, T\}$ then the *complement* of x, denoted by x^C, is the word obtained by replacing each A in x by T and vice versa, and by replacing each C in x by G and vice versa. If x is over the binary alphabet $\{0, 1\}$ then x^C is obtained by replacing each 0 in x by 1 and vice versa. Finally, the Hamming distance $H(x, w)$ between x and word $w = w_1w_2 \ldots w_n$ is the number of indices i for which $w_i \neq x_i$. The following constraints pertain to a set of words, each of length n.

- The **Hamming constraint** with distance parameter d is that for all pairs of distinct words w, x in the set, $H(w, x) \geq d$. A set of words of size M satisfying the Hamming constraint is called a (n, M, d) *code*, or when the parameters are implied, simply a *code*. We let $A_q(n, d)$ denote the maximum size of a code with words of length n over alphabet size q. In most reports on the use of DNA codes, a high Hamming distance is enforced between pairs of code words (see for example [**3, 7, 8, 14, 26, 28, 29, 31**]), in order to limit non-specific hybridization whereby the Watson-Crick complement of a strand x anneals to a distinct strand w.
- The **reverse-complement constraint** with parameter d is that for all pairs of words w, x in the set (where w may equal x), $H(w^C, x^R) \geq d$. In DNA code applications, the reverse-complement constraint is intended to limit hybridization between two strands [**7, 8, 14, 26, 29, 31**]. We call a code that satisfies the Hamming and reverse-complement constraints a reverse-complement code. We let $A_q^{RC}(n, d)$ denote the maximum size

of a reverse-complement code over an alphabet of size q, with parameters n, d defined as for codes. For example, the code words ACG and CGG (representing strands $5' - ACG - 3'$ and $5' - CGG - 3'$ respectively) can be part of a $A_4^{RC}(3,1)$ code but not a $A_4^{RC}(3,2)$ code.
- The **reverse constraint** with parameter d is that, for all pairs of words w, x in the code, $H(w, x^R) \geq d$. We call a code that satisfies the reverse constraint, in addition to the Hamming constraint, a reverse code. We let $A_q^R(n,d)$ denote the maximum size of a reverse code. Our study of this constraint is not motivated directly by the goal of limiting false positives or false negatives in the use of DNA code words, but indirectly by a close relationship between $A_q^{RC}(n,d)$ and $A_q^R(n,d)$. For example, when n is even, one can obtain a reverse complement code from a reverse code simply by complementing the symbols in the second half of each word in the code. Thus, constructions of reverse codes can easily be adapted to obtain constructions of reverse complement codes. We consider the reverse constraint to be simpler than the reverse-complement constraint and so focus on this.
- The final constraint that we consider has a somewhat different flavor. It is motivated by the goal that all code words in the set have similar melting temperature, allowing hybridization of multiple words to proceed simultaneously [**28**]. The melting temperature of a short DNA strand can be accurately estimated using a formula of Wetmur [**30**], which in turn uses estimates of two further parameters of a DNA strand, namely its free energy and enthalpy due to Breslauer et al [**5**]. We use ΔG to denote the Breslauer estimate of the free energy of a DNA word. As explained in Section 6, the free energy constraint is essentially the sum of "nearest neighbor" weights associated with substrings of length 2 in a word. The **free energy constraint** with parameters g, ϵ is that for all words in the code, $g - \epsilon \leq \Delta G \leq g + \epsilon$. A closely related constraint is that the enthalpy of all words in the code lies within a small range.

Section 3 summarizes some well known results on codes. Our new results, presented in Section 4, focus on reverse and reverse-complement codes. Following is a summary of these results.

We first show a close relationship between the maximum sizes of reverse codes and reverse-complement codes. Specifically,

$$A_4^{RC}(n,d) = A_4^R(n,d) \qquad \text{when } n \text{ is even, and}$$
$$A_4^R(n,d+1) \leq A_4^{RC}(n,d) \leq A_4^R(n,d-1) \qquad \text{when } n \text{ is odd.}$$

We then show several methods for obtaining upper and lower bounds on the size of reverse codes with alphabet sizes 2 and 4, including the following.

- **Halving bound:** $A_q^R(n,d) \leq A_q(n,d)/2$. This bound follows from the fact that if S is a $(n, |S|, d)$ reverse code then $S \cup S^R$ is a $(n, 2|S|, d)$ code.
- **Cai's lower bound:** $A_q^R(2n, 2d) \geq \lfloor A_q(n,d)/2 \rfloor$. This result, which constructs a $(2n, 2d)$ reverse code from an optimal (n,d) code, is due to Jin-Yi Cai.
- **Construction for $d = 2$:** We give a simple inductive construction of a reverse code that is optimal for even n, and close to optimal for odd n. For

$q = 2$ or 4,

$$A_q^R(n,2) = q^{n-1}/2 \quad \text{when } n \text{ is even, and}$$

$$(q^{n-1} - q^{\lfloor n/2 \rfloor})/2 \leq A_q^R(n,2) \leq q^{n-1}/2 \quad \text{when } n \text{ is odd.}$$

- **Product bound:** $A_4^R(n,d) \geq A_2^R(n,d)A_2(n,d)$. In particular, reverse codes over an alphabet of size 4 can be obtained by taking the "product" of a reverse code and a code, both over an alphabet of size 2.
- **Doubling construction:** We show how to construct $(2^n, 2^n, 2^{n-1})$ binary reverse codes, which are optimal, i.e. $A_2^R(2^n, 2^{n-1}) = 2^n$.

We apply these results to obtain explicit bounds on $A_4^R(n,d)$ for $2 \leq n \leq 16$ and $2 \leq d \leq 8$. These are presented in section 5. We consider the free energy constraint in section 6.

2. Related work

Deaton et al. [**8, 10**] observe that the well-known sphere-packing bound (see Section 3) can be used to upper bound the size of DNA codes satisfying the Hamming constraint. They describe genetic algorithms for finding DNA codes that satisfy several constraints, including the Hamming and reverse complement constraints. Heuristic methods for finding good DNA encodings have also been described by Garzon et al. [**17**] and by Zhang and Shin [**32**].

In his patent on methods for sorting polynucleotides using DNA tags, Brenner [**3**] gives a greedy algorithm for generation of DNA codes satisfying the Hamming distance constraint. Brenner also considers sets of words over an alphabet of size 3, where one of the 4 possible nucleotides A, C, G, or T is absent. Mir [**24**] also proposed a word design for use in DNA computing over an alphabet representing just 3 of the 4 possible nucleotides; similarly Cukras et al. [**7**] use a three-letter alphabet in designing a library of RNA strands for encoding bits.

For prototype experiments of the Wisconsin DNA computing project, a DNA word design with words of length 16 was developed [**14**]. The internal 8 bases of a word are constrained to satisfy exactly the Hamming and reverse complement constraints with $d = 4$. In addition, the words also satisfy the constraint that 4 out of the 8 bases are from the set $\{C, G\}$. A set of words of size 108 satisfying these constraints was found. Cukras et al. [**7**] design a library of RNA strands which incorporate equi-length "bit encodings" that have pairwise high Hamming distance and have similar melting temperatures, along with other constraints. They use a computer program to find a library with the desired combinatorial properties.

Shoemaker et al. used an algorithm to generate a set of 9,105 20-mers that satisfy the Hamming constraint with $d = 5$ and are predicted to have similar melting temperatures (61 ± 5^oC). In addition, the words in the set are predicted to have no secondary structure. They give no details on their algorithm.

Finally, we list other combinatorial constraints that are relevant to DNA code design but which we do not study in this paper. (i) The first arises, for example, in Brenner and Lerner's work [**4**], where DNA tags and the polymers to be tagged are chemically synthesized in an alternating parallel fashion. Thus each code word (or tag) is the concatenation of "units," one per monomeric chemical unit in the polymer. The units are designed to have the *comma free* or frame-shift constraint: no unit x occurs as a substring in the concatenation of two other distinct units yz. The purpose of imposing this constraint is to limit the possibility of "frame shift errors" in the hybridization process. A greedy algorithm for generating words satisfying a similar frame shift constraint is given by Garzon et al. [**15, 16**]. Roweis et al. [**26**] propose such a frame shift constraint in their proposal for a sticker based model for DNA computation, and point to useful references in the applied mathematics literature on comma free codes and DeBruijn sequences. Smith and Schweitzer [**29**] consider a "modified DeBruijn problem," namely given k, design a DNA strand as long as possible, so that if S is any k-letter substrand of this DNA strand, then S occurs exactly once in the DNA strand and also the reverse of S and its complement do not occur in the strand. (ii) Baum [**2**] developed bounds on the size of DNA word sets satisfying several combinatorial constraints, key among them being the *subword constraint* that for some parameter d, for any pair of distinct words w and x, no subword of w of length d equals a subword of x of length d. (iii) In designing words for surface-based DNA computing, Frutos et al. [**14**] enforced the *GC content constraint* that the fraction of G's and C's in each code word be 1/2. The motivation for this is to limit the range of melting temperatures of the code words. (iv) Seeman [**27**] developed algorithms to design DNA sequences for use in construction of stable DNA structures with branched helix axes, "with the goal of minimizing sequence similarities between segments of molecules." The end product can be thought of as a set of words with similar G-C content, such that if s is any string of some given length d (such strings are referred to as "critons") over $\{A, C, G, T\}$, then s and its reverse complement s^{RC} occur at most once in any word in the set. Additional constraints are imposed on the sequences, for example at junction points in the branched structures. (v) The important *forbidden subwords constraint*, already mentioned in relation to the work of Shoemaker et al. [**28**] above, is that no word in the code set contains as a subword a specified set of undesirable words, such as DNA strands with secondary structure [**26**], strands that are to be used as PCR primers, or strands that are recognized by restriction enzymes. Some of these additional constraints have also been used by other experimental DNA computing projects [**7, 31**].

Finally, Smith and Schweitzer [**29**] propose encoding methods that allow for error correction even when mismatches are assigned weights. A different measure of error potential in DNA code words has been proposed by Rose et al. [**25**].

3. Bounds on the size of a code

In this section, we briefly review previous results on codes that will later be extended or applied in obtaining bounds on reverse codes. Proofs of the bounds in this section can be found in the survey article by Ericson [**13**] and text by MacWilliams and Sloane [**23**], which is a good introduction to the subject. Brouwer et al. [**6**] give a table of upper and lower bounds for $A_2(n, d)$ which we use in obtaining some of our bounds on reverse codes. We use several of these bounds, in conjunction with our halving and product bounds (Theorems 4.4, 4.7) in the construction of Table I.

The following two bounds on $A_q(n, d)$ are described in terms of two quantities. Let S be the set from which words in the code are drawn. Then the first quantity we need is $|S|$, which is clearly q^n. Let $V(s, d')$ be the number of words of S that have distance at most d' from word s, where $s \in S$. $V(s, d')$ is independent of s; denote it by $V(d')$.

THEOREM 3.1. **(Sphere-Packing upper bound)**

$$A_q(n,d) \leq \frac{|S|}{V(\lfloor (d-1)/2 \rfloor)} = \frac{q^n}{\sum_{i=0}^{\lfloor (d-1)/2 \rfloor} \binom{n}{i} (q-1)^i}.$$

THEOREM 3.2. **(Gilbert-Varshamov lower bound)**

$$A_q(n,d) \geq \frac{|S|}{V(d-1)} = \frac{q^n}{\sum_{i=0}^{d-1} \binom{n}{i} (q-1)^i}.$$

THEOREM 3.3. **(Singleton upper bound)**

$$A_q(n,d) \leq q^{n-d+1}.$$

THEOREM 3.4. **(Plotkin upper bound)** *For* $d > \frac{q-1}{q}n$,

$$A_q(n,d) \leq \frac{qd}{qd-(q-1)n}.$$

4. Selected Results on the Size of Reverse and Reverse-Complement Codes

We now present new bounds on the maximum size of reverse codes and reverse complement codes, as defined in the introduction. Recall that for $q \in \{2,4\}$, $A_q^{RC}(n,d)$ denotes the maximum size of a code of length n over an alphabet of size q that satisfies the Hamming and reverse-complement constraints. Similarly, $A_q^R(n,d)$ denotes the maximum size of a code satisfying the Hamming and reverse constraints.

There is a close relationship between the size of reverse and reverse-complement codes for the alphabet $\{A,C,G,T\}$.

THEOREM 4.1.

$$A_4^{RC}(n,d) = A_4^R(n,d), \text{ for } n \text{ even, and}$$

$$A_4^R(n,d+1) \leq A_4^{RC}(n,d) \leq A_4^R(n,d-1), \text{ for } n \text{ odd.}$$

PROOF. First, suppose that n is even. Let $\{x_i\}$ be a (n,M,d) reverse code. Write each $x_i = a_i b_i$, where the lengths of a_i and b_i are equal. Then $\{a_i b_i^C\}$ is a (n,M,d) reverse complement code, and so $A_4^R(n,d) \leq A_4^{RC}(n,d)$. We can prove the other inequality in the same manner.

When n is odd, truncating a $(n, A_4^R(n,d+1), d+1)$ code by removing the middle letter of each code word and then applying the previous result for n even gives $A_4^R(n,d+1) \leq A_4^{RC}(n-1,d)$. Since $A_4^{RC}(n-1,d) \leq A_4^{RC}(n,d)$ for odd n, the first inequality in the case that n is odd follows. The proof of the second inequality is similar. □

The remaining results in this section pertain to reverse codes. Extending the proof of the sphere-packing and Gilbert-Varshamov bounds, i.e. Theorems 3.1 and 3.2, we obtain the following bound for reverse codes with $d = 3$.

THEOREM 4.2. *For* $n \geq 4$,

$$A_q^R(n,3) \leq \frac{q^{\lceil n/2 \rceil} \sum_{i=2}^{\lfloor n/2 \rfloor} \binom{\lfloor n/2 \rfloor}{i} (q-1)^i}{2(1+4(q-2)+(n-4)(q-1))}.$$

PROOF. Let S be the set of all words x (length n, alphabet size q) such that $H(x, x^R) \geq d$. Also, let $V(s, d')$ be the number of words of S that have distance at most d' from word $s \in S$ and let $V^-(d') = \min\{V(s, d')\}$ where the min is taken over all s in S.

Following the proof of the sphere-packing bound for codes, to obtain an upper bound on the size of a code drawn from S we consider the set D_x, consisting of words which are disqualified when a word x from S is chosen to belong to the code. A lower bound on D_x is given by $2V^-(\lfloor (d-1)/2 \rfloor)$. This is because for any word s in S, $H(s, s^R) \geq d$ and hence $V(s, \lfloor (d-1)/2 \rfloor)$ is disjoint from $V(s^R, \lfloor (d-1)/2 \rfloor)$. Therefore,

$$A_q^R(n,d) \leq \frac{|S|}{2V^-(\lfloor (d-1)/2 \rfloor)}.$$

We first calculate the size of S. Note that if $x = x_1 x_2 \ldots x_n$ then $x_j^R = x_{n-j+1}$. We say a mismatch occurs at j if $x_j \neq x_j^R$, i.e. if $x_j \neq x_{n-j+1}$. If a mismatch occurs at j then by symmetry a mismatch also occurs at x_{n-j+1}. In fact, $H(x, x^R)$ is always even.

How many words x have $H(x, x^R) = 2i$? The number of such words is the number of words that have i mismatches at indices $j \leq \lfloor n/2 \rfloor$. There are $\binom{\lfloor n/2 \rfloor}{i}$ choices for these i indices. At each chosen index j, there are q choices for the letter x_j and once this is chosen, there are $q-1$ choices for the letter x_{n-j+1}. Also, there are $\lceil n/2 \rceil - i$ indices j of x at which $x_j = x_{n-j+1}$. There are q choices for the value of x_j at each of these indices. In total, there are

$$\binom{\lfloor n/2 \rfloor}{i} (q(q-1))^i q^{\lceil n/2 \rceil - i}$$

words x such that $H(x, x^R) = 2i$. Therefore,

$$|S| = \sum_{i=\lceil d/2 \rceil}^{\lfloor n/2 \rfloor} \binom{\lfloor n/2 \rfloor}{i} (q(q-1))^i q^{\lceil n/2 \rceil - i} = q^{\lceil n/2 \rceil} \sum_{i=\lceil d/2 \rceil}^{\lfloor n/2 \rfloor} \binom{\lfloor n/2 \rfloor}{i} (q-1)^i.$$

Next, consider a word $x = x_1 x_2 \ldots x_n$ in S. We claim that for $d = 3$ there are at least $4(q-2) + (n-4)(q-1)$ words of distance exactly 1 from x, and therefore $V^-(1) = 1 + 4(q-2) + (n-4)(q-1)$. To show the claim, let j and j' be such that $1 \leq j < j' \leq n/2$, $x_j \neq x_{n-j+1}$ and $x'_j \neq x_{n-j'+1}$ (j, j' exist because $H(x, x^R)$ is even and at least 3). For each of the four possible indices i with $i \in \{j, j', n-j+1, n-j'+1\}$, there are $q-2$ ways to change x_i to obtain a word x' of distance 1 from x, such that $H(x', (x')^R) = H(x, x^R) \geq 3$. For each of the $n-4$ remaining indices i, there are $q-1$ ways to change x_i to obtain a word x' of distance 1 from x, such that $H(x', (x')^R) \geq 3$. Thus, there are at least $4(q-2) + (n-4)(q-1)$ words of distance exactly 1 from x, as required. □

The upper bound of Theorem 4.2 can easily be generalized to $d > 3$ by lower bounding $V^-(\lfloor (d-1)/2 \rfloor)$. Generalizing the argument above, it is not hard to show that $V^-(d') \geq \sum_{i=0}^{d'} \binom{n}{i} (q-2)^i$.

THEOREM 4.3. *Let S be as in Theorem 4.2 and let $V^+(d) = max\ \{V(s,d) | s \in S\}$. Then,*

$$A_q^R(n,d) \geq \frac{|S|}{2V^+(d-1)},$$

PROOF. A simple greedy algorithm yields this lower bound: repeatedly select a word w from S to be in the code, and remove from S all words which differ from w or w^R in less than d places. Each step removes at most $2V^+(d-1)$ words from S and so we end up with a reverse code of at least $|S|/2V^+(d-1)$ words. □

THEOREM 4.4. **(Halving bound)** $A_q^R(n,d) \leq \frac{A_q(n,d)}{2}$.

PROOF. The reverse constraint implies that $H(x, x^R) \geq d$ for every word x in S. Also $x \in S \Rightarrow x^R \notin S$. Thus, starting with a set S satisfying the Hamming and reverse constraints we can get a new set by adding to S the reversals of all words in it. This new set satisfies the Hamming constraint because the new words added are at least distance d apart from each other and from the original words in S. The new set is twice the size of the original. □

THEOREM 4.5. **(Cai's lower bound)**

$$A_q^R(2n, 2d) \geq \left\lfloor \frac{A_q(n,d)}{2} \right\rfloor$$

PROOF. Divide an optimal (n,d) code into two equal parts (after dropping one word if $A_q(n,d)$ is odd). Let these parts be $X = \{x_1, \dots, x_t\}$ and $Y = \{y_1, \dots, y_t\}$, where $t = \lfloor A_q(n,d)/2 \rfloor$. Then $\{x_i y_i^R | i = 1, \dots, t\}$ is a $(2n, 2d)$ reverse code. □

THEOREM 4.6. **(d=2 Construction)**

$$A_q^R(n,2) = \frac{q^{n-1}}{2},\ \textit{for even } n \textit{ and } q \in \{2,4\}, \textit{ and}$$

$$A_q^R(n,2) \geq \frac{q^{n-1} - q^{\lfloor n/2 \rfloor}}{2}, \textit{for odd } n \textit{ and } q \in \{2,4\}.$$

PROOF. The proof builds on the following claim:

CLAIM 4.1. *For even n and $q \in \{2,4\}$, Σ^n can be partitioned into subsets, each containing q^{n-1} words, such that*

1. *any two words from the same subset differ in at least two positions,*
2. *if a word belongs to a subset, its reversal is also in the same subset, and*
3. *all the $q^{n/2}$ palindromes are in the same subset.*

PROOF. (of Claim) The partitions for the base case ($n = 2$) can be $S_1^2 = \{AA, CC, GG, TT\}$, $S_2^2 = \{AC, CA, GT, TG\}$, $S_3^2 = \{AG, GA, CT, TC\}$, $S_4^2 = \{AT, TA, CG, GC\}$ for $q = 4$ and $\{00, 11\}, \{01, 10\}$ for $q = 2$.

For the induction case, when $q = 4$, assume that we have a partition S_i^n of Σ^n, $i \in \{1, 2, 3, 4\}$, with the above properties, with S_1^n containing all of the palindromes. Then S_i^{n+2} for $i \in \{1, 2, 3, 4\}$ can be defined as follows.

$$S_1^{n+2} = S_1^n.S_1^2 \cup S_2^n.S_2^2 \cup S_3^n.S_3^2 \cup S_4^n.S_4^2,$$

$$S_2^{n+2} = S_1^n.S_2^2 \cup S_2^n.S_3^2 \cup S_3^n.S_4^2 \cup S_4^n.S_1^2,$$

$$S_3^{n+2} = S_1^n.S_3^2 \cup S_2^n.S_4^2 \cup S_3^n.S_1^2 \cup S_4^n.S_2^2,$$

$$S_4^{n+2} = S_1^n.S_4^2 \cup S_2^n.S_1^2 \cup S_3^n.S_2^2 \cup S_4^n.S_3^2,$$

where $A.B = \{pwq | w \in A, pq \in B, |p| = |q| = 1\}$.

It is not difficult to verify that this is a partition of Σ^{n+2} having all the three properties, with S_1^{n+2} containing the palindromes. The induction step for $q = 2$ utilizes a similar "product-of-sets" construction. As an example, when $q = 2$, the two subsets S_1^4 and S_2^4 obtained by the above construction are as follows:

S_1^4	S_2^4
0000	0010
0110	0100
1001	1011
1111	1101
0011	0001
0101	0111
1010	1000
1100	1110

□

Now, to complete the proof of Theorem 4.6 when n is even, note that if we take any of the subsets not containing any palindromes and drop half of the words from it (either a word or its reversal), we get a set satisfying the Hamming and reverse constraints (for $d = 2$). The identical upper bound follows from Theorem 4.4 combined with Theorem 3.3.

The construction for odd $n > 1$ and $q \in \{2, 4\}$ uses the sets S_i^{n-1} obtained above as follows:

$$S_1^n = S_1^{n-1}.A \cup S_2^{n-1}.C \cup S_3^{n-1}.G \cup S_4^{n-1}.T,$$

$$S_2^n = S_1^{n-1}.C \cup S_2^{n-1}.G \cup S_3^{n-1}.T \cup S_4^{n-1}.A,$$

$$S_3^n = S_1^{n-1}.G \cup S_2^{n-1}.T \cup S_3^{n-1}.A \cup S_4^{n-1}.C,$$

$$S_4^n = S_1^{n-1}.T \cup S_2^{n-1}.A \cup S_3^{n-1}.C \cup S_4^{n-1}.G,$$

where $B.X = \{w_1 X w_2 | w_1 w_2 \in B, |w_1| = |w_2|\}$. With this construction, each of the four subsets S_i^n has $q^{\lfloor n/2 \rfloor}$ palindromes. By first removing these palindromes from each subset and then dropping half of the remaining words (either a word or its reversal), we obtain four reverse codes with parameter $d = 2$, each of size $(q^{n-1} - q^{\lfloor n/2 \rfloor})/2$. □

THEOREM 4.7. **(Product Bound)** $A_4^R(n,d) \geq A_2^R(n,d)A_2(n,d)$.

PROOF. Let B be a code and let A be a reverse code, both over alphabet $\{0,1\}$, with words of length n. Then each element of the Cartesian product $A \times B$ corresponds to a word over an alphabet of size 4, where a bit in A determines whether A/T or G/C appear in that position, while the corresponding bit in B makes a choice from the 2 remaining possibilities. Moreover, this map is one-to-one and the set of $|A||B|$ words so obtained satisfies the Hamming and reverse constraints. □

THEOREM 4.8. **(Doubling Construction)** *For* $n \geq 2$, $A_2^R(2^n, 2^{n-1}) = 2^n$.

PROOF. Let C be a code with words of length n. Call C a $H^{RC}(n,d)$ code if it has the following property: for all words x, y in C,

$$\begin{aligned} &H(x,y) \geq d, && \text{if } x \neq y \\ &H(x,y^R) \geq d, \\ &H(x,y^C) \geq d, \text{ and} \\ &H(x,y^{RC}) \geq d. \end{aligned}$$

The proof of the following claim is straightforward.

CLAIM 4.2. *Let C be a $H^{RC}(n,d)$ code with $d \leq n/2$. Then the code $C' = CC \cup CC^C$ is a $H^{RC}(2n,2d)$ code, where $CC = \{xx | x \in C\}$ and $CC^C = \{xx^C | x \in C\}$. Moreover, the size of C' is twice that of C.*

We now show that for $n \geq 2$, there is a $H^{RC}(2^n, 2^{n-1})$ code of size at least 2^{n-1}. If C is such a code, then $C \cup C^C$ satisfies the Hamming and reverse constraints and has twice the size of C, and from this the lower bound of the theorem follows. For the upper bound, we have that $A_2(4r, 2r) = 8r$ from [**23**], which by the halving bounds (Theorem 4.4) implies that $A_2^R(2^n, 2^{n-1}) \leq 2^n$ for $n \geq 2$.

Let $n = 2$. It is easy to verify that $\{0111, 0010\}$ is a $H^{RC}(4,2)$ code of size 2. The construction of the claim then inductively yields a $H^{RC}(2^n, 2^{n-1})$ code of size 2^{n-1} for all $n > 2$, as required. □

5. Table of $A_4^R(n,d)$

This section contains a table of the function $A_4^R(n,d)$ for various values of the parameters. The table contains both a lower bound and an upper bound on the true value; whenever the two bounds are equal the entry is a single integer. The superscripts on entries indicate the method by which the bound was obtained. The following chart gives an overview of the superscripts used in the tables.

Superscript	Name of Bound	Relevant Theorem or Reference
s	Sphere-Packing	4.2
h	Halving	4.4
g	Gilbert-Varshamov	4.3
p	Product	4.7
x	$d = 2$ construction	4.6
c	Cai	4.5

Table I
Bounds on $A_4^R(n,d)$

n,d	2	3	4	5	6	7	8
2	2	0	0	0	0	0	0
3	6^x-$(2^3)^h$	0	0	0	0	0	0
4	$(2^5)^{x,h}$	2^p-8^s	2	0	0	0	0
5	120^x-$(2^7)^h$	4^p-26^s	2^p-8^h	0	0	0	0
6	$(2^9)^{x,h}$	16^p-107^h	8^p-107^h	2^p-5^h	2	0	0
7	2016^x-$(2^{11})^h$	33^g-372^h	16^p-372^h	2^p-34^s	2^p-4^h	0	0
8	$(2^{13})^{x,h}$	160^p-1310^h	128^p-1310^h	8^c-118^h	8^c-118^h	2^p-3^h	2
9	32640^x-$(2^{15})^h$	354^g-4681^h	160^p-4681^h	8^g-372^h	8^c-372^h	2^p-14^h	2^p-3^h
10	$(2^{17})^{x,h}$	1184^g-16912^h	320^p-16912^h	32^c-1202^h	32^c-1202^h	8^c-142^h	8^c-8^h
11	523776^x-$(2^{19})^h$	3903^g-61680^h	576^p-61680^h	60^g-3964^h	32^c-3964^h	8^c-420^h	8^c-420^h
12	$(2^{21})^{x,h}$	13233^g-226719^h	2304^p-226719^h	173^g-13294^h	96^p-13294^h	8^p-1276^h	8^p-1276^h
13	2095104^x-$(2^{23})^h$	45012^g-838860^h	4096^p-838860^h	437^g-45221^h	128^p-45221^h	18^g-3964^h	8^p-3964^h
14	$(2^{25})^{x,h}$	155496^g-3121342^h	12539^g-3121342^h	1444^g-155705^h	512^p-155705^h	64^p-12555^h	32^p-12555^h
15	134209536^x-$(2^{27})^h$	541020^g-11671106^h	40385^g-11671106^h	4280^g-541746^h	1024^p-541746^h	128^p-40439^h	64^p-40439^h
16	$(2^{29})^{x,h}$	1901386^g-43826196^h	132111^g-43826196^h	13066[illegible]-1902111^h	4096^p-1902111^h	576^p-132160^h	512^p-132160^h

6. The Free Energy Constraint

In this section, we present an algorithm to calculate the number of DNA strands (of a certain length) whose free energy equals a given value. The algorithm relies on a heuristic proposed by Breslauer et. al. [5] to approximate the free energy of any DNA strand. The data produced by the above algorithms can be used to efficiently generate a random strand with free energy close to a given value. This data could be used, for example, by a simulated annealing algorithm for finding a set of DNA strands with similar melting temperatures.

6.1. Algorithm Outline. The free energy of the DNA strand $u_1 u_2 \dots u_L$ is approximated by the following formula from Breslauer et. al. [5]

$$\Delta G_{total} = correction\ factor + \sum_{i=1}^{L-1} w(u_i, u_{i+1})$$

where $w(x, y)$ is the observed free energy of the 2-mer xy. Thus, the free energy is hypothesized to depend only on the nearest-neighbor interactions of nucleotides in the strand.

Let $N(l, u, e)$ be the number of DNA strands of length l, beginning with nucleotide u which have free energy e. Then, with the convention that 1-mers have free energy 0 and that $N(l, u, e)$ equals 0 when $e < 0$, $N(l, u, e)$ can be calculated for $l > 1$ from "previous" values as follows.

$$N(l, u, e) = \sum_{v \in \{A,C,G,T\}} N(l-1, v, e - w(u, v)).$$

The running time of this algorithm is $O(L^2 S^2 M 10^P)$ where L is the length of the DNA strands, S is the alphabet size (4 in our case), M is the maximum free energy of a 2-mer and P is the precision to which 2-mer free-energies are calculated (in number of decimal places). The space used is $O(L^2 S M 10^P)$.

Once all the N values have been calculated by this procedure we can find the number of strands whose free energies lie within the range $[P, Q]$ in $O(S(Q - P))$ time. A plot showing the number of strands corresponding to a free energy value/range can also be produced in $O(LSM)$ time.

6.2. Random Generation Algorithm. This section shows how the data obtained from the dynamic programming algorithm above can be used to randomly select a strand from all strands of a given length and free energy, using an ordering implied by the N-array. Let $S = \{w_1, \dots, w_N\}$ be the set of all strands of length L and free energy E. To randomly select a strands from S we generate a random number r in the range $[1, N]$. By the dynamic programming algorithm, $N = N_A + N_C + N_G + N_T$, where N_j is the size of S_j, the set of strands of S which begin with nucleotide j. Since N_A, N_C, N_G, N_T are known (they are entries in the N-array) we can fix the first nucleotide in the random strand to be generated, depending on whether r is in the range $[1, N_A]$, $[N_A + 1, N_A + N_C]$, $[N_A + N_C + 1, N_A + N_C + N_G]$ or $[N_A + N_C + N_G + 1, N]$.

Applying this process iteratively we can generate the entire strand. i.e. if r is in the range $[N_A + 1, N_A + N_C]$ (say), we fix the first nucleotide to be C. We then consider the set S_C and then choose a strand at random from this set by using the random number $r - N_A$, which will be uniformly distributed over the range $[1, N_C]$.

7. Conclusions

Design of good DNA codes is a hard and important problem for DNA computing and other biotechnology applications. Current approaches focus on DNA codes that satisfy natural combinatorial constraints. Theoretical work on this problem can build on classical results from coding theory, thereby providing useful design tools, and also providing limits on the attainable size of certain codes. The theoretical results of this paper are a first step in this direction, focusing on codes that satisfy the Hamming and reverse complement constraints.

Additional combinatorial constraints need to be considered, particularly the frame shift constraint (see Section 2). Codes that satisfy the frame shift constraint with Hamming distance equal to 1 have been extensively studied (see for example [**11, 18**]), but to our knowledge there is little known when the Hamming distance is greater than 1.

Perhaps more difficult theoretically, but well motivated from a practical point of view, would be to derive good methods for generating codes that satisfy the free energy constraint *in addition to* one or more other constraints such as the Hamming, reverse complement, or frame shift constraints.

Finally, more experimental work is badly needed to validate the use of combinatorial constraints in design of DNA codes in the first place. Related to this, it is desirable to have a method for predicting the free energy of arbitrary duplexes – that is, a method that generalizes formulas based on "nearest neighbor" interactions in the special case of a duplex formed from a strand and its Watson-Crick complement [**5, 30**].

Acknowledgements

We thank Jin-Yi Cai for allowing us to include his result, Theorem 4.5, in this paper. We also thank Lloyd Smith and the anonymous referees for their valuable comments on the paper.

This research was supported by NSF and DARPA via grants CCR-96-13799 and CCR-97-25021.

References

[1] L. M. Adleman, "Molecular Computation of Solutions to Combinatorial Problems," Science, **266**, 11 November, 1994, pages 1021–1024.

[2] E. B. Baum, "DNA Sequences Useful for Computation," Proc. of DNA Based Computers II, DIMACS Workshop June 10-12, 1996, L. F. Landweber and E. B. Baum, Editors, DIMACS Series in Discrete Mathematics and Theoretical Computer Science, Vol. 44, 1999, pages 235-242.

[3] S. Brenner, "Methods for Sorting Polynucleotides using Oligonucleotide Tags," U.S. Patent Number 5,604,097, Feb 18, 1997.

[4] S. Brenner and R. A. Lerner, "Encoded combinatorial chemistry," Proc. Natl. Acad. Sci. USA, Vol 89, pp 5381-5383, June 1992.

[5] K. Breslauer,R. Frank, H. Blocker, L. Marky, "Predicting DNA duplex stability from the base sequence," Proc. Natl. Acad. Sci. USA 83 (1986), pages 3746-3750.

[6] A.E.Brouwer, J.B. Shearer, N.J.A. Sloane, W.D. Smith, "A New Table of Constant Weight Codes," IEEE Transactions on Information Theory, Vol. 36, No. 6, November 1990, pages 1334-1380.

[7] A. R. Cukras, D. Faulhammer, R. J. Lipton, and L. F. Landweber, "Chess Games: A model for RNA Based Computation," Preliminary Proc. of the Fourth International DIMACS Meeting on DNA Based Computers, U. Pennsylvania, 1998, pages 27-37.

[8] R. Deaton, R. C. Murphy, M. Garzon, D. R. Franceschetti, and S. E. Stevens, Jr., "Good Encodings for DNA-Based Solutions to Combinatorial Problems," Proc. of DNA Based Computers II, DIMACS Workshop June 10-12, 1996, L. F. Landweber and E. B. Baum, Editors, DIMACS Series in Discrete Mathematics and Theoretical Computer Science, Vol. 44, 1999, pages 247-258.

[9] M. Garzon, R. Deaton, P. Neathery, D. R. Franceschetti, and S. E. Stevens, Jr., "On the Encoding Problem for DNA Computing," Preliminary Proc. 3rd DIMACS Workshop on DNA Based Computers, June 23-25, 1997, pages 230- 237.

[10] R. Deaton, M. Garzon, R. C. Murphy, J. A. Rose, D. R. Franceschetti, and S. E. Stevens, Jr., "Genetic Search of Reliable Encodings for DNA-Based Computation," Koza, John R., Goldberg, David E., Fogel, David B., and Riolo, Rick L. (editors), Proceedings of the First Annual Conference on Genetic Programming 1996.

[11] W. L. Eastman, "On the construction of comma-free codes," IEEE Transactions on Information Theory, Vol. 11, 1965, pages 263–267.

[12] A. A. El Gamal, L. A. Hemachandra, I. Shperling, and V. K. Wei, "Using Simulated Annealing to Design Good Codes," IEEE Transactions on Information Theory, Vol. IT-33, No. 1, January 1987.

[13] T. Ericson, "Bounds on the Size of a Code," Topics in Coding Theory, Springer-Verlag, 1989.

[14] A. G. Frutos, Q. Liu, A. J. Thiel, A. M. W. Sanner, A. E. Condon, L. M. Smith, and R. M. Corn, "Demonstration of a Word Design Strategy for DNA Computing on Surfaces," Nucleic Acids Research, Vol. 25, No. 23, December 1997, pages 4748-4757.

[15] M. Garzon, R. Deaton, P. Neathery, D.R. Franceschetti, and R.C. Murphy, "A New Metric for DNA Computing," Proc. 2nd Genetic Programming Conference, Morgan Kaufman, 1997, pages 472-478.

[16] M. Garzon, R. Deaton, L.F. Nino, S.E. Stevens Jr., and M. Wittner, "Encoding Genomes for DNA Computing," Proc. 3rd Genetic Programming Conference, Madison, WI, 1998.

[17] M. Garzon, R. J. Deaton, J. A. Rose, and D. R. Franceschetti, "Soft Molecular Computing," Preliminary Proc. Fifth International Meeting on DNA Based Computers, June 14-15, MIT, 1999, pages 89-98.

[18] S. W. Golomb, B. Gordon, and L. R. Welch, "Comma-Free Codes," Can. J. Math., Vol 10, 1958, pages 202–209.

[19] S. Litsyn, A. Vardy, "The Uniqueness of the Best Code," IEEE Transactions on Information Theory, Vol. 40, No. 5, November 1994, pages 1693-1698.

[20] Y. Klein, S. Litsyn, A. Vardy, "Two New Bounds on the Size of Binary Codes with a Minimum Distance of Three," Designs, Codes and Cryptography, 6, 1995, pages 219-227.

[21] Klaus-Uwe Koschnick, "Some New Constant Weight Codes," IEEE Transactions on Information Theory, Vol. 37, No. 2, November 1991, pages 370-371.

[22] F.R. Kschischang, S. Pasupathy, "Some ternary and Quaternary Codes and Associated Sphere Packings," IEEE Transactions on Information Theory, Vol. 38, No. 2, March 1992, pages 227-246.

[23] F. J. MacWilliams and N. J. A. Sloane, "The Theory of Error-Correcting Codes," North-Holland, 1977.

[24] K. U. Mir, "A Restricted Genetic Alphabet for DNA Computing," Proc. of DNA Based Computers II, DIMACS Workshop June 10-12, 1996, L. F. Landweber and E. B. Baum, Editors, DIMACS Series in Discrete Mathematics and Theoretical Computer Science, Vol. 44, 1999, pages 243-246.

[25] J. A. Rose, R. Deaton, D. R. Franceschetti, M. Garzon, and S. E. Stevens, Jr., "A Statistical Mechanical Treatment of Error in the Annealing Biostep of DNA Computation," Special program in DNA and Molecular Computing at the Genetic and Evolutionary Computation Conference (GECCO-99), Orlando, FL., July 13-17, 1999, Morgan Kaufmann.

[26] S. Roweis, E. Winfree, R. Burgoyne, N. V. Chelyapov, M. F. Goodman, P.W.K. Rothemund, and L. M. Adleman, "A Sticker-Based Model for DNA Computation," Proc. of DNA Based Computers II, DIMACS Workshop June 10-12, 1996, L. F. Landweber and E. B. Baum, Editors, DIMACS Series in Discrete Mathematics and Theoretical Computer Science, Vol. 44, 1999, pages 1-29.

[27] N. C. Seeman, "DeNovo design of sequences for nucleic acid structural engineering," J. Biomolecular Structure and Dynamics, ISSN 0739-1102, Vol. 8, No. 3 (1990), pages 573-581.

[28] D. D. Shoemaker, D. A. Lashkari, D. Morris, M. Mittman, and R. W. Davis, "Quantitative phenotypic analysis of yeast deletion mutants using a highly parallel molecular bar-coding strategy," Nature Genetics, Vol 16, December 1996, pages 450–456.

[29] W. D. Smith and A. Schweitzer, "DNA Computers in Vitro and in Vivo," NECI Technical Report, March 20, 1995.

[30] J. G. Wetmur. "DNA Probes: Applications of the Principles of Nucleic Acid Hybridization," Critical Reviews in Biochemistry and Molecular Biology 26(3/4):227-259, 1991, pages 227–259.

[31] E. Winfree, F. Liu, L. A. Wenzler, and N. Seeman, "Design and Self-Assembly of Two-Dimensional DNA Crystals," Nature Vol. 394, No. 6, August 1998, pages 539-544.

[32] B-T. Zhang and S-Y. Shin, "Molecular Algorithms for Efficient and Reliable DNA Computing," Proc. 3rd Annual Genetic Programming Conference, Edited by J. R. Koza, K. Deb, M. Doringo, D.B. Fogel, M. Garzon, H. Iba, and R. L. Riolo, Morgan Kaufmann, 1998, pages 735-742.

(A. Marathe) Computer Sciences Department, 1210 West Dayton Street, University of Wisconsin, Madison, WI 53706

E-mail address: `marathe@cs.wisc.edu`

(A. Condon) Computer Sciences Department, 1210 West Dayton Street, University of Wisconsin, Madison, WI 53706

E-mail address: `condon@cs.wisc.edu`

(R. Corn) Chemistry Department, 1101 University Avenue, University of Wisconsin, Madison, WI 53706

E-mail address: `corn@chem.wisc.edu`

DIMACS Series in Discrete Mathematics
and Theoretical Computer Science
Volume **54**, 2000

Soft Molecular Computing

Max Garzon, Russell J. Deaton, and John A. Rose

ABSTRACT. Molecular computing (MC) utilizes the complex interaction of biomolecules and molecular biology protocols to effect computation. Lab experiments in MC are unreliable, inefficient, unscalable, and expensive compared to conventional computing standards. A critical issue in MC is therefore to test protocols to minimize errors and mishaps that can thwart experiments when actually run *in vitro*. The purpose of this paper is to describe `Edna`, an integrated software platform developed to address this problem. The platform will allow MC practicioners to use digital computers to gain insight on the performance of a protocol before it actually unfolds in the tube. Currently, `Edna` provides tools to find good encodings for a given set of hybridization conditions, by design or evolution, tools to visualize the quality of these encodings, tools to estimate the complexity of given protocols based on bounded complexity, and a virtual test tube simulator based on local interactions between electronic DNA molecules. The virtual tube has allowed us to reproduce Adleman's experiment *in silico*. `Edna` includes graphical interfaces, click-and-drag facilities, is object-oriented, extensible, and so that it can easily evolve as the field progresses.

1. Introduction

Adleman's ground-breaking work [**2**] demonstrated the way to use molecules for computational purposes. Five years later, substantial obstacles remain to bring the potential of molecular computing into the realm of practical and efficient computation, at least in a niche of core applications. Molecular Computing problems are solved in three phases: *encoding* that maps the problem onto DNA strands, *molecular operations* that perform the basic core processing, and *extraction/detection* that makes the results visible to the naked eye. A typical set up consists of a multiset of molecules in a tube that encode the initial conditions of the computation. An algorithmic problem is solved by using a DNA program, herein referred to as a *protocol*, which is defined as a time-ordered sequence of elementary steps given by basic bio-operations (`BIOPS`) that change the physico-chemical state of the molecules inside the tube.

The decisive impact of errors generated by improper hybridizations during a DNA computation has now been recognized. Because the physical processes involved in molecular computing are much more uncertain, involved, and expensive than digital operations, simulating those processes and protocols beforehand would

2000 *Mathematics Subject Classification.* Primary 68Nxx, 92–04, 92D20, 37–04, 37Mxx.

© 2000 American Mathematical Society

be advantageous. The situation is analogous to the fabrication of solid-state integrated circuits (IC). IC fabrication facilities are extremely costly to set up. In order to insure success, new chips are commonly simulated from the physical fabrication processes, through the device physics, to logic and circuit level verification. Likewise molecular computing devices could be simulated from the low-level physical processes, such as hybridization reactions, through the molecular biology protocols, to a final verification of the algorithms that they implement. Preliminary work in this direction has centered on the encoding problem (see [**6**], [**10, 11**] and [**26**], for example), because of its decisive influence on the outcome of a computation. Once the encoding of the inputs of a problem have been chosen, the results of a DNA experiment are essentially determined up to laboratory execution. If the encodings are prone to errors, the experiment can be repeated any number of times and always provide an erroneous result. Focus on the massive parallelism and complexity of interactions in the tube, an important source of computational power in molecular computing, however, has prevented other important aspects of the problem from being addressed, such as reliability, efficiency and scalability of the protocols.

The purpose of this paper is to describe `Edna`, an integrated software platform that our group has developed in the last few years to address this problem. The basic assumption is that molecular computing will be effective to address problems beyond the reach of conventional computing. Ideally, one would like to have some sort of virtual tube, much like a wind tunnel for aerodynamics, in which some electronic DNA-like molecules could be tested under variable reaction conditions, and at a much lower price than actual lab experiments. Since such an object is probably nonexistant (more below in the discussion in the conclusion), the next best thing is to create a piece of software, running on solid-state computers, that could provide insight on the actual test tube runs before they are carried out in the lab. In general, the software platform will allow practicioners to determine good encoding molecules for the data in the problem, the best experimental conditions for their protocols, some sense of the reliability of the answers produced, and even perhaps some indication of the complexity of the task as it will unfold in the tube. Currently, `Edna` contains tools to find good encodings for a given set of hybridization conditions, by design or evolution, tools to visualize the quality of these encodings, tools to estimate the reliability of the protocols, and tools to estimate the complexity of given protocols based on bounded complexity [**9**]. These tools are justified and validated on various analyses of encodings and reactions conditions that are briefly described in the following sections. It is clear that the hypotheses on which the tools have been built will not be agreed upon by all researchers in the field and will necessarily change overtime. Therefore, the platform has been made object-oriented and flexible enough that it can be easily modified and/or extended to incorporate further research on these issues and be upgraded as the field progresses. Although not emphasized below, `Edna` includes graphical interfaces and click-and-drag facilities to enable easy use. Preliminary results indicate that `Edna` may even be potentially useful to solve *in silico* instances of some problems which would be challenging enough for *in vitro* experiments.

2. Encoding Tools

Computational algorithms rely upon physical systems for their implementation. As a result of inherent physical uncertainty, no architecture can ever be entirely error-free. This applies to solid-state electronics, quantum computational devices, and DNA-based computers. A quantitative characterization of the impact of physical uncertainties upon the computational reliability is therefore necessary, whenever a high degree of confidence in the computational result is desired. This problem has been resolved for conventional computation after half a century of solid-state eletronics, but remains unsolved for DNA computing. Specifically, the encoding problem for DNA computing consists of mapping the instances of an algorithmic problem in a systematic manner onto specific molecules and chemical protocols so that the resulting reaction products are very likely to contain the answers to the problem's instances. A good solution of the problem prevents unwanted hybridization errors and enables easy retrieval of the answer(s) in the extraction steps.

In the past, two kinds of approaches have been taken to frame and address the encoding problem: heuristic and combinatorial. Usually, heuristic methods to minimize error have been used to encode the problem instances in a set of oligonucleotide codewords [**18, 25, 6, 30, 26**]. In each case, the encoding strategy may be useful for minimizing the potential for a particular kind of hybridization error, or for a given specific problem instance and computational architecture. They lack, however, a desirable generalizability to arbitrary computational protocols. A more general, physical measure of error potential has been recently introduced by [**21**], called the computational incoherence, ξ. Being physically based, ξ should generalize encoding to any problem instance or computational architecture. Although a physically based encoding strategy, such as ξ, may be closer to the ground truth as far as hybridization error is concerned, its determination can be excessively expensive computationally.

In the combinatorial approach, a measure of hybridization affinity is used to quantify the likelihood that two strands will form a double strand. These measures include Hamming distance [**6**], [**26**], and a new hybridization distance introduced in [**8**]. The latter h-metric appears to capture some of the physical constraints on an encoding in a purely combinatorial measure. Therefore, a combinatorial measure of the tendency to hybridize may have an advantage if computational short-cuts from coding theory or computational biology can be applied, or the structure of the space of all possible encodings (the so-called DNA hypercubes) is known. A tool for finding good encodings in the h-metric space is included in `Edna`.

We present some details on each of these two approaches. Since the Hamming distance ignores frame-shift errors and is not generally appropriate except in special circumstances (such as the Madison group's codewords), we restrict ourselves to the computational incoherence and the h-metric.

2.1. Computational Incoherence. One physically-based measure that has been used to characterize the hybridization potential of a particular species is the duplex melting temperature, T_m. An examination of the T_m's of planned versus unplanned hybrids has been demonstrated to be a useful diagnostic method for the screening [**10**] and design [**7**] of computational sets of molecules. However, T_m is not entirely satisfactory for a fundamental characterization of hybridization error. T_m properly refers to the propensity of stacking for a stochastically observed pair of adjacent base pairs, and not to the overall state of a given duplex. In addition, as the melting curve of any finite length DNA duplex will necessarily have a substantial width, knowledge of T_m is insufficient to predict the state (bound or unbound) of an observed instance of the duplex at arbitrary reaction temperatures. Furthermore, relative rankings in the melting temperatures of two duplexes need not correspond to relative rankings in energetic stability at a given temperature [**5**].

An alternative physical measure that addresses some of these drawbacks, the computational incoherence (CI), has been proposed in [**21**]. The annealed ensemble consists thus of numerous instances of many distinct duplexes, in equilibrium with a set of single-strand DNA species. The computational incoherence, ξ, uses the principles of equilibrium statistical mechanics to estimate the ensemble average probability of an error hybridization per hybridization event, given a specified set of planned hybridizations, a set of encodings, and a reaction temperature. Basically, ξ gives the probability of an unplanned hybridization in a collection DNA molecules, and therefore, is a measure of error potential for the ensemble, rather than for pairs of oligonucleotides, as previously has been the case for T_m, Hamming distance, and other error measures. Details of the derivation can be found in [**21**]. What follows is the gist of that argument.

The principles of statistical thermodynamics have been used to charactize the melting behavior of DNA oligonucleotides and polynucleotides [**28, 19**]. Together with ξ, they are applied to the problem of unplanned hybridizations in a DNA computation. The basic quantity to calculate is the partition function of a duplex that is randomly selected from a tube containing a DNA computation. The partition function, Z, describes the equilibrium distribution of the quantum states of the duplex among the macroscopic configurations of that duplex. The number of quantum states at a given energy of the duplex is given by the Gibbs factor, $Z_j = \exp(-\Delta G_j/RT)$, where ΔG_j is the Gibbs free energy of formation for the duplex, R is the molar gas constant, and T is the absolute temperature. In a DNA computation, the randomly selected duplex will be observed to lie in exactly one of two distinct states: it will be hybridized in a manner either consistent (state $\overline{e}$), or inconsistent (state e) with the hybridization rules of the computation. The probability that this duplex is in state e is then given by the ratio of the sum of the Gibbs factors for all possible "error" configurations of the duplex and the partition function for the duplex. Therefore, as an ensemble measure of hybridization error, a measure ξ can systematically group all unplanned hybridizations into a single error category. In accordance with the principles of equilibrium statistical mechanics, however, the mixture must be allowed to reach equilibrium prior to the addition of any secondary enzymes (i.e., ligase). This ξ is completely general. The possible configurations of a duplex include the details of both primary and secondary structure. such as bulges, mismatches, and other secondary structures that contribute to a specific configuration [**5**]. In addition, concentrations of the reacting species are included through the normal equilibrium expression [**5**]. Therefore, the general form for ξ accounts for all possible states of both single- and double-stranded species.

The problem, however, is that in the general form, calculation of ξ is intractable. This is due to several reasons. First, calculation of ξ is only as good as the thermodynamic data available. Measurements of the energetic factors contributing to ΔG_j for base-pair stacking exist [**24**], as well as for some details of the secondary structure [**17**]. Complete data for all possible structural details under all possible reaction conditions, however, is not available. In addition, all possible configurations of the species present in the reaction must be considered. This number is huge, and analogously to the problem of computing the number of alignments between two sequences in computational biology, it is computationally intractable.

Lastly, direct calculation of ξ requires that the concentrations of all species be calculated, which is a difficult problem involving coupled quadratic equations of arbitrary molecularity. Given these factors, approximations must be made in order to reduce ξ to a tractable form.

In [**21**], an approximation ξ' of the computational incoherence ξ has been computed as follows. ΔG_j is defined relative to a zero energy state, which is equal to that of the unstacked, dissociated, single strands [**5**], contributions from single stranded regions of the duplex are considered negligible (thus reducing Z to the conformal partition function, Z_c [**5**]), and attention is restricted to short polygos (roughly, length $\leq$ 200 bps). With these assumptions, the model becomes the staggered zipper model [**4, 5**], in which configurations containing internal loops are discarded. The Gibb's factor of each duplex configuration is then related to the total difference in the free energies between the stacked and unstacked states of the duplex. The ΔG_j of duplex j may be estimated by summing the free energy changes of stacking each successive pair of adjacent hydrogen-bonded base pairs $\Delta G_{i,i+1}$ during formation of the duplex from the totally unstacked state, in accordance with the nearest-neighbor model of duplex formation. An estimate of the nearest-neighbor parameters $(\Delta H^0_{i,i'}, \Delta S^0_{i,i'})$ for each of the 10 WC nearest-neighbor pairs has been reported in [**24, 1**]. The total free energy of formation of a duplex j is given by

$$(1) \qquad \Delta G_j = \sum_i \Delta G_{i,i+1} + \Delta G_{init} + \Delta G_{sym},$$

where the sum is perfomed over adjacent pairs of hydrogen bonded base pairs i and $i+1$, ΔG_{init} is the free energy of initiation at each end, and ΔG_{sym} is a symmetry correction. For the annealing reaction in which dissociation is negligible and the initial concentrations of single stranded species are equal, the ensemble average probability that a duplex will be found in an error binding mode is computed as

$$(2) \qquad \xi \approx \frac{\sum_{i,j\geq i} Z_e^{i,j}}{\sum_{i,j\geq i} Z_c^{i,j}},$$

where $Z_e^{i,j}$ is the sum of the Gibbs factors of all error configurations and $Z_c^{i,j}$ is the conformal partition function between single-stranded DNA species i and j.

`Edna` includes a tool to compute this form of ξ of given encodings. Reaction specifications required from the user include the set of graph characteristics (vertex number and edge identities) and the length of encoded signel-stranded DNA molecules. Stochastic generation of edges is also supported to allow the assessment of the characteristics of random graphs. Reaction conditions may be set by the user or set automatically to the calculated $T^{max}_{reaction}$. `Edna` can calculate the following information for each assessed encoding:

- The set of T_m and ΔT_m for planned modes of annealings;
- A maximum suggested reaction temperature appropriate for annealing, defined as equal to the minimum member of the set $T_m - \frac{1}{2}\Delta T_m$ over planned modes of annealing;
- ξ' computed for the selected set of reaction conditions;
- An error diagnostic matrix consisting of the set of conditional hybridization error probabilities for each pair of encoding oligos.

In addition, Edna implements a standard genetic algorithm for evolving encodings using $-log\xi'$ as a fitness function. The user can select various parameters of the genetic algorithm, including mutation and crossover rates, population size, crossover point number, and parental number. The initial population of encodings may be seeded by the user, or may be generated partially or entirely at random. Convergence occurs upon the achievement of either a set maximum number of evolved generations or the emergence of an encoding having a minimally acceptable fitness. The built-in genetic algorithm facility has allowed us to evolve much more coherent encodings for Adleman's original problem.

2.2. The h-metric. The physico-chemical reality of the tube, particularly frameshifts in hybridization, makes it clear that the Hamming distance is not an adequate measure of hybridization likelihood. Frame shifts appear to be accountable for, nevertheless, by a generalization of the Hamming metric, the so-called *h-metric* [**8**]. The h-metric seems to capture enough of the reality of reactions conditions and the complexity of test tube hybridizations to frame and adequately solve the encoding problem. Moreover, it is combinatorially-based and computable in quadratic time. The h-measure between two oligos x and y is defined as the minimum of all Hamming distances obtained by successively shifting and lining up the WC-complement of y against x. the h-metric is defined for so-called poligos, namely equivalence classes of $n-$mers at an h-measure 0 of each other. (This is a technical point required by the fact that the h-measure does not give, strictly speaking, a metric space.) If some shift of one strand somewhat or other matches a segment of the other, the measure is reduced from n by the value of the Hamming distance between the shifted strands. Thus oligos in the same class are WC complementary and will most certainly hybridize. A small measure indicates that the two oligos are likely to stick to each other one way or another; a large measure indicates that *under whatever physico-chemical conditions* y finds itself in the proximity of x, they are far from containing many WC complementary pairs (let alone segments), and are therefore less likely to hybridize, i.e., they are more likely to avoid an erroneous unwanted hybridization. The maximum length of these segments in controlled by a parameter τ, that is a fairly coarse expression of the reaction conditions. Therefore the h-metric can be regarded as measuring the degree of inertia against hybridization.

The resulting metric space of all possible encoding $n-$mers is called the $\mathtt{DNA}_n$ cube of dimension n. A solution of the encoding problem thus consists of finding good codes, in the sense of conding theory, in the higher-dimensional DNA hypercubes, in lieu of the ordinary Hamming boolean hypercubes. They furnish encodings that are, by definition, capable of *preventing* errors them from occurring, largely independently of the type of experiment they are used for. Error-*correcting* approaches have been suggested by several authors (see, for example, [**3, 12**]), but they attempt to detect and correct errors (as is common in information theory), that have been allowed to occur. It appears possible, therefore, to *avoid* as many errors as possible, and perhaps use error-correcting methods to handle errors that cannot be prevented with the encodings generated through the h-metric.

Likewise, Edna includes a tool to compute, visualize, manipulate, and evolve encodings for molecular protocols based on the h-metric criterion. The results are integrated so that encodings obtained using one criterion (e.g., h-distance) can be compared against other criteria (e.g. computational incoherence). Edna has been instrumental in ellucidating the structure of the DNA cubes, which will give information on optimal encodings, and in determining how much of the relevant physical-chemistry is effectively captured by the h-metric.

2.3. Complexity Estimates of Molecular protocols. Preparing for an experiment in molecular computation includes gaining some insight into how the computation will actually go. That means *efficiency* of the reactions (are there enough molecules to produce the desired products? Are they too numerous to stretch reaction times to equilibrium too long? Are the concentrations of each reactant appropriate?) and some indication of the *reliability* and *scalability* of the results obtained from the experiments. Encodings discussed in the previous section provide a minimum guarantee of reliability. Efficiency and scalability have more to do with the actual reactions that are supposed to take place inside the tube, i.e., with the complexity of the molecular computation.

Various concepts have been used in the field to quantify the complexity of a computational protocol in molecular computing. Earlier ones have readily borrowed from Turing complexity and used as resource a count of tube manipulations to be performed. The goodness of a protocol is measured in the traditional way, by the asymptotic behavior of this count as the size of the problems (and hence molecules) approaches infinity. Since this approach emphasizes events outside the tube and since oligos larger than 20−mers have scarcely been handled easily in the lab for DNA computing, it does not appear to be a very realistic or useful abstraction for analysis of molecular protocols.

An alternative metric, *bounded complexity* (BC), more focused on reactions conditions and the events happening in the interior of the tube, has been introduced in [**9**]. Due to the fact that there are many different factors involved in each protocol, the complexity is measured in a so-called TVn unit, which is a common unit that permits meaningful comparisons between different solutions to the same or even different problems. One TVn unit $1h$ gives the

$$1h \equiv \textit{unit cost per } 1^oC \textit{ change of temperature}$$
$$\textit{per one nucleotide per } 1\ \mu l \textit{ volume reaction.}$$

In order to measure the *complexity* of a protocol, one associates a cost to each elementary operation χ. Each operation is implemented as a laboratory procedure, and therefore this cost depends in general on the full state of the tube, i.e. on many factors like temperature, volume of the test tube, number of different DNA molecules present, enzymes used and their cost, and number of nucleotides that are expected to interact within the tube, to name a few. In general, however, we will make the simplifying assumption that the bulk of the cost depends only on a few of these parameters, i.e., we have assumed that operations only depend explicitly on V, T, N and n. The efficiency of a protocol π can be measured by

$$C(\pi) = \sum_{\oplus \in \pi} C_{\oplus}(n(\Psi), N(\Psi), V(\Psi), T(\Psi)),$$

where it is understood that Ψ denotes the successive tubes that arise from the application of preceding operations. `Edna` will include a tool to specify molecular protocols and estimate and compare their efficiency through bounded complexity.

2.4. Virtual Test Tubes. The foregoing sections have discussed events that are expected to occur outside the test tube in preparation for the actual experiments to be carried out in the tube. At this point, it is only natural to ask, why stop here?

Why not provide a simulation of the reactions that might actually happen, full with all the random brownian motion and chains of complex molecular interactions that undoubtedly take place in actual test tubes? The sneak preview is worth watching, particularly since it is much less costly than the actual test tube run.

`Edna` includes tools to perform a simulation and analysis of test tube experiments based on the measures presented in the previous sections. The simulation is based on purely local molecular interactions between cells in a cellular automaton-like space. The cells represent quanta of 2D or 3D space that may be empty or occupied by nucleotides, solution molecules, and/or other reactants. Each cell is also characterized by associated parameters that render the tube conditions in a realistic way, such as temperature, salinity, covalent bonds, etc. The cellular space can be set up to boundary conditions that reflect tube walls. The transitions of the automaton implement the massive random motion of single stranded molecules compounded over space and time, their multiple attempts and eventual success at hybridizing into double stranded molecules according to various hybridization rules, under various temperatures and reaction conditions and for a variety of input encodings. The virtual tube comes to equilibrium under the Gibbs free energy function, showing a possible outcome of the actual experiment. That this asymptotic behavior has something to do with the actual test tube run can be justified on several grounds. After all, the underlying physical chemistry is based on fairly well known *local* interactions among the molecules and between them and their tube environment. What is unknown, and presumably very powerful, is the compounded effect of all the massive number of interactions in the real tube. A virtual test tube run represents a Montecarlo sample of that effect. Second, experiments that have been actually performed in the lab, such as Adleman's experiment, have been run on `Edna`. Their outcomes may be compared with the known outcomes to benchmark the reliability of the virtual tube and the analyses on which is has been built. The degree of reliability of the virtual test tube can thus be appropriately validated. Preliminary results on the HPP show that `Edna` can be a powerful tool to debug, and in many cases even perform, molecular protocols with high reliability. A systematic study in under way and will be reported elsewhere.

2.5. Discussion and Conclusions. The approach in this paper has been to explore the contribution that silicon-based computers can make to the success of biomolecule-based computing. This approach is justified for several reasons. First, as a new computing paradigm, molecular computing must be evaluated against the performance of conventional solid-state based computing. The impressive progress rates and proven reliability of the latter has led many practicioners to believe that molecular computing can only compete advantageously in a niche of 'killer' applications. Second, basic obstacles remain in the way of bringing the potential of molecular computing to the realm of practical, efficient and reliable computations. A critical one is that molecular computing (MC) utilizes the complex interaction of biomolecules and protocols based on biotechnological developments to perform computation. Although biotechnology to manipulate molecules has been developed to satisfactory standards for biological purposes, they are far from meeting analogous standards in conventional computing media. In particular, molecular computing lab runs in the test tube are too complicated, unreliable, inefficient, and expensive when compared to ordinary runs of conventional software. A re-evaluation of the models on which such a teachnology is based is required.

Edna, a software platform, has been developed to address the basic problems of encoding, reliability, efficiency and scalability for molecular protocols using DNA molecules. The platform will allow MC practicioners to take advantage of digital computers to gain realistic insights on actual test tube performance of a protocol before it actually unfolds in the tube. Edna is object-oriented and extensible, so that it can easily evolve as the field progresses. Edna is therefore a research tool that makes it possible to use the advantages of conventional computing to bring to molecular computing comparable levels of reliability and efficiency.

Finally, Edna poses a number of interesting questions directly addressing what is perhaps the critical issue in MC: is it the case that there exist elemental unknown computations happening deep inside biomolecule ensembles, powerful enough to propel molecular computing beyond the confines of conventional computers? Or, is it the case that the results of molecular computing could be achieved, in principle, by some sort of electronic DNA, but at such an enormous computational expense to be impossible in real life due to the sheer massive number/volume of molecules involved and/or the inability of current electronics to make more efficient use of the much smaller electrons? Preliminary estimates based on extrapolations of a distributed version of Edna running over computer networks (which in fact would create a virtual test tube the size of the internet) give some evidence that even in the latter case, Adleman's experiment would be somewhat of a challenge.

References

[1] H. T. Allawi, J. SantaLucia, Jr., Thermodynamics and NMR of Internal G*T Mismatches in DNA. Biochemistry **36**, 10581 (1997).

[2] L. M. Adleman, Science, **266**, 1021 (1994).

[3] L.M. Adleman (1996) On Constructing a Molecular Computer. In [**16**], 1-21.

[4] A. Benight, R. Wartell, and D. Howell, Theory agrees with experimental thermal denaturation of short DNA restriction fragments. *Nature* **289**, 15 (1981).

[5] C. R. Cantor, P. R. Schimmel, *Biophysical Chemistry, Part III: The Behavior of Biological Macromolecules* (Freeman, New York, 1980).

[6] R. Deaton, M. Garzon, R. E. Murphy, J. A. Rose, D. R. Franceschetti, S. E. Stevens, Jr., Physical Review Letters, **80**, 417 (1998).

[7] A. Frutos, Q. Liu, A. Thiel, A. Sanner, A. Condon, L. Smith, R. Corn, Demonstration of a Word Design Strategy for DNA Computing on Surfaces. *Nucleic Acids Research* **25**, 4748 (1997).

[8] M. Garzon, P. Neathery, R. Deaton, R.C. Murphy, D.R. Franceschetti, S.E. Stevens, Jr.. A New Metric for DNA Computing. In [**13**], 472-478.

[9] M, Garzon, N. Jonoska, S. Karl. The Bounded Complexity of DNA Computing. Proc. of the 4th workshop, Princeton University, 1998. In [**23**] and special issue of *Biosystems*.

[10] A. J. Hartemink, D. K. Gifford, Thermodynamic Simulation of Deoxyoligonucleotide Hybridization of DNA Computation, in [**15**].

[11] J. Khodor, D.K. Gifford, A. Hartemink (1998), Design and Implementation of Computational Systems Based on Programmed mutagenesis. In [**23**], 101-107; 287-297.

[12] R. Karp, C. Kenyon, O. Waarts (1996), Error-resilient DNA Computation. Proc. 7th Annual Symposium on Discrete Algorithms, SODA 458-467.

[13] J.R. Koza, K. Deb, M. Dorigo, D.B. Fogel, M. Garzon, H. Iba, R.L. Riolo, eds. (1997). *Proc. 2nd Annual Genetic Programming Conference*, Morgan Kaufmann.

[14] J.R. Koza, K. Deb, M. Dorigo, D.B. Fogel, M. Garzon, H. Iba, R.L. Riolo, eds. (1998). *Proc. 3rd Annual Genetic Programming Conference*, Morgan Kaufmann.

[15] L.F. Landweber, E.B. Baum (eds.), *DNA Based Computers II*, Proc. of the 2nd workshop, Princeton University, 1996. DIMACS series of the American Mathematical Society, vol. 44 (1999), 247-258, Providence RI.

[16] R. Lipton, E. Baum (eds.) *DNA Based Computers*. DIMACS Series in Discrete Mathematics and Theoretical Computer Science, vol. **27**. American Mathematical Society, Providence RI.

[17] L. Marky, K. Blumenfeld, S. Kozlowski, and K. J. Breslauer, Salt-Dependent Conformational Transitions in the Self-Complementary Deoxydodecanucleotide d(CGCAATTCGCG): Evidence for Hairpin Formation. *Biopolymers*, **22**, 1247-1257 (1983).

[18] K. U. Mir, *DNA Based Computers II, Princeton University, 1999*, DIMACS Proc. Series (American Mathematical Society, Providence, RI, 1997).

[19] T. Paner, M. Amaratunga, M. Doktycz, A. Benight, Analysis of Melting Transitions of the DNA Hairpins formed from the Oligomer Sequences d[GGATAC$(X)_4$GTATCC](X = A,T,G,C). *Biopolymers* **29**, 1715 (1990).

[20] M. T. Record, T. Lohman, A Semiempirical Extension of Polyelectrolyte Theory to the Treatment of Oligoelectrolytes: Application to Oligonucleotide Helix-Coil Transitions. *Biopolymers* **17**, 159 (1978).

[21] J. A. Rose, R. Deaton, D. R. Franceschetti, M. Garzon, S. E. Stevens, Jr., A Statistical Mechanical Treatment of Error in the Annealing Biostep of DNA Computation. Special program in DNA and Molecular Computing at the Genetic and Evolutionary Computation Conference (GECCO-99), Orlando FL, July 13-17, 1999. Morgan-Kaufmann, 1829-1834.

[22] P. Ross, F. Howard, M. Lewis, Thermodynamics of Antiparallel Hairpin-Double Helix Equilibria in DNA Oligonucleotides from Equilibrium Ultracentrifugation. *Biochemistry* **30**, 6269 (1991).

[23] H. Rubin. D. Wood (eds.), 4th DIMACS workshop on DNA Computers. Proc. of the 4th workshop, University of Pennsylvania, 1998. DIMACS Series in Discrete Mathematics and Theoretical Computer Science, vol. **48**. American Mathematical Society, Providence RI.

[24] J. SantaLucia, Jr., A unified view of polymer, dumbbell, and oligonucleotide DNA nearest-neighbor thermodynamics *Proc. Natl. Acad. Sci.* **95**, 1460 (1998).

[25] W. D. Smith, *DNA Based Computers, Princeton University, 1996*, DIMACS Proc. Series (American Mathematical Society, Providence, RI, 1996).

[26] Progress toward Demonstration of a Surface-based DNA Computation. In [**14**], 709-717.

[27] G. Vesnaver, K. Breslauer, The contribution of DNA single-stranded order to the thermodynamics of duplex formation. *Proc. Natl. Acad. Sci.* **88**, 3569 (1991).

[28] R. M. Wartell, A. S. Benight, Thermal Denaturation of DNA Molecules: A Comparison of Theory with Experiment. PHYSICS REPORTS (Review Section of Physics Letters) **126**, 67 (1985).

[29] J. G. Wetmur, *Preliminary Proceedings of the Third Annual Meeting on DNA Based Computers, University of Pennsylvania, 1997*, DIMACS Proc. Series (American Mathematical Society, Providence, RI, 1997).

[30] B-T Zhang, S-Y Shin, Molecular Algorithms for Efficient and Realiable DNA Computing, In [**14**], 735-742.

(M. Garzon) Computer Science, The University of Memphis, Memphis, TN 38152
E-mail address: `mgarzon@memphis.edu`

(R. Deaton) Electrical Engineering, The University of Memphis, Memphis, TN 38152
E-mail address: `rjdeaton@memphis.edu`

(J. Rose) Electrical Engineering, The University of Memphis, Memphis, TN 38152
E-mail address: `johnrose@memphis.edu`

DIMACS Series in Discrete Mathematics
and Theoretical Computer Science
Volume **54**, 2000

A Study on the Hybridization Process in DNA Computing

Masahito Yamamoto, Jin Yamashita, Toshikazu Shiba, Takuo Hirayama, Shigeharu Takiya, Keiji Suzuki, Masanobu Munekata, and Azuma Ohuchi

ABSTRACT. A hybridization process is a basic and significant process in DNA computing. Therefore, in order to clarify the problems that may be occurred in the process, we have to analyze the process in detail. In this paper, we focus on the hybridization process. First, we introduce a simulation model of the process and then discuss the usefulness of the simulator. Next, we propose a new encoding method for avoiding miss-annealings (errors). In order to show the validity of the simulator and the effectiveness of our encoding method, we introduce a Hamiltonian path problem with 4 vertices and discuss the results of laboratory experiments.

1. Introduction

Adleman has shown that the instance of directed Hamiltonian Path Problems (HPPs) can be solved by biological experiments using DNA [**1**]. Adleman's experiment was based on the "generate and test" paradigm, which first randomly generates all candidate solutions in parallel and then chooses the correct solution by biological operations. Following Adleman's experiment, other NP-complete problems, such as the satisfiability problem (SAT) and the maximal clique problem, have been solved[**2**][**3**]. However, few authors have described experiments in detail or discussed practical problems in DNA computing. Even with regards to HPPs, there have been no detailed reports on what kinds of path are generated or what kinds of error occur in the DNA computing process. DNA computing for solving an HPP essentially consists of the following three processes. One process is the encoding of the problem to DNA oligonucleotides. To implement DNA computation, we have to design DNA sequences of the oligonucleotides to be used. The vertices and edges of the graph are encoded in oligonucleotides, or single strands, of DNA. Another process is hybridization, which is the process of generating many paths in the directed graph by using the feature for fusing two single stranded molecules by Watson-Click complementary base pairing. The third process is the detection of the DNA path encoded as the Hamiltonian path in the graph. However, in order to detect the desired DNA path in this process, a sufficient amount of DNA paths must be generated in the first two processes. As Kurtz et al. pointed out, in simple HPPs, such as the line digraph, the rate of production of target DNAs reduces

2000 *Mathematics Subject Classification.* Primary 92C40; Secondary 05C45.

exponentially in problem size [**4**]. Therefore, it is clearly important to know the expected amount of target DNAs in advance for more difficult problems.

In this paper, in order to discuss these features of the DNA computing process, we focus on the hybridization process. Since the hybridization process is a basic and significant process in DNA computing, analyzing this process would provide useful information for future works on DNA computing. First, we present our simulation model of the hybridization process based on the concentration dynamics model. This simulator takes the concentration of each of the DNA oligonucleotides and computes the concentration of the resultant DNA paths. By simulating the hybridization process in DNA computing, we can estimate the concentration of the desired DNA paths in advance. Next, we propose a new encoding method that enables HPP to be encoded into some 20 mer DNA oligonucleotides. Previous works have revealed that the method used to encode the HPP into DNA oligonucleotides affects the reliability of computation [**5**][**6**][**7**]. Our proposed method is based on the degree of binding between a pair of DNA oligonucleotides.

In order to achieve the above two objectives, we introduce an HPP with 4 vertices. This problem is a simple feedback digraph F_4 in Kurtz's paper, and seems to be nonsense to solve in itself, but it is appropriate to discuss some practical problems that occur in DNA computing [**4**]. We confirm that the simulator computed the correct concentration of DNA paths by determining the quantity of these paths practically. Furthermore, we introduce a "stopper", which stops the generation of paths in the hybridization process, and by using some "stoppers" with various degrees of similarity between themselves and any other DNA oligonucleotides, we show the effectiveness of our proposed encoding method. Based on the results of experiments, we show the effectiveness of our simulation model of DNA computing and clarify the practical problems that occur in the DNA computing process. Furthermore, we verify that our proposed degree of binding between two oligonucleotides is effective for avoiding errors in computation.

2. A Simulation Model of the Hybridization Process

In this section, we propose a simulation model of the hybridization process based on the concentration dynamics model. First, some definitions are given and then the simulation model is described.

2.1. Definitions.

DEFINITION 2.1. Let p be a nucleic acid base, i.e., $p \in \{\text{A,G,C,T}\}$, then the Watson-Crick complement $\overline{p}$ of p is denoted as follows:

$$\overline{p} = \begin{cases} \text{A} & (p = \text{T}) \\ \text{G} & (p = \text{C}) \\ \text{C} & (p = \text{G}) \\ \text{T} & (p = \text{A}) \end{cases}$$

DEFINITION 2.2. A $5'$ to $3'$ oligonucleotide of length n is denoted as $s \stackrel{def}{=} |p_1p_2p_3\cdots p_n >$, where $p_i \in \{\text{A,G,C,T}\}$ $(i = 1, \cdots, n)$, while a $3'$ to $5'$ oligonucleotide is represented as $s^R \stackrel{def}{=} < p_1p_2p_3\cdots p_n|$. Then, the complementary pair of s is represented as follows: $\overline{s}^R \stackrel{def}{=} < \overline{p_1p_2p_3}\cdots\overline{p_n}|$.

DEFINITION 2.3. Let s be a single-strand DNA sequences. Given a Watson-Click complementary pair $\overline{s}^R$ of s. The degree of binding $\mathcal{F}(s)$ is defined as follows:

$$\mathcal{F}(s) \stackrel{def}{=} n(3 \cdot \%(\mathrm{G}+\mathrm{C}) + 2 \cdot \%(\mathrm{A}+\mathrm{T})) \tag{2.1}$$

where %(G+C) and %(A+T) are the GC content and the AT content, respectively. The degree of binding represents the number of hydrogen bonds between s and $\overline{s}^R$.

From now on, we assume that we treat only linear DNAs, neither branched DNAs nor secondary structure DNAs, in our simulation model.

DEFINITION 2.4. We define the DNA complex as follows:

1. A single-strand DNA is a DNA complex.
2. A linear double-stranded DNA that has at most two sticky ends is a DNA complex.

DEFINITION 2.5. Let x_i and x_j be the DNA complexes. When x_i and x_j include subsequences $s_1, s_2, \cdots, s_m$ and $\overline{s_1}^R, \overline{s_2}^R, \cdots, \overline{s_m}^R$ in exposed regions (sticky ends), respectively, and a threshold non-negative integer θ is given, the degree of binding $\mathcal{F}_\theta$ between x_i and x_j is defined as follows:

$$\mathcal{F}_\theta(x_i, x_j) \stackrel{def}{=} \begin{cases} \max_k\{\mathcal{F}(s_k)\} & (\text{if } \max_k\{\mathcal{F}(s_k)\} \geq \theta) \\ 0 & (\text{otherwise}). \end{cases} \tag{2.2}$$

DEFINITION 2.6. Let x_i and x_j be the DNA complexes and a threshold non-negative integer θ is given. If $\mathcal{F}_\theta(x_i, x_j) > 0$, then we call that a new DNA complex x_m can be generated by DNA complexes x_i and x_j, and denote the situation by $x_i * x_j \to x_m$.

Note that the different DNA complexes may be generated from the same pair of DNA complexes.

DEFINITION 2.7. Let $\mathcal{O}$ be a set of single-strand DNAs. The DNA complex on $\mathcal{O}$ is defined recursively as follows:

1. A single-strand DNA $O_i \in \mathcal{O}$ is a DNA complex on $\mathcal{O}$.
2. Let x_i, x_j be the DNA complexes. Then a DNA complex x_m so that $x_i * x_j \to x_m$ is a DNA complex on $\mathcal{O}$.

DEFINITION 2.8. From above definitions, we can define a set B_i of indices of DNA complexes on $\mathcal{O}$ and a set M_i of the pair of indices of DNA complexes on $\mathcal{O}$ as follows:

$$B_i \stackrel{def}{=} \{j | \mathcal{F}_\theta(x_i, x_j) > 0\}. \tag{2.3}$$

$$M_i \stackrel{def}{=} \{(j,k) | x_j * x_k \to x_i, \;\; \mathcal{F}_\theta(x_j, x_k) > 0\}. \tag{2.4}$$

2.2. Simulation Model. Using the previous definitions, we represent the concentration dynamics model of DNA complexes as shown in (2.6).

$$\begin{aligned} \frac{d[x_i]}{dt} &= \sum_{(j,k)\in M_i} (Va_{jk}[x_j][x_k] - Vd_i[x_i]) \\ &\quad - \sum_{j \in B_i} (Va_{ij}[x_i][x_j] - Vd_m[x_m]) \\ &\qquad\qquad (x_i * x_j \to x_m). \end{aligned} \tag{2.5}$$

where $[x_i]$ represents the concentration of the DNA complex x_i at time t and Va_{ij} is the reaction rate constant of annealing between two DNA complexes x_i and x_j, and Vd_i is that of denaturing of the DNA complex x_i.

Our hybridization simulator takes the initial concentration of each DNA oligonucleotide and computes the concentration of the resultant DNA paths. In the process of computation, annealing with low degree of binding (less than θ) are ignored. The computation is performed by the 4-dimensional Runge-Kutta method.

In this paper, for the simplicity, we set the reaction rate constants are all equal, i.e., $Va_{ij} = Va$ (constant) and $Vd_i = Vd$ (constant) similarly to Kurtz's analysis [**4**]. In our simulator, we can compute the final concentration of DNA complexes for arbitrary graphs, depending on the sequences of initial DNAs.

2.3. Encoding Method. In this subsection, we propose a new encoding method based on the above degree of binding. We define the degree of similarity between two DNA oligonucleotides as follows.

DEFINITION 2.9. Let O_i and O_j be single-strand DNAs and a threshold non-negative integer θ is given. When both have common subsequences $s_1, s_2, \cdots, s_m$, the degree of similarity Φ between O_i and O_j is defined as follows:

$$\Phi(O_i, O_j) \stackrel{def}{=} \begin{cases} \max_k\{\mathcal{F}(s_k)\} & (\text{if } \max_k\{\mathcal{F}(s_k)\} \geq \theta) \\ 0 & (\text{otherwise}). \end{cases} \tag{2.6}$$

In order to design the DNA sequences, we start from the initial sequences generated randomly, and tried to minimize the summation of the above Φ for all pairs of vertices and edges. Furthermore, we considered some complemented sequences and intermediate sequences (subsequences that are generated by ligation between two DNA sequences). Actually, designed sequences were calculated by a procedure based on the simulated annealing method. As a result, we obtained the sequences shown in Tab. 1. Moreover, we prepared some stoppers mentioned above for the experiments as shown in Tab. 2.

TABLE 1. Designed sequences of vertices and edges

Name	Sequences($5' \to 3'$)
O_0	GCCTA CCTTT CCGTT CTGCG
O_1	TCTTG CTGTG GTGTC GGTTG
O_2	TGAGG ATGGG ACTTC GTCAG
O_3	TAATG CCCCT GGAGC CAAAA
O_{01}	CACAG CAAGA CGCAG AACGG
O_{12}	CCCAT CCTCA CAACC GACAC
O_{21}	CACAG CAAGA CTGAC GAAGT
O_{23}	AGGGG CATTG CTGAC GAAGT

3. Experimental Settings

In order to confirm that our proposed simulation model is accurate and to show the effectiveness of our encoding method, we introduced an HPP with 4 vertices as shown in Fig. 1.

TABLE 2. Designed sequences of stoppers

Name	Sequences	Degree of binding with O_1	Note
S_1	CAACC GACAC	26	complement
S_2	GACCC GACAC	19	
S_3	TGTAA GACAC	13	
S_4	TGTAA TTCAC	8	
S_5	TATGT AAGCG	7	minimum

FIGURE 1. The directed graph with 4 vertices.

This HPP is the problem of deciding whether or not there exist Hamiltonian paths from the vertex 0 to the vertex 3. This problem is very simple and nonsense to solve in itself, but it is appropriate to discuss some practical problems that occur in DNA computing. Since the graph has a cycle between the two vertices 1 and 2, there exist infinite paths from the vertex 0 to the vertex 3 in the graph, i.e., 0123, 012123, 01212123, $\cdots$, etc.. Therefore, we can confirm that the simulator computed the correct concentrations of DNA paths by comparing them with the quantity of these paths practically. In this paper, we set $Va = 0.01$, $Vd = 0$, i.e., $Va/Vd = \infty$, thus denaturing can be ignored. Also, we set the maximun number of steps in Runge-Kutta method to 100,000.

Furthermore, we introduce a "stopper", which stops the generation of paths in the hybridization process, as shown in Fig 2. Using some "stoppers" with various degrees of similarity between themselves and any other DNA oligonucleotides, we show the effectiveness of our proposed encoding method.

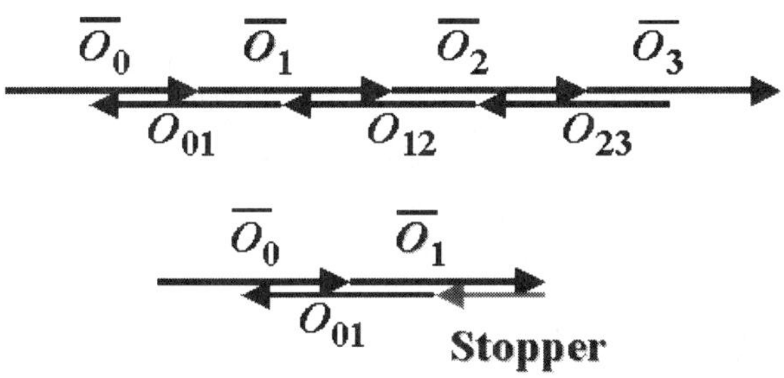

FIGURE 2. Stopper

4. Materials and Methods

4.1. Construction of random paths (Step 1). To construct random DNA paths with 4 vertices, a mixture containing 50 pmols of each oligonucleotide was made. The reaction contains 8 oligonucleotides encoding 4 vertices and 4 edges, and an additional oligonucleotide was also added when the stopper was used. The 5' end of each oligonucleotide was phosphorylated by using 30 units of T4 polynucleotide kinase in 270 μl of reaction mixture containing 50 mM Tris-HCl pH 7.5, 10 mM MgCl2, 5 mM dithiothreitol (DTT), and 0.2 mM ATP. The reaction was performed at 37 °C for 30 min, and then the reaction mixture was heated to 70 °C, kept at that temperature for 10 min, and gradually cooled to construct random paths by the annealing of oligonucleotides. ATP and DTT were added at the final concentrations of 0.5 mM and 9 mM, respectively, to the reaction mixture, and then 350 units of T4 DNA ligase was added. The ligation reaction was performed at 16 °C for 12 hr.

4.2. Amplification by PCR of DNA paths that begin with V_0 and end with V_3. Amplification of DNA paths that begin with V_0 and end with V_3 was done using 1 μl of ligation mixture as a template and 2.5 units of Taq DNA polymerase (GIBCO BRL) by appropriate cycles of PCR incubation at 94 °C for 45 sec, at 62 °C for 30 sec, and 72 °C for 45 sec. Two specific primers that can anneal with V_0 [SP0 (5'-GCCTA CCTTT CCGTT CTGCG-3')] and V_3 [SP3 (5'-TTTTG GCTCC AGGGG CATTG- 3')] were added to the PCR reaction.

4.3. Analyses of DNA Paths Amplified by PCR.. Amplified DNA paths were applied to 6 % polyacrylamide gels, and the DNA bands were visualized by ethidium bromide. The image of the gel electrophoresis was scanned by a Bio Image Analyzer (FA1000, Aisin Cosmos Co., Japan). Estimation of the amount of DNA in each DNA band was done by measuring the intensity of the bands using an image-analyzing software (NIH image).

5. Results and Discussions

TABLE 3. Quantification of the relative amount of each DNA path by measuring the intensity of DNA bands

Path	Number of Cycles					
	0	7	10	15	20	25
0123	0	0	0.24	0.41	0.96	1.00
012123	0	0	0	0.1	0.36	0.36
01212123	0	0	0	0	0.21	0.23

Tab. 3 shows the relative amount of DNA paths, after PCR amplification, measured by the above method and these values are plotted in Fig. 3. To estimate the relative amount of each DNA path, cycles of PCR amplification were varied, and the relative intensity of each DNA band of each cycle was plotted on a graph, as shown in Fig. 4. While the intensity of DNA bands linearly increased from 7 to 15 cycles of PCR, no efficient amplification was observed after 20 cycles of amplification. This could be because either primers or dNTPs for PCR amplification were used up after 20 cycles. In order to measure the relative amount of DNA paths precisely, the intensity of DNA bands of 15 cycles of PCR amplification (before shortage of primers or dNTPs) was estimated. A comparison of the relative populations of DNA paths after hybridization obtained from experiments and from simulation is shown in Fig. 5. The relative population of each DNA path obtained from experiments and from simulation was well matched.

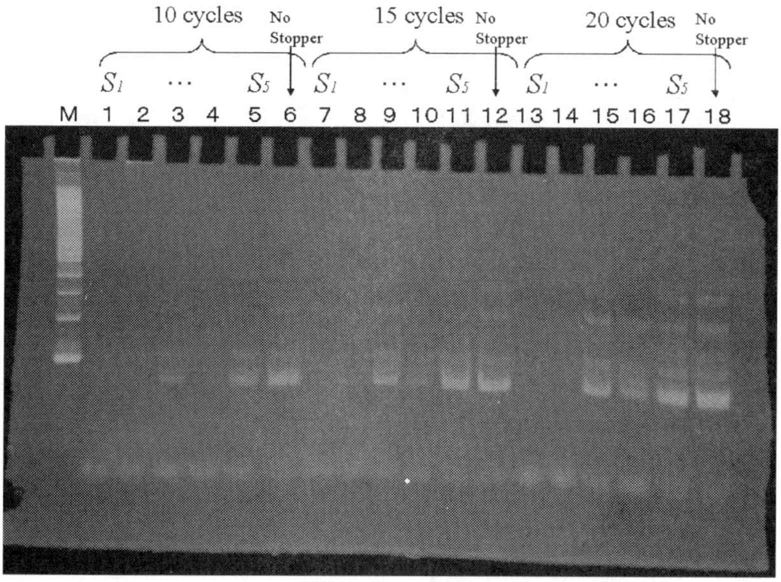

FIGURE 3. Formation and quantification of DNA paths after various cycles of PCR amplification. Amplified DNA paths after various cycles of PCR were analyzed by 6 % polyacrylamide gel electrophoresis [**8**]. DNA bands were visualized by ethidium bromide staining. Lane M, DNA size marker (100 bp ladder); lanes 1-6, amplified DNA paths after 10 cycles of PCR; lanes 7-12, 15 cycles of PCR; lanes 13-18, 20 cycles of PCR. In lanes 1, 7 and 13, S_1 was added to the hybridization reaction as a stopper DNA; in lanes 2, 8 and 14, S_2 was added; in lanes 3, 9 and 15, S_3 was added; in lanes 4, 10 and 16, S_4 was added; in lanes 5, 11 and 17, S_5 was added; and in lanes 6, 12 and 18, no stopper DNA was added.

Fig. 6 shows that a comparison of the relative amount of DNA paths when the degree of similarity of stoppers was varied. From this figure, the number of generated DNA paths varied according to the value of the degree of similarity of stoppers. This means that our proposed encoding strategy is effective for avoiding miss-annealing.

6. Conclusions

In this paper, we focus on the hybridization process in DNA computing and propose a simulation model of hybridization. Moreover, we present a new encoding method for avoiding errors. In order to discuss the validity of our simulation model and the effectiveness of our proposed encoding method, a simple HPP with 4 vertices is introduced. The results of laboratory experiments show that our simulation model is useful and that our proposed encoding method is relatively effective for avoiding errors. In the experiments, we used native gels, instead of denaturing gels, because the applied graph is very small and simple, but our analysis may become to be more accurate if we use denaturing gels [**9**].

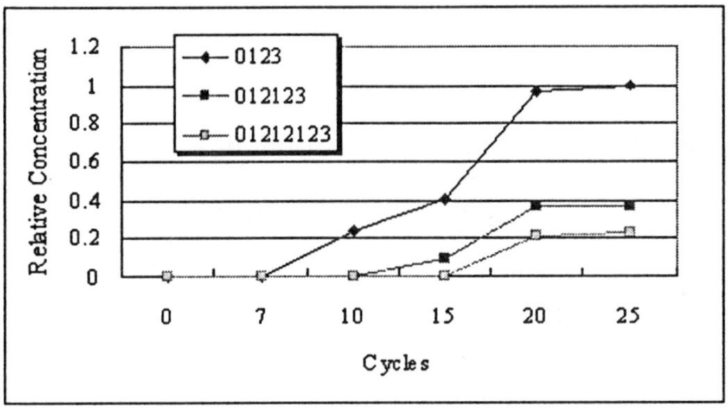

FIGURE 4. Quantification of the amount of each DNA path by measuring the intensity of DNA bands. The image of the gel electrophoresis was scanned by a Bio Image Analyzer (FA1000, Aisin Cosmos Co., Japan), and the relative intensity of each DNA band was quantitated by an image-analyzing software (NIH image). A DNA path 0→1→2→3 is represented by closed triangles, 0→1→2→1→2→3 by closed rectangles, and 0→1→2→1→2→1→2→3 by open rectangles.

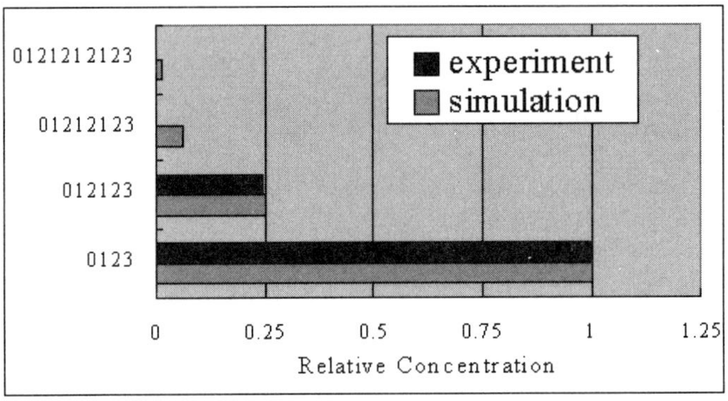

FIGURE 5. Comparison of the relative amounts of DNA paths obtained from experiments and from simulation.

References

[1] L. Adleman. Molecular Computation of Solutions to Combinatorial Problems *Science*, 266:1021-1024, 1994.
[2] R. J. Lipton. DNA Solution of Hard Computational Problems. *Science*, 268:542-545, 1995.
[3] Q. Ouyang, P. Kaplan, S. Liu and A Libchaber. DNA Solution of the Maximal Clique Problem. *Science*, 278:446-449, 1997.
[4] S. A. Kurtz, S. R. Mahaney, J. S. Royer, J. Simon. Active Transport in Biological Computing. *DNA Based Computers II, DIMACS Series in Discreate Mathematics and Theoretical Computer Science*, 44:171-179, 1999.
[5] M. Amos, A. Gibbons and D. Hodgeson. Error-Resistant Implementation of DNA Computations. Proc. of 2nd DIMACS Int. Conf. Evol. Comp., Indiannapolis, 243-248, 1996.

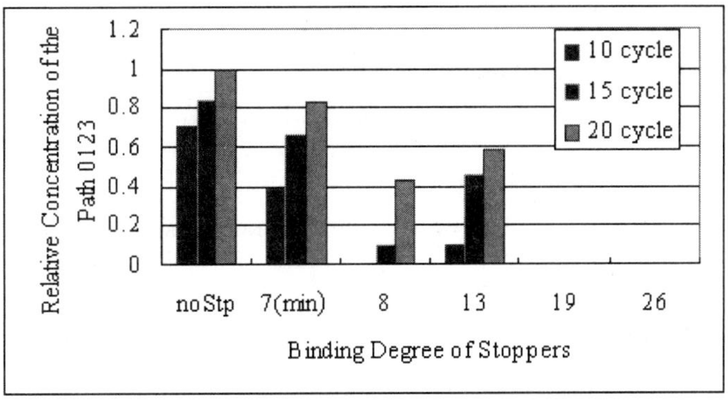

FIGURE 6. Comparison of the relative amounts of DNA paths when the degree of similarity of stoppers was varied.

[6] R. Deaton, R. C. Murphy, J. A. Rose, M. H. Garzon, D. R. Franceschetti and S. E. Stevens. Good Encodings for DNA-based Solutions to Combinatorial Problems. *DNA Based Computers II, DIMACS Series in Discreate Mathematics and Theoretical Computer Science*, 44:247-258, 1999.

[7] E. Baum. DNA Sequences Useful for Computation. *DNA Based Computers II, DIMACS Series in Discreate Mathematics and Theoretical Computer Science*, 44:235-241, 1999.

[8] J. Sambrook, E. F. Fritsh and T. Maniatis. Molecular Cloning: A Laboratory Manual, *2nd ed. Cold Spring Harbor Laboratory Press*, 1989.

[9] P. D. Kaplan, G. Cecchi, A. Libchaber. DNA based molecular computation: Template-template interaction in PCR. *DNA Based Computers II, DIMACS Series in Discreate Mathematics and Theoretical Computer Science*, 44:97-104, 1999.

(M. YAMAMOTO) GRADUATE SCHOOL OF ENGINEERING, HOKKAIDO UNIVERSITY, SAPPORO 060-8628, JAPAN

E-mail address: masahito@complex.eng.hokudai.ac.jp

(J. YAMASHITA) KYUSHU UNIVERSITY, FUKUOKA 612-53, JAPAN

E-mail address: jin@bio-math10.biology.kyushu-u.ac.jp

(T. SHIBA) GRADUATE SCHOOL OF ENGINEERING, HOKKAIDO UNIVERSITY, SAPPORO 060-8628, JAPAN

E-mail address: shiba@dove-mc.eng.hokudai.ac.jp

(T. HIRAYAMA) GRADUATE SCHOOL OF ENGINEERING, HOKKAIDO UNIVERSITY, SAPPORO 060-8628, JAPAN

E-mail address: caesar@complex.eng.hokudai.ac.jp

(S. TAKIYA) RESEARCH CENTER FOR MOLECULAR GENETICS, HOKKAIDO UNIVERSITY, SAPPORO 060-0810, JAPAN

Graduate School of Environment Earth Science, Hokkaido University, Sapporo 060-0810, Japan

E-mail address: takiya@rmg.hokudai.ac.jp

(K. Suzuki) Graduate School of Engineering, Hokkaido University, Sapporo 060-8628, Japan

E-mail address: suzuki@complex.eng.hokudai.ac.jp

(M. Munekata) Graduate School of Engineering, Hokkaido University, Sapporo 060-8628, Japan

E-mail address: munekata@dove-mc.eng.hokudai.ac.jp

(A. Ohuchi) Graduate School of Engineering, Hokkaido University, Sapporo 060-8628, Japan

E-mail address: ohuchi@complex.eng.hokudai.ac.jp

DIMACS Series in Discrete Mathematics
and Theoretical Computer Science
Volume **54**, 2000

Simulating Biological Reactions: A Modular Approach

Alexander J. Hartemink, Tarjei S. Mikkelsen, and David K. Gifford

ABSTRACT. We develop a general framework for simulating a sequence of biological reactions using small simulation modules. We demonstrate the usefulness of such a framework by implementing a simulator called the CYBERCYCLER. The CYBERCYCLER contains DNA hybridization, polymerization, ligation, and melting modules linked together to simulate a thermocycling process. This simulator enables us to interpret the behavior of our own programmed mutagenic unary counter in the laboratory. We describe the modules we implemented and then present a comparison of output from the CYBERCYCLER with the results of laboratory experiments to evaluate the effectiveness of the CYBERCYCLER simulator. We define a concrete specification for transformation modules and for populations of molecules passed between modules. This specification enables the construction of tools for simulating an arbitrarily complex sequence of biological reactions. We provide Java interfaces and classes for building such tools.

1. Introduction

Biological reactions implementing any non-trivial form of DNA-based computation are likely to be extremely complex: computations intended to solve large problems will require large populations of molecules, and the biological reactions used to carry out the computations will transform these populations in complicated ways. The prospect of reasoning about all the products being formed in these reactions is quite daunting.

In our laboratory, we have constructed a simple unary counter to experimentally demonstrate the feasibility of DNA-based computation using programmed mutagenesis [**12, 14**]. While the counter has been engineered to operate cleanly [**13**], the repeated thermocycling that increments the counter causes a wealth of product species to be formed after just a few cycles. This reaction complexity makes it difficult to predict or even interpret the results of frequent thermocycling; even when the results are predictable and interpretable, simply keeping track of the myriad interactions taking place during the reaction remains a significant challenge.

2000 *Mathematics Subject Classification.* Primary 68U20; Secondary 92C45, 92C40, 92B99, 80A30.

Hartemink is partially funded by a National Science Foundation Graduate Research Fellowship.

© 2000 American Mathematical Society

Tools that can predict, analyze, interpret, and keep track of the products being formed in biological reactions would be of great benefit. In that context, the contribution of this work is three-fold: first, we develop a general framework for building tools to simulate biological reactions; second, we present the CYBERCYCLER, a tool built within this framework and able to simulate the reactions necessary for understanding our laboratory unary counter implementation; and third, we provide Java interfaces and classes that define a concrete specification for this framework, enabling the construction of tools for simulating an arbitrarily complex sequence of biological reactions. The CYBERCYCLER demonstrates that this general framework is useful for building simulation tools, and the Java interfaces and classes instantiate this framework, producing a flexible and extensible environment for building simulation tools more easily.

Overview of this Paper. We begin by surveying some related work to motivate our research and then develop a general framework for building tools to simulate biological reactions. Once the basic elements of this framework have been elucidated, we present the CYBERCYCLER, a tool for simulating thermocycled reactions such as those incrementing our unary counter [**12, 14, 13**]. We briefly outline the various modules we implemented for the CYBERCYCLER, and then compare the output of the CYBERCYCLER with the results of laboratory experiments to evaluate the effectiveness of the CYBERCYCLER simulator. Finally, we offer a collection of Java interfaces and classes as a specification for building general simulation tools. We close by suggesting some directions for future work.

2. Related Work

The need for rational design and analysis of biological reactions has spawned a variety of software tools. These tools can generally be divided into three categories: thermodynamic prediction of secondary structure, thermodynamic analysis of hybridization (including optimal PCR primer selection), and simulation of specific models of DNA-based computation.

The *mfold* algorithm developed by Zuker, *et al.* [**31**] is perhaps the most notable program using thermodynamics to predict secondary structure of single-stranded DNA and RNA.

Tools in the second category include BIND by Hartemink, *et al.* [**12**], Oligo from MBI, Inc. [**15**], HYBsimulator from AGCT, Inc. [**2**], and the recent HYTHER program developed at the SantaLucia laboratory [**18**]. PCRsim by Rubin, *et al.* [**22**] and the Amplify program by Engels [**9**] simulate standard PCR reactions.

Among the various tools developed for the purpose of simulating specific models of DNA-based computation, Reif, *et al.* [**21**] have developed a graph-theoretic model for simulating recombinant DNA reactions, and Hagiya, *et al.* [**10**] have developed an applet for analyzing their DNA computing algorithms. Hartemink, *et al.* [**13**] wrote SCAN to optimally select nucleotide sequences according to given design constraints.

We are unaware of previous work describing the simulation of arbitrary sequences of DNA hybridization, polymerization, and ligation reactions. Though each of these reactions has been studied and simulated separately, we were unable to combine available tools so as to use them in concert.

3. General Framework for Simulation

While a plethora of software tools exist to simulate specific reactions, there is a profound lack of tools designed to simulate arbitrary sequences of biological reactions. A framework that effectively facilitates such simulation must be flexible, extensible, and general. By developing an "open source" framework with these properties, we hope to do more than construct a simulation tool; we hope to offer a means for easily building and expanding simulation tools.

3.1. Modular Design. In our framework, we represent a complicated sequence of reactions as a sequence of transformations on molecular populations. Populations are the material of a reaction, while transformations are the agents of change in a reaction. Transformations can represent active agents (*e.g.*, enzymes), environmental changes (*e.g.*, temperature increase), mechanical operations (*e.g.*, sampling or splitting populations), or filters (*e.g.*, micropore).

Transformation modules are implemented according to a concrete specification which allows them to be composed with one another. As a result, complex reactions can be simulated by building appropriate transformation modules and then linking them together in a sensible way.

This framework provides a tool-building environment which is flexible and completely general. It encourages code reuse, but remains fully extensible: researchers can tailor or tweak modules according to their needs without having to rewrite entire simulators. As new data or models for various transformations become available, modules can be updated and redistributed easily.

3.2. Taggable Data. We permit every species within a population to be associated with an arbitrary number of tags. Tags on species can be used to facilitate reaction simulation (*e.g.*, a tag indicating that a particular species is immobilized, allowing it to be separated from other species that are not), experimental analysis (*e.g.*, the history tag, discussed below), or output functionality (*e.g.*, the radio-label tag, also discussed below). Transformations may read or write tags in performing their respective duties, and are responsible for updating them as necessary.

4. CYBERCYCLER Design and Implementation

The CYBERCYCLER accomplishes two purposes. Primarily, it helps us better understand our unary counter by predicting the products being formed during the reaction as it is thermocycled. Secondarily, it serves as a "proof of concept" for our general simulation framework.

To build the CYBERCYCLER within our general simulation framework, we make a few simplifying assumptions. We assume that polymerase and ligase can be modeled as though they are acting serially even though they are both present simultaneously in the reaction. In the case of polymerase and ligase, this seems reasonable. Similarly, we assume that hybridization and polymerization can be modeled in a serial fashion. This seems justified by the fact that PCR frequently consists of a low temperature "annealing" step followed by a raised temperature "extension" step for polymerization, segregating the two operations so as to minimize non-specific binding.

In the context of our general framework, we implemented transformation modules representing melting, hybridization, polymerization, ligation, and CyberGel output. As shown in Figure 1, the first four of these modules are linked together

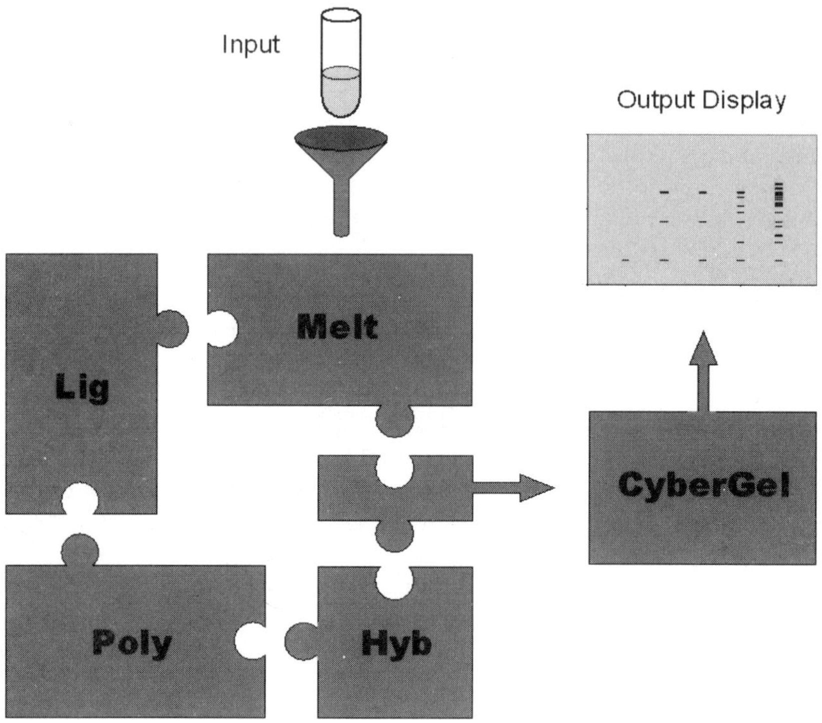

FIGURE 1. *A schematic look at the* CYBERCYCLER *simulator: In this depiction, various modules are composed to form the* CYBERCYCLER. *The initial population enters the melting module as input, and the resulting population is passed in turn to the hybridization module. The polymerization module receives its input from the hybridization module and passes its result to the ligation module, which cycles its result back to the melting module. The population is sampled after each time through the melting module, with the sample going to the CyberGel module for display.*

in a loop, and the population products are sampled and passed to the CyberGel module after each cycle.

Furthermore, we tag each strand in the population with a radio-label tag and a history tag. The radio-label tag is a simple Boolean tag indicating if a strand has been labeled with radioactive ^{32}P. Any labeled strand that is polymerized or ligated yields a labeled strand. In the CyberGel module, only labeled strands produce bands, enabling the output to be compared with autoradiographs from laboratory experiments.

The history tag describes the reaction in which a strand was formed. Strands present in the initial population have empty history fields, but any strand formed during a reaction is tagged with a string describing how it was formed. This allows every strand's pedigree to be tracked throughout the simulation, a form of analysis that is impossible in a laboratory setting.

In the following subsections, we describe the various CYBERCYCLER modules and the modeling choices we made in their implementation.

4.1. The Hybridization Module. The hybridization module takes a population as input and numbers each single-stranded species of DNA from 1 to n, where n is the total number of single-stranded species of DNA in the population. Then for each $i \in \{1, \ldots, n\}$ and each $j \in \{i, \ldots, n\}$, the module considers the possible hybridization of strand i with strand j at every offset position between the two strands[1]. For each of these strand×strand×position possibilities, the module calculates the predicted ΔH° and ΔS° for the hybridization reaction.

Predictions of enthalpy and entropy are calculated using the nearest neighbor stacking model [**12, 23, 24, 26, 20, 8**]. Thermodynamic parameters used in the calculation of ΔH° and ΔS° are the most recent published, taken from [**23, 17, 3, 4, 5, 6**]. Hybrid melting temperature is computed from ΔH°, ΔS°, strand concentrations, and cation concentration, as described in [**12, 29**]. The hybridization module is parameterized over the annealing temperature, and it considers as valid any binding whose melting temperature exceeds the annealing temperature.

For our purposes, a conservative (overstated) estimate of the possible products in the unary counter reaction is desirable. Therefore, for the sake of computational efficiency, our module implementation does not employ a chemical kinetic model to determine product concentrations, but instead uses only the simple heuristic that any new strand is formed at a concentration equal to the lowest concentration of any of the reactants involved in its formation. This tends to overstate strand concentrations, but the effect of overstated concentrations on predicted melting temperature is almost negligible and errs in the direction of conservative product estimation.

Different applications may require simulations that compute more accurate values of product concentrations, but this is easily achieved as a result of our modular framework: a re-implemented hybridization module could incorporate reaction dynamics to determine how much of each reactant will be used in forming each possible product[2].

4.2. The Polymerization Module. The polymerization module checks each strand in the population for bound strands present after the hybridization step. For each of the m bound strands, the module checks for a stable $3'$ end and if present, extends the bound strand until it reaches the end of the template, or another bound strand blocking its path. As before, our implementation is conservative in that it predicts extension products representing polymerization to each of these possible blockage lengths.

We assume that the polymerase exhibits perfect fidelity (it introduces no mismatched nucleotides, insertions, or deletions), full extension of the strand product

[1] This includes the cases where one strand extends past the end of the other strand; in total, the module considers length(i) + length(j) - 2 possible hybridization positions. Note also that the module considers the case where j is equal to i in order to check for possible dimer formation. Though the module does not currently check for monomeric secondary structures such as hairpins, our population specification is perfectly capable of representing such structures.

[2] See [**11, 25, 28**] for a more detailed discussion of reaction dynamics and chemical kinetics. In general, accurate prediction of product concentrations may be quite complicated since there are competitive reactions and multi-strand hybridization possibilities. For a modestly-sized population, however, it should not be too difficult to achieve a reasonable approximation of product concentrations.

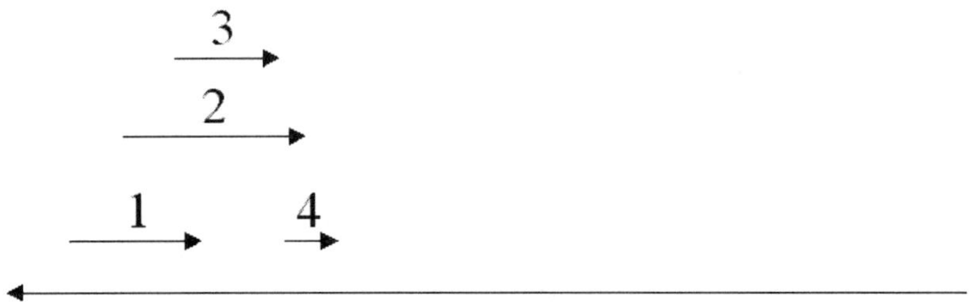

FIGURE 2. *Four binding strands may overlap: In this depiction, some of the binding strands overlap with one another in such a way that certain strand combinations preclude simultaneous binding. A naïve implementation might consider all 15 non-empty strand combinations, but our recursive implementation considers only the 6 that do not result in strand overlap.*

(until it encounters either a blocking strand or the end of the template), and no exonuclease activity (either $3' \to 5'$ or $5' \to 3'$).

In generating extension products corresponding to all possible blockage lengths, our module implementation incorporates a recursive polymerization procedure. Starting from the strand located most proximate to the $3'$ end of the template, the procedure calls itself again twice, once incorporating the strand, and once without it. In the recursive call where the strand is incorporated, the next strand considered is the first one whose $5'$ end is beyond the $3'$ end of the current strand. In the recursive call without the strand, the next strand considered is the next one in the $5'$-sorted list. This recursive procedure allows large portions of the binary subset tree to be pruned.

In the case shown in Figure 2 for example, the number of considered combinations is reduced from $2^m - 1 = 15$ to 6. As m grows and the degree of overlap increases[3], the computational savings becomes even more significant. Anecdotally, running times for simulations of five-cycle reactions with this recursive procedure were around a minute, whereas running times for an iterative version of the module were around an hour.

4.3. The Ligation Module. The ligation module receives the intermediate population[4] from the polymerization module and looks for possible ligation sites:

[3]Overlap becomes increasingly common in later cycles as polymerase generates new species from subsequences of existing species in every cycle.

[4]The population of strands that is passed between the two modules has no real significance and exists only because of our modeling assumption that we can treat the two as separable modules and apply them serially. We refer to such a population as an *intermediate pseudo-population*, since it is purely an artifact of the modeling. Our claim in modeling the reaction this way is that the population that results from applying the polymerization and ligation modules in series is a reasonably accurate approximation of the one produced if the modules were to be applied simultaneously, but of course, an artifactual pseudo-population may be produced along the way. This pseudo-population should not be considered as existing in a real experimental setting.

adjoining strands bound to a single template. If the $3'$ end of the upstream strand and the $5'$ end of the adjacent downstream strand are both stable, the module ligates the pair together to produce a longer strand.

Once again, the ligation module is conservative in that it assumes that any putative ligation can either fail or succeed and reports both possibilities as products. For our purposes, it suffices to assume that half the putative ligations fail and half succeed, so as to allow the CYBERCYCLER to consider both possibilities. Applications needing a more accurate measure of ligation efficiency are easily accommodated by a simple re-implementation of the module.

4.4. The Melting Module. Like the hybridization module, this module is parameterized over temperature. However, when the temperature is higher than the highest temperature permissive for nucleic acid hybridization, the actual implementation of this module is trivial. When that assumption holds, as is the case in our unary counter, the module simply examines the various species in the population and, for each one that is double-stranded, it produces two single-stranded species. The resulting population is then passed to the next module in the chain. In the case of the CYBERCYCLER, the population is sampled and then given to the hybridization module.

4.5. The CyberGel Module. The CyberGel module receives a sampled population and displays the nucleic acid species contained therein in the format of an electrophoretic gel. This enables a researcher to examine the contents of a population using an output modality that offers the benefits of familiarity and comparability with laboratory-produced gels. Furthermore, it permits the researcher to examine annotations associated with bands in the gel, which lends it a comparative advantage over laboratory gels. For example, for each band in the CyberGel, the researcher can learn the exact contents of the band: what strand species (or combination of species) is present in the band, what its sequence is, and how the strand was formed (courtesy of its history tag). Though the CYBERCYCLER maintains product concentrations for each strand in a population, the CyberGel module currently displays uniformly-sized bands. A better implementation would render the bands with a size and intensity proportional to the concentration of the corresponding strand species.

5. Using the CYBERCYCLER

We now turn to a discussion of how we use the CYBERCYCLER to predict the products that are formed during the operation of our unary counter. For verification purposes, we have also used the CYBERCYCLER to examine standard PCR, assembly PCR, and library assembly with ligase (as in Adleman's Hamiltonian Path experiment [**1**]), but those results are not included here.

The initial population in our CYBERCYCLER simulation consists of all the pieces of DNA necessary for implementing the programmed mutagenic unary counter (for a more detailed description of this counter see [**12, 14, 13**]). We radio-label one of the mutagenic rule strands (length 21bp) and then pass the population of five strands to the CYBERCYCLER. The CYBERCYCLER generates an extensive textual description of the products, consisting of the complete population, concentration, sequence content, bindings, reactions, and strand histories after each cycle. The CyberGel module creates a more intuitive graphical representation of the products.

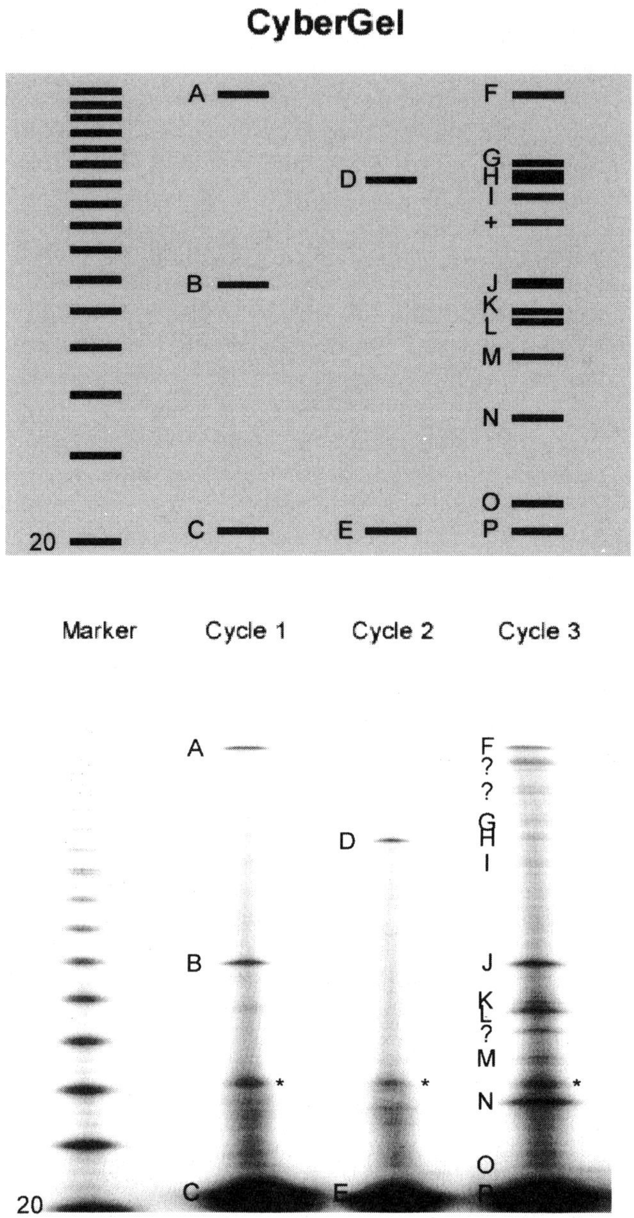

FIGURE 3. *The CyberGel output of a* CYBERCYCLER *simulation run at the same conditions as a set of reactions carried out in the laboratory. Bands in the CyberGel are shown uniformly-sized, independent of concentration. Bands labeled A through P correspond in each gel, while bands labeled with a + or ? are discrepancies. Bands labeled with a * are artifactual and can be discounted.*

In Figure 3, we present the CyberGel output of a CYBERCYCLER simulation run at the same experimental conditions as a set of reactions carried out in the laboratory. We see that the CYBERCYCLER is quite good at predicting and interpreting bands appearing in the laboratory gel. In particular, each of the bands labeled A through P in the laboratory gel has a corresponding band in the CyberGel.

The 42bp band in each cycle of the laboratory gel (labeled with a *) is an artifact and can be discounted. We have observed a band of exactly twice the length of the radio-labeled rule strand in hundreds of unary counter experiments. While the exact origin of this band is not known, it has been reproducibly observed by other researchers as well.

Nevertheless, a few discrepancies between the two gels still remain. One band in the CyberGel (labeled with a +) is not present in the laboratory gel. This is likely due to our conservative modeling: while the product represented by this band might potentially be able to form, it may not be at all likely.

Additionally, three third cycle bands in the laboratory gel (labeled with a ?) are missing in the CyberGel. We believe these bands represent strands that are shortened due to monomeric secondary structure in the unary counter template molecule. They are not predicted by the CYBERCYCLER because our hybridization module does not currently screen for monomeric secondary structure within the population.

Despite these discrepancies, one powerful feature of the CYBERCYCLER is its ability to track the histories of strands throughout the reaction. For example, if we examine the band in the CyberGel that did not appear in the laboratory gel (labeled with a +), we can see that it is 92bp long and is formed when strand C binds to strand D and is extended until it is blocked by strand B, to which it is ligated.

6. Specification of the General Framework

To provide a concrete specification of the framework developed in Section 3, we have developed Java interfaces for populations, population transformations, structures, and sequences, along with abstract classes implementing each of those interfaces. Furthermore, we have created a number of classes extending the abstract classes so that other researchers can begin to use these interfaces with minimal initial investment of time. Classes implementing these interfaces can be composed together as long as they adhere to the interface specification, enabling classes written at disparate locations to be distributed and used elsewhere. A complete description of the Java interfaces and classes, as well as all the source code for this specification are available from *http://www.psrg.lcs.mit.edu/biosim/*.

7. Future Work

We continue to develop the biological simulator specification and are in the process of creating more transformation modules. The CYBERCYCLER was originally written in C++ and its transformation modules are in the process of being converted to Java classes. As they are completed, they will be posted to *http://www.psrg.lcs.mit.edu/biosim/*.

Our modules could be extended as needed for different simulation applications. For example, the hybridization module might be re-implemented to check for hairpins and other monomeric secondary structures.

Our modules could also be supplemented as needed by implementing additional transformation modules. Modules that seem straightforward to implement include split, merge, radio-label (phosphatase and kinase), filter, immobilize, restriction cut, transcribe, reverse transcribe, and protein translate.

This large collection of modules could then be composed to represent reactions like PCR, RT-PCR, LCR, library construction, inserting a fragment of DNA into a plasmid vector, sticker bit-setting, affinity purification, etc.

An integrated simulation desktop environment would simplify tool-building immensely. A simulation desktop environment would permit a researcher to load a set of available transformation modules into his tool box, and then draw a flowchart-like representation of the experiment he wished to simulate, inserting the desired modules from his tool box as needed.

8. Conclusion

We believe our general framework for biological simulation is a useful one for a variety of reasons. Its modular design allows for code reuse and distributed code development, enables researchers to compose modules simulating arbitrarily complex sequences of reactions in the laboratory, provides an easy way for researchers to tailor modules to their desired precision or update modules as more data become available, and establishes a standard interface for future modules, yet to be developed.

The presence of taggable data is also useful in that it permits the modules to track various aspects of the reactions and provide the researcher with more information about what is happening to the various species during each reaction, much more than could be gleaned from laboratory experimentation and with dramatically less work. For instance, we implemented history tags to track when and how strands are formed during the thermocycling process, and radio-label tags to assist in generating an appropriate CyberGel output image.

We are eager to implement new modules and see what modules other groups decide to implement or improve. The collaborative possibilities of this simulation venture are extremely exciting.

Acknowledgments. The authors wish to thank Julia Khodor for her helpful comments and suggestions, and also for the provision of the electrophoretic gel displayed in Figure 3. Hartemink also gratefully acknowledges the support of the National Science Foundation.

References

[1] Adleman, L. (1994) *Science* **266**, 1021-1024.
[2] Advanced Gene Computing Technologies, Inc. *http://www.hybsimulator.com/.*
[3] Allawi, H. T. & SantaLucia, J. (1997) *Biochemistry* **36**, 10581-10594.
[4] Allawi, H. T. & SantaLucia, J. (1998) *Biochemistry* **37**, 2170-2179.
[5] Allawi, H. T. & SantaLucia, J. (1998) *Biochemistry* **37**, 9435-9444.
[6] Allawi, H. T. & SantaLucia, J. (1998) *Nucleic Acids Res.* **26**, 2694-2701.
[7] Anato, V. P. & Tinoco, I. (1992) *Nucleic Acids Res.* **20**, 819-824.
[8] Breslauer, K., Frank, R., Blöcker, H. & Marky, L. (1986) *Proc. Natl. Acad. Sci. USA* **83**, 3746-3750.
[9] Engels, W. R. (1993) *Trends in Biochemical Sciences* **18**, 448-450.
[10] Hagiya, M. *http://nicosia.is.s.u-tokyo.ac.jp/MCP/eng/index.html.*
[11] Hammes, G. (1978) *Principles of Chemical Kinetics.* Academic Press, New York.

[12] Hartemink, A. & Gifford, D. (1997) Thermodynamic Simulation of Deoxyoligonucleotide Hybridization for DNA Computation. *Proceedings of the 3rd DIMACS Workshop on DNA Based Computers*, June.

[13] Hartemink, A., Gifford, D. & Khodor, J. (1998) Automated Constraint-Based Nucleotide Sequence Selection for DNA Computing. *Proceedings of the 4rd DIMACS Workshop on DNA Based Computers*, June.

[14] Khodor, J. & Gifford, D. (1998) Design and Implementation of Computational Systems Based on Programmed Mutagenesis. *Proceedings of 4th DIMACS Workshop on DNA Based Computers*, June.

[15] Molecular Biology Insights, Inc. *http://oligo.net/*.

[16] Petruska, J., Goodman, M. F., Boosalis, M. S., Sowers, L. C., Cheong, C. & Tinoco, I. (1988) *Proc. Natl. Acad. Sci. USA* **85**, 6252-6256.

[17] Peyret, N., Anada, P., Allawi, H. T. & SantaLucia, J. (1999) *Biochemistry* **38**, 3468-3477.

[18] Peyret, N. & SantaLucia, J. *http://sun2.science.wayne.edu/~jslsun2/hyther.htm*.

[19] Pritchard, C. E. & Southern, E. M. (1997) *Nucleic Acids Res.* **25**, 3403-3407.

[20] Quartin, R. & Wetmur, J. (1989) *Biochemistry* **28**, 1040-1047.

[21] Reif, J. H. *http://bmc.cs.duke.edu/*.

[22] Rubin, E. & Levy, A. (1996) *Nucleic Acids Res.* **24**, 3538-3545.

[23] SantaLucia, J. (1998) *Proc. Natl. Acad. Sci. USA* **95**, 1460-1465.

[24] SantaLucia, J., Allawi, H. & Seneviratne, P. A. (1996) *Biochemistry* **35**, 3555-3562.

[25] Steinfeld, J., Francisco, J. & Hase, W. (1989) *Chemical Kinetics and Dynamics*. Prentice-Hall, Englewood Cliffs, New Jersey.

[26] Sugimoto, N., Nakano, S., Yoneyama, M. & Honda, K. (1996) *Nucleic Acids Res.* **24**, 4501-4505.

[27] Werntges, H., Steger, G., Riesner, D. & Fritz, H. (1986) *Nucleic Acids Res.* **14**, 3773-3790.

[28] Weston, R. & Schwartz, H. (1972) *Chemical Kinetics*. Prentice-Hall, Englewood Cliffs, New Jersey.

[29] Wetmur, J. (1991) *Crit. Rev. in Biochem. and Mol. Biol.* **26**, 227-259.

[30] Zenkova, M. A. & Karpove, G. G. (1993) *Uspekhi Khimii* **62**, 414-435.

[31] Zuker, M., Mathews, D. H. & Turner, D. H. (1999) *Algorithms and Thermodynamics for RNA Secondary Structure Prediction: A Practical Guide in RNA Biochemistry and Biotechnology*. J. Barciszewski & B. F. C. Clark, eds. NATO ASI Series, Kluwer Academic Publishers.

(A. Hartemink) Massachusetts Institute of Technology, Laboratory for Computer Science, 545 Technology Square, Cambridge, Massachusetts 02139
E-mail address: `amink@mit.edu`

(T. Mikkelsen) Massachusetts Institute of Technology, Laboratory for Computer Science, 545 Technology Square, Cambridge, Massachusetts 02139
E-mail address: `tarjei@mit.edu`

(D. Gifford) Massachusetts Institute of Technology, Laboratory for Computer Science, 545 Technology Square, Cambridge, Massachusetts 02139
E-mail address: `gifford@mit.edu`

DIMACS Series in Discrete Mathematics
and Theoretical Computer Science
Volume **54**, 2000

Experimental Progress in Computation by Self-Assembly of DNA Tilings

Thomas H. LaBean, Erik Winfree, and John H. Reif

ABSTRACT. Approaches to DNA-based computing by self-assembly require the use of DNA nanostructures, called tiles, that have efficient chemistries, expressive computational power, and convenient input and output (I/O) mechanisms. We have designed two new classes of DNA tiles, TAO and TAE, both of which contain three double-helices linked by strand exchange. Structural analysis of a TAO molecule has shown that the molecule assembles efficiently from its four component strands. Here we demonstrate a novel method for I/O whereby multiple tiles assemble around a single-stranded (input) scaffold strand. Computation by tiling theoretically results in the formation of structures that contain single-stranded (output) reported strands, which can then be isolated for subsequent steps of computation if necessary. We illustrate the advantages of TAO and TAE designs by detailing two examples of massively parallel arithmetic: construction of complete XOR and addition tables by linear assemblies of DNA tiles. The three helix structures provide flexibility for topological routing of strands in the computation, allowing the implementation of string tile models.

1. Introduction

In his seminal paper on molecular computation [**Adl94**], Adleman demonstrated a generate-and-separate algorithm for solving Directed Hamiltonian Path Problems (DHPP). In the first phase of his algorithm, a combinatorial library of DNA molecules representing potential solutions to the problem is generated by random hybridization and ligation of a small number of oligonucleotide species. In the second phase, a sequence of affinity separation and other laboratory procedures isolates the DNA molecules representing the actual solution; the sequence of this molecule, and thus the answer to the mathematical problem, can then be determined. However, the generate-and-separate strategy has two difficulties. First, for moderately large problems, a naïve generation strategy must generate an astronomical number of DNA strands to represent all potential solutions. Second, the number of laboratory procedures, each time consuming and error-prone, grows with the size of the problem.

The first difficulty can be avoided by interspersing the generate procedures and the separation procedures [**FB97, FB99**]. Simple examples of this approach include DNA implementations of dynamic programming [**BB98**] and breadth-first

2000 *Mathematics Subject Classification.* Primary 68Q05.

© 2000 American Mathematical Society

search [**MAS99, YS99**]. However, this approach usually comes at the cost of making the second difficulty more severe, by multiplying the number of procedures required.

Our work has focused on another approach: computation by self-assembly. Self-assembly is the spontaneous self-ordering of substructures into superstructures driven by annealing of, in this case, Watson-Crick base-pairing DNA sequences. Computation by self-assembly entails the building up of superstructures from starting units such that the assembly process itself performs the actual computation. Adleman made use of a simple form of computation by self-assembly in his original experiment [**Adl94**]: his algorithm does not blindly generate all possible sequences of vertices; instead, the oligonucleotide sequences and the logic of Watson-Crick complementarity guide the self-assembly processes so that only valid paths are generated. We have generalized this approach to two-dimensional (2D) self-assembly processes, where it can be seen that computation by self-assembly is Turing-universal [**Win96**]. The building blocks in these constructions are branched DNA complexes, which we call *DNA tiles*, consisting of several individual DNA oligonucleotides that associate with well-defined secondary structure. DNA tiles are designed to contain several short sections of unpaired, single-strand DNA (ssDNA) extending from the ends of selected helices that function as binding domains, which we call *pads*. Individual tiles interact by annealing with other specific tiles via their ssDNA pads to self-assemble into desired superstructures. The use of pads with complementary base sequences allows the neighbor relations of tiles in the final assembly to be intimately controlled; thus the only large-scale superstructures formed during assembly are those that encode valid mappings of input to output. Consequently, the second difficulty mentioned previously has been addressed: rather than implementing a DNA computing algorithm using a sequence of multiple laboratory procedures, our approach essentially uses only four: mixing the input oligonucleotides to form the DNA tiles, allowing the tiles to self-assemble into superstructures, ligating strands that have been co-localized, and then performing a single separation to identify the correct output.

We now give a brief description of the mathematical tiling and DNA construction techniques underlying our work.

1.0.1. *Computation by Tiling.* A class of domino tiling problems were defined by Wang [**Wan61**]. One is given a finite set of unit size square tiles, each of whose sides are labeled with symbols over a finite alphabet (the pads). Additional restrictions may include the initial placement of a subset of these tiles, and the dimensions of the region where tiles must be placed. The problem is to place the tiles, chosen with replacement and without rotation, to completely fill the given region so that each pair of abutting tiles have identical symbols on their contacting sides. Proofs of the undecidability of the infinite tiling problem rely on constructions wherein tiling patterns simulate single-tape Turing Machines [**Buc62, Wan63, Ber66, Rob71, Wan75**]. Other results include reductions of NP-complete problems to finite-size tiling problems [**LP81**]; tilings and patterns in general are surveyed in [**GS87**].

1.0.2. *DNA as a Construction Material.* Nano-fabrication of structures in DNA was pioneered by Seeman starting in the 1980s (e.g., see [**See82**]). Seeman and his students have built a multitude of ingenious constructions including: DNA branched junctions [**SCK89, WMKS91, DZS92**], knots [**SCD$^+$93**], Borromean rings [**MSS97**], a cube [**CS91**], and a truncated octahedron [**ZS94**] (reviewed in e.g.

[**See98**]). Recently, they have reported the construction from DNA of a nanomechanical device capable of controlled movement [**MSSS99**].

The most relevant constructs for our current purposes are the double-crossover (DX) complexes, which consist of two double-helices interlocked by exchange of oligonucleotide strands at two separate crossover points [**FS93**]. In DX complexes, a particular oligonucleotide strand base-pairs with its complement in one helix then crosses over to continue base-pairing with its complementary section on a second, adjacent helix; the multiple crossover points constrain the helix axes to be co-planar. Because DX complexes can be used as DNA tiles in self-assembly reactions, we will often refer to them as such.

DX complexes come in five varieties that differ from one another in the geometry of strand exchange and the topology of the strand paths through the tile. Anti-parallel crossovers cause a reversal in direction of strand propagation through the tile following exchange of strand to a new helix. We will primarily concern ourselves with tiles containing anti-parallel crossovers (DAO and DAE) rather than those with parallel crossovers (DPE, DPOW, DPON), because laboratory characterizations of parallel tiles showed them to be less stable structurally than anti-parallel tiles [**FS93**]. DX tiles provide up to four ssDNA pads for encoding associations with neighboring tiles. DX complexes have been shown to be sufficiently rigid to avoid formation of circular structures as seen with the more flexible dsDNA [**LYQS96**]. Recently, 2D lattices consisting of hundreds of thousands of DX units have been constructed and observed by atomic force microscopy [**WLWS98**], and DX complexes and lattices have been used successfully as substrate for enzymatic reactions including cleavage and ligation [**LSS99**].

1.0.3. *Parallelism, Speed and Tiling Depth.* The massive parallelism inherent in DNA-based computers has, since its inception, driven thinking in the field. In computation by self-assembly, parallelism reveals itself in many ways. Each superstructure may contain information representing a different calculation (*global parallelism*). Growth on each individual superstructure may occur at many locations simultaneously (*local parallelism*). Self-assembly may be restricted such that certain assembly reactions can proceed only after others have been completed (*serial self-assembly*). Alternatively, self-assembly reactions may be limited by no such restrictions (*free self-assembly*). As examples, BCA tiles utilize local parallelism and serial self-assembly [**Win96**]; DHHP tiles utilize both local and global parallism and serial self-assembly [**WYS98**]; and self-assembly of linear, hairpin, and branched DNA molecules to generate regular, bilinear, and context-free languages makes use of global parallelism and free self-assembly [**WYS98, Eng99**], as do the proposals of Jonoska et al [**JKS99, JKS98**].

It it expected, but unproven, that free self-assembly might be faster and more robust than serial self-assembly. Three measures assist discussion: the *depth* of a superstructure is the maximum number of self-assembly reactions experienced by any substructure (the depth of the graph of reaction events), and the *size* of a superstructure is simply the number of tiles it contains. A superstructure consisting of $n \times m$ tiles, where $n > m$, is said to have m *layers*. The depth (size) of a tiling system is the maximum depth (size) of any superstructure it generates. Likewise for the number of layers. For example, in a tiling system where $A+B \to AB, C+D \to CD, AB+CD \to ABCD$, we know that the depth is ≥ 2 and the size is ≥ 4. Again, although it needs further study, tiling systems with low depth, small size, and few layers are considered desirable, motivating the search for efficient computations performed by such systems. Tiling systems that produce only superstructures with k layers, for some constant k, are said to use *linear self-assembly*. As an example, the two tiling systems for addition discussed here produce $3 \times n$ and $1 \times n$ superstructures for n-bit sums and hence are linear, but known tiling systems for multiplication produce $n \times n$ for n-bit products [**WER99**], and hence are not linear. The BCA and DHHP tile systems also are not linear.

Reif [**Rei99**] proposed DNA self-assembly methods of linear size and small depth to solve a number of fundamental problems (e.g., prefix computation, permutation, integer addition and subtraction, multiplication by a constant number, finite state automata simulation, and string fingerprinting) that form the basis for the design of many parallel algorithms. In particular, he gave linear self-assemblies for cumulative XOR and integer addition with some similarities to those described here. Furthermore, Reif showed how these elementary operations can be combined to perform more complex computations, such as bitonic sorting and general circuit evaluation in $O(\log n)$ experimental steps. To increase the likelihood of successful assembly, Reif also proposed step-wise assembly, which controls the assembly in distinct stages.

A second approach for small-depth computations is to compress several tile layers into single tiles, so that the simplest form of linear self-assembly suffices. This approach works particularly well when the topology and routing of the strands in the DNA tiles is carefully considered, leading to the notion of *string tiles* [**WER99**]. The TAO and TAE tiles reported here are particularly useful as string tiles, and this approach will be discussed in Section 3.

1.0.4. *Previous Methods and New Improvements in I/O.* Input and output are critical to the practical use of DNA-based computing. Winfree [**Win96**] used the first and last layers of the assembly for input and output, respectively. Reif [**Rei99**] proposed the use of DNA nano-structures known as *frames* for input and output; this allows the input and output to be ssDNA strands rather than assemblies of tiles. Precedent for Reif's proposal for output can be found in techniques used to confirm experimental constructions of exotic 3-dimensional structures, wherein oligonucleotides were ligated together to form new, longer strands which, if present at the end of the construction process, reported on the success of formation of the desired structure (for example see [**LYQS96**]). We also make use of reporter strands, both to monitor proper assembly and to output the results of a computational assembly. Further, we advance a similar concept for input by describing and testing scaffold strands, which are long DNA strands capable of serving as nucleation points for assembly. Preformed, multimeric scaffold strands are added to the hybridization/annealing mixture in place of the monomeric oligo corresponding to the tile's reporter segment. Tiles assemble around the scaffold strand, automatically forming a chain of connected tiles which can subsequently be used as the input layer in a computational assembly (see Subsection 2.2 below).

2. TAO Tile and 2D Assemblies

Besides expanding the repertoire of DNA tiles, our specific goals in prototyping a new tile type were to provide four ssDNA pads per tile and simultaneously to allow a reporter strand to trace its way diagonally through the tile. This latter property is essential for several of the computational constructions we will consider later, wherein the diagonal strand, when ligated to its counterparts in neighboring

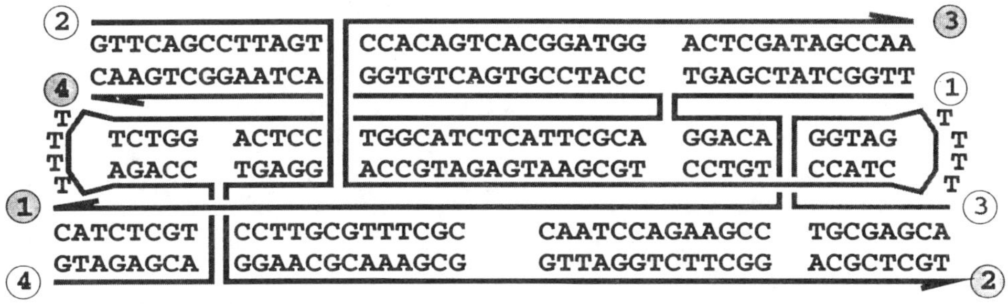

FIGURE 1. **Strand and Sequence Trace through TAO Tile.** The four oligonucleotides are numbered and labeled with white circles on their 5′ ends and dark circles on the 3′ ends. Base-pairing partners are shown one above the other for each of the three helices. The two TTTT segments are hairpins on the ends of the middle helix. Horizontal lines show chain direction along the helices but do not show the helical twist, therefore, position of the strands around the circumference of the helices is not accurately represented. Vertical lines represent the strand crossover points. Oligos 1 and 2 are 52 bases in length; oligos 3 and 4 are 72 bases long. Through most of our discussion oligo-1 is considered the reporter strand, however, oligo-2 is symmetrical to oligo-1 and can also be used in a reporter role. Approximate tile dimensions estimated from modeling are $65 \times 140 \times 20$Å.

tiles, produces a reporter strand that traces through a diagonal arrangement of tiles and records the values of the input and output strings from all the tiles.

Building on previous nomenclature [**FS93**], we refer to our tile design as TAO: "Triple" helix, with Anti-parallel crossovers, and an Odd number of helical half turns between crossovers (see Figure 1). Its three double-helices are designed to be co-planar. Anti-parallel crossovers indicate a reversal of strand direction following exchange to a new helix. An odd number of half-turns between crossovers ensures the desired reporter strand path through the tile (compare, for example, with the strand trace for TAE tiles in Section 3).

The prototype TAO, at 42 bases in length, goes through 4 full helical turns. Spacing between crossover points on the top helix (as represented in Figure 1) is 16 bases, for 3 helical half-turns; on the middle helix, 5 bases for 1 half-turn; and on the bottom helix, 26 bases for 5 half-turns. The shortest dsDNA arms between crossover and tile edge are 8 bases on the bottom helix. The prototype TAO was designed without ssDNA sticky ends for ease of structural analysis. Tiles functional for superstructure assembly will be equipped with 5′ ssDNA pads of about 5 bases in length. In the 2D assemblies, described below, only 4 pads will be required, therefore the TAO as shown contains hairpins on both ends of the middle helix. These hairpins reduce the number of oligonucleotides needed for construction of the tile; however, they can be eliminated if more pads are required or if altered strand traces through the tile are desired.

The specific sequences used for the prototype TAO tile were designed using stochastic hill-climbing to search for nucleotide sequences that would provide the desired base-pairing while avoiding undesired pairings. The particular sequences chosen are drawn from a huge potential pool of sequences; we can, therefore, maintain the desired TAO structure while using different internal sequences to represent different digit values for tiles in the computational superstructure. The computational assemblies described below make use of binary encoding schemes (tiles hold values 1 or 0); however, for some calculations it may be useful to utilize expanded encodings (for example base-10). The tile assemblies are completely compatible with expanded alphabets, and would simply require a larger number of starting tile types.

2.1. TAO Tile Construction Results. TAO tiles have been successfully built and analyzed previously [**LYK+99**]. Summarizing this work, four oligonucleotides (with sequences given in Figure 1) were chemically synthesized and purified by polyacrylamide gel electrophoresis (PAGE) under denaturing conditions. DNA concentrations of the pure stocks were measured by UV-absorption at 260 nm. Stoichiometric and non-stoichiometric mixtures of oligos (buffered at pH 7.9 - 8.0 and containing 10 mM $MgCl_2$) were heated to 95°C for 5 minutes, then annealed by cooling slowly to ambient temperature over the course of 1 to 1.5 hours. Following annealing the tiles were subjected to various tests of structural integrity.

First, formation of individual tiles of designed strand stoichiometry and molecular weight was determined by PAGE under non-denaturing conditions. Further, complex was subjected to random backbone cleavage analysis (using hydroxyl radical chemistry), showing cleavage and protection patterns on the nucleotide level as predicted from the tile design. That is, bases predicted to be buried within the tile, either in crossover points or helix faces occluded by neighboring helix, showed protection and decreased sensitivity to backbone scission. As expected, bases predicted to be exposed to solvent water showed the highest rates of scission.

Finally, Ferguson analysis (measurement of electrophoretic migration rates in gels of various polyacrylamide concentrations) was used to compare hydrodynamic properties of TAO complexes with those of two related complexes, a DX tile and a dsDNA of the same overall length. These results showed a decrease in electrophoretic mobility for the TAO tile compared with other complexes, as would be expected for a molecule with a more compact shape.

The thermal stability of the tile was measured by monitoring UV absorption during heat induced melting of tile structure. The particular TAO tile we examined appears to be stable at temperatures below approximately 50°C.

In summary, the prototype TAO has been shown to form with the proper oligonucleotide stoichiometry, desired molecular weight, and expected structural signatures on the whole complex and individual nucleotide levels. The tile also displays acceptable thermal stability for use as a component in the construction of computational tilings.

2.2. Results of Lattice Assembly on Preformed Scaffold Strands. In the 2D assembly experimental designs described below, we will make use of random formation of input tile layers. These input layers are capable of providing a reporter strand to enable read-out of the specific inputs from each computational complex. For many applications it may prove useful to encode specific inputs rather than randomly assembling large populations of input layers. Additionally, it may be necessasry to use the output of one self-assembly step as the input to a subsequent step. If preformed reporter strands of desired structure could be used as a scaffold on which to assemble a tile layer, they could be used to input specific values into computational assemblies. To test tile assembly onto preformed inputs, we constructed scaffold strands. The scaffold strand is the structural equivalent of a reporter strand, but in our test experiments it is formed by conventional, double-strand nucleic acid techniques rather than resulting from a tile assembly.

Scaffold strands from oligos-1 and -2 of the TAO tile (as shown in Figure 1, above) were prepared by the following procedure: anneal tile strand with a 24-base bridge oligo, ligate to form concatamers, and gel purify ssDNA concatamers of desired length. The tile oligo sequences and bridge annealing positions are as follows:

```
         5'                                                                                                   3'
         ----------------------------------------------------
Oligo-1:TTGGCTATCGAGTGGACACCGAAGACCTAACCGCTTTGCGTTCCTGCTCTACTTGGCTATCGAGTGGACACCGAAGACCTAACCGCTTTGCGTTCCTGCTCTAC
Bridge 1:                                       AAGGACGAGATGAACCGATAGCTC
                                              3'                        5'

         5'                                                                                                   3'
         ----------------------------------------------------
Oligo-2:GTTCAGCCTTAGTGGAGTGGAACGCAAAGCGGTTAGGTCTTCGGACGCTCGTGTTCAGCCTTAGTGGAGTGGAACGCAAAGCGGTTAGGTCTTCGGACGCTCGT
Bridge 2:                                       AGCCTGCGAGCACAAGTCGGAATC
                                              3'                        5'
```

The overlined sequence represents a single copy of tile oligo which is then repeated (without overlining) to show how two copies of a tile oligo are held together by a bridge oligo functioning as a splint. The bridge oligo (24 bases in length) anneals to the 5′ end of one copy of tile oligo and to the 3′ end of another copy, to make long multimers of the following form:

In the annealing/ligation reaction for scaffold strand formation, we used approximately 400 pmoles of phosporylated strand-1 (or strand-2), 100 pmoles unphosphorylated strand, and 400 pmoles bridge strand. These ratios gave acceptable concentrations of concatamers of desired sizes for purification by denaturing PAGE. We constructed and purified scaffold strands from oligo-1 and oligo-2 in lengths ranging from $2x$ to $10x$ (i.e. $10x$ strands have 10 tandem copies of the sequence required for a TAO tile). Following preparation and purification of desired scaffold strands, simple tile assemblies were formed by replacing monomeric tile oligos with multimeric scaffolds in the tile formation annealing step. Tile assemblies form around these scaffold strands, which act as tethers connecting adjacent tiles. These assemblies were analyzed by PAGE and atomic force microscopy (AFM); results are shown in Figure 2. The gel results (panel **A**) are exactly as expected. Scaffold strands $2x$-oligo-1 and $3x$-oligo-1 are 104 and 156 bases, respectively. They are ssDNA and run faster than dsDNA markers of the same length. The $2x$-TAO and $3x$-TAO assemblies should contain 252 and 378 base-pairs, respectively. The bands run slightly faster than dsDNA of the same molecular weight due to their more compact shape.

The AFM study (panel **B**) suggests that the assemblies are linear chains in preference over branched or linked structures. In separate experiments, we examined assemblies formed around scaffolds of four different sizes. Monomer concentration was 1.2 to 3.8 uM. Because we used a 2-fold excess of tile strands, relative to the number of tile positions supported by the scaffold strands, many three-strand partial tile complexes were also present in the solution, appearing as dots in the images. Average lengths of linear chain assemblies, measured by hand from several images,

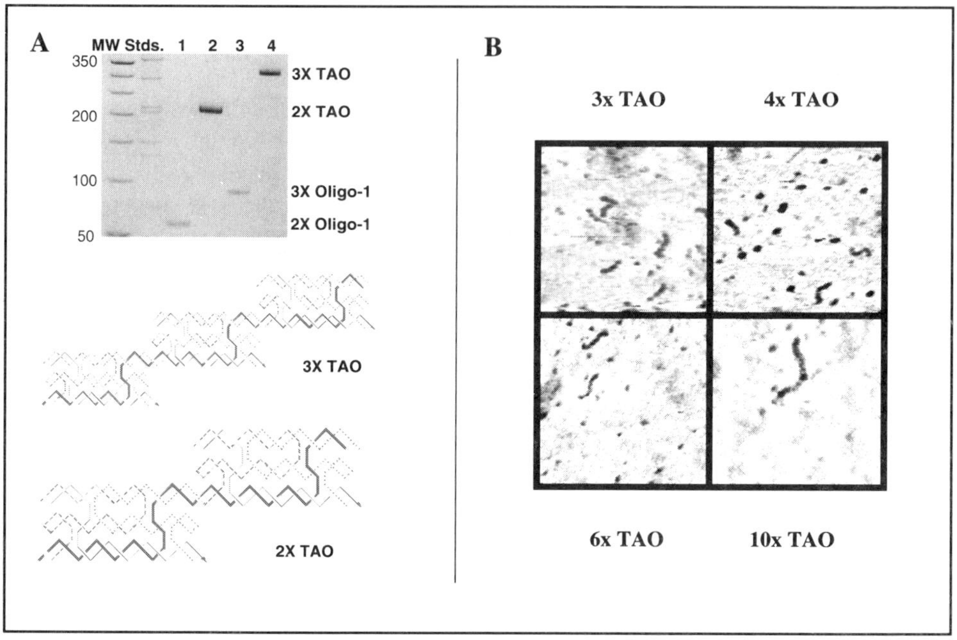

FIGURE 2. **Use of scaffold strands in small non-computational assemblies.** Panel **A** shows a polyacrylamide gel (run under non-denaturing conditions) which demonstrates assembly of TAO tile layers around $2x$ and $3x$ scaffold strands; below are diagrams of three-tile ($3x$ TAO) and two-tile ($2x$ TAO) assemblies constructed around $3x$-oligo-1 and $2x$-oligo-1 scaffold strands. Lanes 1 and 3 of the gel are scaffold strand alone; lanes 2 and 4 are scaffold strands annealed with oligos 2, 3, and 4 to form tile assemblies. Molecular weight markers (MW Stds.) are dsDNA with sizes indicated in basepairs. Panel **B** shows AFM images (clockwise from upper left) of three-, four-, six- and ten-tile assemblies constructed around scaffold strands $3x$-oligo-2, $4x$-oligo-2, $6x$-oligo-1 and $10x$-oligo-1. Each image is 500×500 nm. Height above the mica surface is indicated by increasing darkness.

were $3x$ TAO: 54 nm ($N = 19$, $\sigma = 6.3$); $4x$ TAO: 66 nm ($N = 9$, $\sigma = 1.3$); $6x$ TAO: 95 nm ($N = 15$, $\sigma = 3.5$); and $10x$ TAO: 175 nm ($N = 1$). This compares favorably to the expected lengths of 45, 60, 90, and 150 nm respectively, based on a diagonal length of 15 nm per tile. However, the AFM study by itself is not conclusive evidence that all tile positions on a scaffold strand are fully formed.

We have demonstrated the use of scaffold strands as a means of assembling a DNA lattice. We might wish to answer another question regarding tile assembly around scaffold strands: is there a maximum length beyond which scaffold strands would function poorly due to, for example, tangling of partially formed lattices?

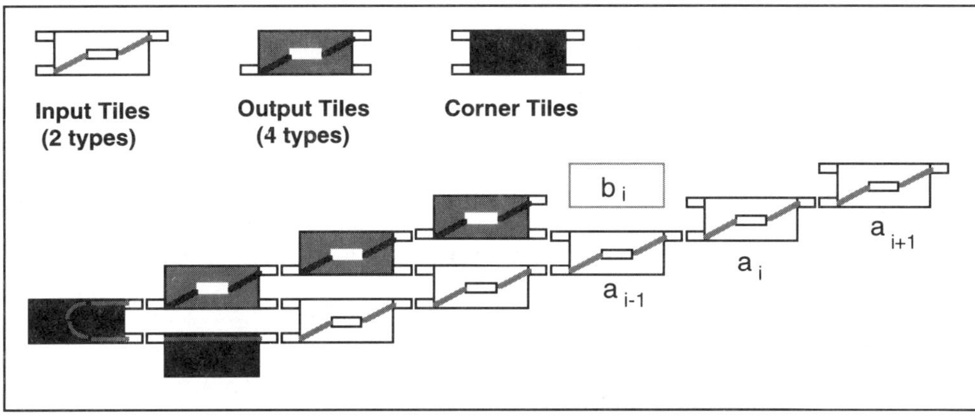

FIGURE 3. **2D Assembly for Cumulative XOR.** The tiles schematics represent TAO tiles. ssDNA pads are shown sticking off the body of the tiles. The small central box on the input and output tiles represents a nucleotide sequence encoding the value (0 or 1) of the tile. The reporter strand is shown as a diagonal line passing through each tile. The figure shows a partially assembled superstructure in which the next step is the binding of an output tile into the slot labeled b_i, followed by continued assembly of the output layer up and to the right.

As of yet, we have not tested the upper length limit; it would be useful to test tile formation on scaffolds of hundreds or thousands of tandem repeats.

2.3. 2D Assemblies for XOR and addition. We will now outline our procedures for using the TAO tile described above to perform massively parallel arithmetic. The simplest 2D assembly computer will perform cumulative XOR computation using a 2-layer architecture: one input tile layer and one output layer. We will also describe the slightly more complicated 2D assemblies required to execute bit-wise XOR and addition of two binary inputs, using 3-layer superstructures. For computations on specific inputs, these procedures will make use of the scaffold strands mentioned in the previous section. Otherwise, the input tiles will randomly assemble and thereby generate a molecular look-up table in which each reporter strand encodes the inputs and outputs of a random calculation. A sufficient number of DNA tile molecules will provide full coverage of all possible n-bit input strings. Such look-up tables may be useful as input for further computations as they represent a unique library of sequences with a complex structural theme.

Recall that the exclusive-or (XOR) operation is a Boolean function that, given two inputs, yields 0 if the inputs hold the same value, and returns 1 if the inputs disagree in value. The input to a cumulative XOR computation is a sequence of Booleans $a_1, a_2, a_3 ...$, and the output is a sequence of Booleans $b_1, b_2, b_3 ...$ where $b_1 = a_1$ and for $i > 1$, $b_i = b_{i-1}$ XOR a_i. Thus the ith output is the cumulative XOR of the 1st through the ith inputs.

A 2-layer tile system for computing cumulative XOR is shown in Figure 3 and Figure 4. As shown, it consists of 8 tile types: 2 input tiles (with value x and pads clockwise from lower left: C, A_x, and C, where $x \in \{0,1\}$), 4 output tiles (with value (x XOR y) and pads B_y, $B_{(x \text{ XOR } y)}$, and A_x, where $x, y \in \{0,1\}$), and 2 corner tiles (with pads C' and C, and B_0 and C'). The two types of input tile differ only in the sequences for the upper-left pad and the internal sequence encoding the tile's value. The other two pads (upper-right and lower-left) are complementary on all input tiles, so that assembly of the input layer can randomly generate all possible n-bit inputs. Output tiles use the bottom two pads to read inputs (one from the next input tile and one from the previous output tile), while the upper-right pad encode outputs. Four output tile types are required, one for each possible pair of input values in the XOR truth table. The reporter strand is shown as a diagonal line passing through each tile by the following path: in one pad, through the sequence encoding the tile's value, and out through a pad on the opposite side of the tile. The corner tiles function primarily to route the reporter strand and thus connect its input and output halves.

In addition to connecting the input and output halves of the reporter strand, the corner tiles set the boundary conditions for the computation. The two corner tiles put the first input bit (a_1) into position for formation of the binding slot for the first output bit (b_1). Of the two corner tiles depicted in Figure 3 and Figure 4, the one farthest left (two pads in one side and reporter strand looped between them) can be a TAO. For the other corner tile (two pads on the top helix connected by reporter strand) we require a different strand topology which can be provided by a TAE tile as described below in Section 3 or by a DAE tile [**FS93**].

Figure 3 shows a partially assembled superstructure in which the next step is the binding of an output tile into the slot labeled b_i, followed by continued assembly of the output layer up and to the right. To properly bind, the output tile must match its lower-right pad with the upper-left pad of tile a_i and its lower-left pad with the upper-right pad of the previous output tile (b_{i-1}). Following the annealing of tile b_i, a new slot b_{i+1} will be formed with one pad from b_i and one pad from a_{i+1}.

The 2-layer 2D assemblies for cumulative XOR calculation, as proposed, will function with TAO tile structures, but could alternatively use simpler double-crossover tiles containing a hairpin on one helix [**WER99**]. We chose to use TAOs in our design in order to test the tiles and assembly techniques that will be required for 3-layer superstructures for bit-wise XOR and addition.

We now move to discussion of the 3-layer assemblies as shown in Figure 5. The components and methodology for 3-layer superstructures closely follows that for the 2-layer cumulative XOR assembly. Again, we rely on random assembly of input layers, but this time we need two input layers per superstructure, and we need corner tiles on both ends of the superstructure in order to properly join both the input and the output portions of the reporter strand into a continuous piece of DNA.

The pads on input and corner strands are essentially as before, except that the second set of input tiles uses its own labels. For addition, the eight output tiles make use of all four pads as well as the (internal) value encoded on the reporter strand: the value is (x XOR y XOR c) and, clockwise from lower left, the pads are C_c, A'_y, $C_{c'}$, and A_x, where $x, y, c, c' \in \{0,1\}$ and $c' = 1$ iff $x+y+c \geq 2$. Thus, the

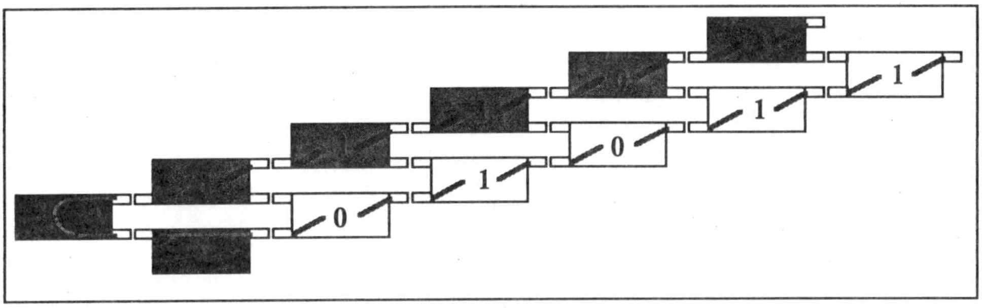

FIGURE 4. **Example I/O for cumulative XOR.** This pictured assembly has input 01011 and output 01101. In examining this cartoon of the assembly, we read both input and output strings from left to right. However, the reporter strand starts ($5'$ end) in the input tile most distant from the corner tiles and proceeds ($5'$ to $3'$) through the input tiles in reverse order, then loops through the corner tiles and continues through the output tiles in forward order. Therefore, the ordering of information on the reporter is: $a_n, a_{n-1}, a_{n-2}, ..., a_1, b_1, b_2, b_3, ..., b_n$.

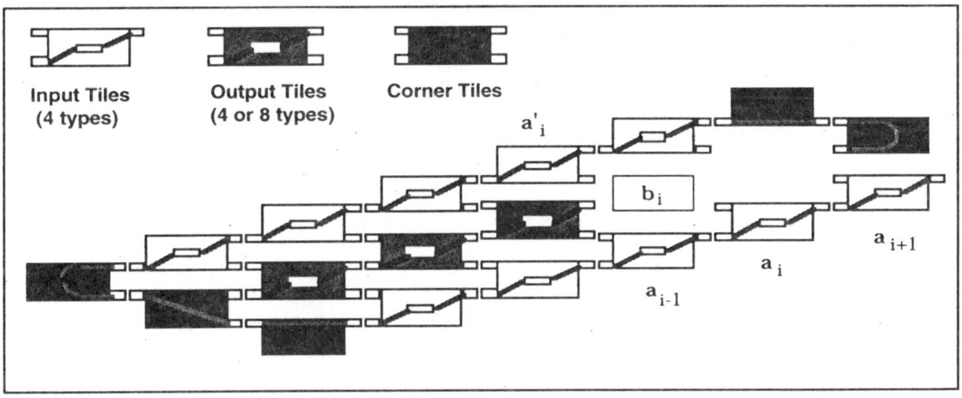

FIGURE 5. **Bit-wise XOR and Addition.** This schematic representation of a 3-layer assembly maintains the conventions used in the previous figures. The superstructure will assemble such that $b = a + a'$ or $b_i = a_i$ XOR a'_i depending on which set of output tiles is used. The reporter strand traces through the entire assembly in a spiral, starting in the nth tile of the lower input layer and down through the 1st, through the corner tiles, through the upper input layer (1st to n^{th}), through distal corner tiles and down through the output layer (n^{th} to 1st).

pads along the middle diagonal represent the carry bit during the addition. For bit-wise XOR, there is no need to pass a carry bit, so lower-left pads are complementary to upper-right pads on all output tiles, therefore only four tile types are required.

The total number of tile types required for the cumulative XOR assembly is at least eight: 2 inputs, 4 outputs, and 2 corner tiles for the lower-left end of the superstructure. It may also prove useful to include corner tiles which terminate the upper-right end of the assembly, thus "bookending" the computation and perhaps increasing accuracy by ensuring that input and output strings correspond in length. The bit-wise XOR and addition assemblies require at least 13 and 17 distinct tile types, respectively. Again expansion of these repertoires by an increase in the number of corner tiles may be desirable to ensure proper nucleation and termination of the superstructures. The 1-layer assembly for addition described below in Section 3 requires exactly 10 tile types.

The 2D assemblies described here allow read-in of specific inputs by scaffold strands (Section 2.2, above) and read-out of input and output strings on reporter strands. The reporters can be purified following assembly and ligation, since they have a large, characteristic size and can be separated from other DNA in the test tube by denaturing PAGE. To determine a particular output string, a reporter strand can be used as template for PCR (polymerase chain reaction). Adleman [**Adl94**] made use of graduated PCR using six different pairs of primers to gauge the distance between nodes on the Hamiltonian path encoded on his DNA. For the binary representation we use, we need only two primer pairs (2 PCR reactions) for our read-out, both containing a constant primers whose binding site is located in a corner tile, and each containing one of two primers corresponding to the value encoding sequence words (0 or 1) located in each tile. The reaction containing corner-tile primer plus value=1 primer, will provide distances from the corner tile to each tile whose value is set to 1. Since these distance increments are uniform, values for each digit in the I/O strings can be read directly. Likewise, using primer corresponding to the sequence word for bit value=0, we will obtain the digit locations in the input and output for bits whose value is 0.

Two potential problems may prevent the 2D assembly computations from proceeding as described. First, the scheme requires that assembly progress in two distinct stages: random input assembly, followed by specific output assembly. If output tiles begin to anneal with each other or with partially formed input chains, there exists the potential for improper assembly. The ssDNA pad sequences must be designed such that input tiles associate with each other at a slightly lower temperature than that which allows output tile annealing. While this temperature separation is possible in theory it may be difficult to achieve in practice. The second potential problem involves the "slot filling" association of output tiles into the forming computational assembly. The proper annealing of output tile requires matching with two (or 3 for addition) ssDNA pads, however, with some probability an improper output tile may fill a slot based on matching with a single pad (or 2 pads for addition). If the incorrect output tile fails to dissociate it will block the processor, returning either a truncated answer or an answer which no longer corresponds to the input. Results of an initial experiment have shown that dsDNA with two matching sticky-end achieves preferential binding over single match DNA [**WYS98**], however this critical result has yet to be demonstrated in more complex assemblies involve larger tile structures. Although linear assemblies with few layers are likely to be less affected by these problems than fully 2D assemblies, these potential problems have induced interest in alternative strategies for computation by self-assembly, such as the one given in the next section.

3. String Tile Assemblies

The concept of string tile assemblies derives from Eng's observation that allowing neighboring tiles in an assembly to associate by *two* sticky ends on each side, he could increase the computational complexity of languages generated by linear self-assembly [**WER99**]. By allowing contiguous strings of DNA to trace through individual tiles and the entire assembly multiple times, surprisingly sophisticated calculations can be performed with 1-layer linear assemblies [**WER99**]. For example, we describe below a string tile system that computes addition or bit-wise XOR.

String tiling, in essence, is the collapse of multi-layer assemblies into simpler superstructures by allowing individual tiles to carry multiple segments of the reporter strand. The 2-layer lattice for cumulative XOR can be reduced to unbranched dsDNA. The 3-layer addition assembly collapses to a linear row of three-helix (or perhaps even two-helix) tiles. By allowing a reporter strand to wind through the assembly, passing through each tile multiple times, we can encode in each tile an entire row of the truth table. Input and output values are recorded on words internal to the tile, while information passed to neighboring tiles (e.g. carry bits) are encoded on the ssDNA pads.

String tile arithmetic implementations have a number of advantageous properties.

1. Input and output strings assemble simultaneously.
2. Each row in the truth table for the function being calculated is represented as a single tile type, where all inputs and outputs are encoded on that tile. Each bit-wise operation is directly encoded in the structure of a tile. Contrast this with the 2D systems described above, in which input bits and output bits each had their own tile types.
3. Adjacent tiles only have to agree on the value of the carry bit between them; one tile passes and one tile receives the bit. The set of pads for carry=0 is distinct from the carry=1 pads.
4. Adjacent tiles associate via multiple sticky ends, all of which either agree or disagree, and therefore there is no need to differentiate between single and double pad matches (as required in the 2D assemblies).

The TAE tile shown in Figure 6 has four half-turns of DNA helix between the crossover points connecting the top helix with the middle helix and between the crossover points connecting the middle helix with the bottom helix. This could be trimmed to two half-turns without loss of the desired strand routing; however, the longer spacing provides an increased number of base positions for the internal sequences that encode the (input and output) values of the tile. Longer sequences also may be useful for improving primer binding for graduated PCR read-out of the reporter strand. The longest oligonucleotide in the tile (shown in light gray in Figure 6) may be useful as a diagonal reporter strand in certain constructions. It is not required for the assembly described below and may make purification of the desired reporter (black) more difficult, since it winds around reporter segments repeatedly. It should be possible to introduce a strand break in the long oligo, perhaps on the middle helix, thus increasing the number of oligos in the tile but

FIGURE 6. **TAE Tile.** The structure of the TAE resembles the TAO in that it is constructed of three double-helices linked by strand exchange, however, it contains an Even (rather than Odd) number of helical half-turns between crossover points linking each pair of helices. Even spacing of crossovers allows reporter strands (shown in black) to stretch straight through each helix from one side of the tile to the other. These 3 reporter segments will be used for building up a long strand which records inputs and outputs for the entire assembly.

decreasing the strand tangling. At the nick, the strands should be lacking a 5′ phosphate, so that the nick does not get sealed by ligase.

Figure 7 presents details for a linear assembly capable of calculating addition or bit-wise XOR. The scheme is illustrated using TAE tiles, however it is not particular to TAEs. The system is a variation of one outlined previously which utilized DPE tiles (Double-crossover (2 helices), Parallel crossovers, with an Even number of half-turns between) to execute a similar strand trace through the assembly [**WER99**]. TAE tiles may be preferable to DPE because early studies on DX complexes containing parallel crossovers showed them to be less structurally stable than their anti-parallel counterparts [**FS93**]. It is possible that DPE tiles could be stabilized by increasing the lengths of dsDNA arms flanking the strand exchange points; however, this has not yet been demonstrated experimentally.

I_{A_i} and I_{B_i} indicate sequences encoding binary values for the two input bits for the current operation. O_i encodes the value of the output bit. The sticky end pads on the right encode the value for the carry bit coming into this operation, while the left-hand pads pass the new carry value on to the next bit operation. Each of the nine tile locations has two possible sequence words; one each for values 1 and 0. One tile type is required for each of the eight rows in the truth table, plus one tile type for each of the two ends of the assembly. The truth table provides the allowed pairings of values for each of the nine tile locations. The c_i column provides the value for all three right-hand pads $(c1_i, {}^\sim c2_i, c3i)$ while column c_{i+1} gives the value for the three left-hand pads $({}^\sim c1_i, c2_i, {}^\sim c3_i)$. Sequence words for $c1$, $c2$, and $c3$ are distinct from one another so that each tile binding reaction involves annealing of three separate pairs of pads. Note that ${}^\sim c1$ represents the complement of $c1$. When two tiles bump into each other, they either agree at all 3 pads and bind or they disagree at all three pads and fail to bind. Binding that is not in register

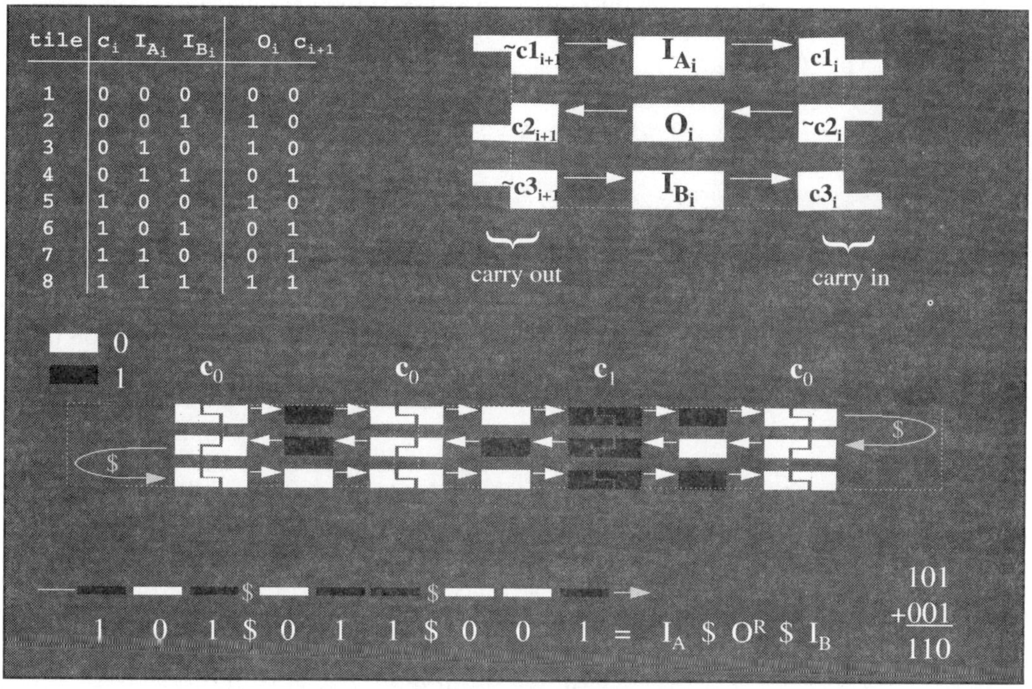

FIGURE 7. **String Tile Addition with TAE Building Blocks.** An example tile is shown schematically in the upper right corner; it is of the TAE form with the trace of the three reporter segments indicated by arrows. The truth table of values for c_i, I_{A_i}, I_{B_i}, and O_i is shown in the upper left; each row corresponds to a type of tile. An example of an assembled structure is given at the bottom, along with its interpretation. The reporter strand follows a Z-shaped path from the upper left to the lower right.

(e.g. $c1$ bound to $\sim c3$) is thereby disallowed. An example computational assembly showing addition of two 3-bit numbers is shown at the bottom. The linear assembly is composed of five tiles, starting on the right: an end or terminal tile, then a type 4 tile, then a type 5, then a type 3, then a left-end tile. Shown below the assembly is a summary of the bit values in the order they are recorded on the reporter strand (carry values, although also written on the reporter strand, are not shown in this simplified drawing). The '$' symbols indicate marker sequences from the end tiles which mark the boundaries between input and output strings and will be useful as primer binding sites during read-out by graduated PCR.

The ordering of I/O digit values on the reporter strand is depicted in Figure 7. Following the strand from $5'$ to $3'$ we encounter: I_A (n^{th} bit through 1st bit), O (1st through n^{th}), I_B (n^{th} through 1st). This orientation allows the inputs to be read with the highest digit first (on the left) but may be undesirable since the output digits are listed on the reporter strand in the opposite direction from the inputs. Other digit value orderings are possible, and with the use of slightly more complex tiles (4 or 5 helices) we can encode outputs in the same orientation as inputs [**WER99**].

The linear assembly for addition and bit-wise XOR presented in Figure 7 displays many benefits over 2D assembly methods: simpler superstructure (1 layer), fewer tile types (exactly ten types are sufficient), shorter reporter strands, neighbor tile matching by 3 pads instead of one, etc. The string tile addition scheme outlined in Figure 7 may have the greatest chance of success of all the DNA computational assembly procedures discussed. We are currently working to experimentally test this system.

4. Discussion and Conclusions

We have described the design of two novel DNA tile types, TAO and TAE, as well as summarizing the experimental analysis of a TAO. We have outlined the use of reporter strands and scaffold strands for I/O in computational assemblies, and presented experimental data demonstrating formation of lattice structures around scaffold strands. We have also detailed three computational assemblies, including a new string tile system which simplifies many aspects of DNA tilings and shows promise for experimental construction. We have been working primarily on the implementation of arithmetic for prototyping functional DNA-based computers. It is our intention that these experiments and prototypes will enable development of materials and techniques that may advance DNA-based computing beyond the realm of demonstration and into useful applications.

Acknowledgments

We wish to sincerely thank: Ned Seeman and Hao Yan for their considerable and invaluable roles in construction and analysis of the TAO tile; Tony Eng and Grzegorz Rosenberg for development and enlightening discussion of string tile concepts; and Wolfgang Frey and Ashutosh Chilkoti for assistance with atomic force microscopy. This work was supported in part by Grants NSF/DARPA CCR-9725021, CCR-96-33567, NSF IRI- 9619647, ARO contract DAAH-04-96-1-0448, and ONR contract N00014-99-1-0406.

References

[Adl94] Leonard M. Adleman. Molecular computation of solutions to combinatorial problems. *Science*, 266:1021–1024, November 11, 1994.

[BB98] Eric B. Baum and Dan Boneh. Running dynamic programming algorithms on a DNA computer. In Landweber and Baum [**LB98**], pages 77–85.

[Ber66] Robert Berger. The undecidability of the domino problem. *Memiors of the AMS*, 66:1–72, 1966.

[Buc62] J. Richard Buchi. Turing machines and the entscheidungsproblem. *Mathematiche Annalen*, 148:201–213, 1962.

[CS91] Junghuei Chen and Nadrian C. Seeman. The synthesis from DNA of a molecule with the connectivity of a cube. *Nature*, 350:631–633, 1991.

[DZS92] Shou Ming Du, Siwei Zhang, , and Nadrian C. Seeman. DNA junctions, antijunctions, and mesojunctions. *Biochemistry*, 31:10955–10963, 1992.

[Eng99] Tony Eng. Linear DNA self-assembly with hairpins generates the equivalent of linear context-free grammars. In Rubin and Wood [**RW99**].

[FB97] Bin Fu and Richard Beigel. Molecular computing, bounded nondeterminism, and efficient recursion. *Automata, Languages and Programming*, 1256:816–826, 1997.

[FB99] Bin Fu and Richard Beigel. A comparison of resource-bounded molecular computation models. *Algorithmica*, 24(2):87–95, 1999.

[FS93] Tsu-Ju Fu and Nadrian C. Seeman. DNA double-crossover molecules. *Biochemistry*, 32:3211–3220, 1993.

[GS87] Branko Grünbaum and G. C. Shephard. *Tilings and Patterns.* W. H. Freeman and Company, New York, 1987.

[JKS98] Nataša Jonoska, Stephen A. Karl, and Masahico Saito. Three dimensional DNA structures in computing. In Lila Kari, Harvey Rubin, and David H. Wood, editors, *Proceedings of the 4th DIMACS Meeting on DNA Based Computers, held at the University of Pennsylvania, June 16-19, 1998*, pages 189–200, preliminary, 1998.

[JKS99] Nataša Jonoska, Stephen A. Karl, and Masahico Saito. Creating 3-dimensional graph structures with DNA. In Rubin and Wood [**RW99**], pages 123–135.

[LB98] Laura F. Landweber and Eric B. Baum, editors. *DNA Based Computers II: DIMACS Workshop, June 10-12, 1996*, volume 44, Providence, RI, 1998. American Mathematical Society.

[LP81] H. R. Lewis and C. H. Padadimitriou. *Elements of the theory of computation.* Prentice-Hall, 1981.

[LSS99] Furong Liu, Ruojie Sha, and Nadrian C. Seeman. Modifying the surface features of two-dimensional DNA crystals. *Journal of the American Chemical Society*, 121(5):917–922, 1999.

[LYK+99] Thomas H. LaBean, Hao Yan, Jens Kopatsch, Furong Liu, Erik Winfree, John H. Reif, and Nadrian C. Seeman. Construction, analysis, ligation, and self-assembly of DNA triple crossover complexes. *in preparation*, 1999.

[LYQS96] Xiaojun Li, Xiaoping Yang, Jing Qi, and Nadrian C. Seeman. Antiparallel DNA double crossover molecules as components for nanoconstruction. *Journal of the American Chemical Society*, 118(26):6131–6140, 1996.

[MAS99] Nobuhiko Morimoto, Masanori Arita, and Akira Suyama. Solid phase DNA solution to the Hamiltonian Path Problem. In Rubin and Wood [**RW99**].

[MSS97] C. D. Mao, W. Q. Sun, and N. C. Seeman. Assembly of Borromean rings from DNA. *Nature*, 368:137–138, March 13, 1997.

[MSSS99] C. D. Mao, W. Q. Sun, Z. Y. Shen, and N. C. Seeman. A nanomechanical device based on the B-Z transition of DNA. *Nature*, 397:144–146, January 14, 1999.

[Rei99] John Reif. Local parallel biomolecular computing. In Rubin and Wood [**RW99**], pages 217–254.

[Rob71] Raphael M. Robinson. Undecidability and nonperiodicity of tilings of the plane. *Inventiones Math.*, 12:177–909, 1971.

[RW99] Harvey Rubin and David Harlan Wood, editors. *DNA Based Computers III: DIMACS Workshop, June 23-25, 1997*, volume 48 of *DIMACS: Series in Discrete Mathematics and Theoretical Computer Science*, Providence, RI, 1999. American Mathematical Society.

[SCD+93] Nadrian C. Seeman, Junghuei Chen, Shou Ming Du, John E. Mueller, Yuwen Zhang, Tsu Ju Fu, Yinli Wang, Hui Wang, and Siwei Zhang. Synthetic DNA knots and catenanes. *New Journal of Chemistry*, 17:739–755, 1993.

[SCK89] Nadrian C. Seeman, Junghuei Chen, and Neville R. Kallenbach. Gel electrophoretic analysis of DNA branched junctions. *Electrophoresis*, 10:345–354, 1989.

[See82] Nadrian C. Seeman. Nucleic-acid junctions and lattices. *Journal of Theoretical Biology*, 99(2):237–247, 1982.

[See98] Nadrian C. Seeman. DNA nanotechnology: novel DNA constructions. *Annual Review of Biophysics and Biomolecular Structure*, 27:225–248, 1998.

[Wan61] Hao Wang. Proving theorems by pattern recognition. II. *Bell System Technical Journal*, 40:1–42, 1961.

[Wan63] Hao Wang. Dominoes and the AEA case of the decision problem. In *Proc. Symp. Math. Theory of Automata*, pages 23–55, New York, 1963. Polytechnic Press.

[Wan75] Hao Wang. Notes on a class of tiling problems. *Fundamenta Mathematicae*, 82:295–305, December 1975.

[WER99] Erik Winfree, Tony Eng, and Grzegorz Rozenberg. String tile models for DNA computing by self-assembly. *in preparation*, 1999.

[Win96] Erik Winfree. On the computational power of DNA annealing and ligation. In Richard J. Lipton and Eric B. Baum, editors, *DNA Based Computers: DIMACS workshop, April 4, 1995*, volume 27, pages 199–221, Providence, RI, 1996. American Mathematical Society.

[WLWS98] Erik Winfree, Furong Liu, Lisa A. Wenzler, and Nadrian C. Seeman. Design and self-assembly of two-dimensional DNA crystals. *Nature*, 394:539–544, 1998.

[WMKS91] Yinli Wang, John E. Mueller, Börries Kemper, and Nadrian C. Seeman. Assembly and characterization of five-arm and six-arm DNA branched junctions. *Biochemistry*, 30:5667–5674, 1991.

[WYS98] Erik Winfree, Xiaoping Yang, and Nadrian C. Seeman. Universal computation via self-assembly of DNA: Some theory and experiments. In Landweber and Baum [**LB98**].

[YS99] Hiroshi Yoshida and Akira Suyama. Solution to 3-SAT by breadth first search. In Erik Winfree and David K. Gifford, editors, *Proceedings of the* 5th *DIMACS Meeting on DNA Based Computers, held at MIT , June 14-15, 1999*, preliminary, 1999.

[ZS94] Yuwen Zhang and Nadrian C. Seeman. The construction of a DNA truncated octahedron. *Journal of the American Chemical Society*, 116:1661–1669, 1994.

(T. LaBean) Department of Computer Science, Duke University, Durham, NC 27708,
E-mail address: thl@cs.duke.edu

(E. Winfree) Department of Molecular Biology, Princeton University, Princeton, NJ 08544

Departments of Computer Science and Computation and Neural Systems, California Institute of Technology, Pasadena, CA 91125
E-mail address: winfree@hope.caltech.edu

(J. Reif) Department of Computer Science, Duke University, Durham, NC 27708,
E-mail address: reif@cs.duke.edu

DIMACS Series in Discrete Mathematics
and Theoretical Computer Science
Volume **54**, 2000

2D DNA Self-Assembly for Satisfiability

Michail G. Lagoudakis and Thomas H. LaBean

ABSTRACT. DNA self-assembly has been proposed as a way to cope with huge combinatorial NP-HARD problems, such as satisfiability. However, the algorithmic designs proposed so far either involve many biosteps or are highly dependent on the particular instance to be solved. This paper presents an algorithmic design for solving satisfiability problems using two-dimensional DNA self-assembly (tiling). The main driving factor in this work was the design and encoding of the algorithm in a general way that separates the algorithm from the data and minimizes the dependency on particular instances. In effect, a large amount of work and preparation can be done in advance as a batch process. In practice, it is likely that the total time for computation will be decreased significantly and laboratory procedures will be simplified.

1. The Satisfiability (SAT) Problem

The *Boolean Satisfiability* (SAT) problem is the most well known representative of the NP-HARD and NP-COMPLETE classes of problems. These problems require non-deterministic polynomial computation time for optimal solutions. As a result, they are theoretically 'intractable' and, as the instance size grows, they very quickly become impossible to solve on conventional digital computers.

An instance of the SAT problem consists of a number of *Boolean variables* $x_1, x_2, ..., x_m$ and a number of *clauses* $C_1, C_2, ..., C_n$. Each clause is a disjunction of distinct literals, whereby a *literal* is a single variable x_i or its negation $\bar{x}_i$. A solution to the problem is a *satisfying assignment*, that is an assignment of binary truth values to the variables x_i such that the conjunction of all clauses is satisfied. Boolean formulas represented in this particular format are said to be in *Conjunctive Normal Form* (CNF); a conjunction of disjunctions.

An example with 5 variables and 8 clauses is given below (+ indicates disjunction; clauses are enclosed by a set of parentheses):

$$(x_1 + x_2 + \bar{x}_3)(\bar{x}_2 + x_4)(\bar{x}_1 + \bar{x}_5)(\bar{x}_1 + x_2 + x_3 + \bar{x}_4 + x_5)(\bar{x}_3)(x_2 + x_5)(\bar{x}_5)(x_1)$$

In this case, the assignment $(x_1, x_2, x_3, x_4, x_5) = (1, 1, 0, 1, 0)$ is satisfying and, therefore, a solution to the problem.

In general, the number of literals in a single clause is limited only by the total number of variables. However, even if it is required that each clause has exactly k literals ($k \geq 3$), the problem (known as kCNF-SAT) is still intractable. Moreover, it is proven that any SAT instance can be turned into an equivalent kCNF-SAT

2000 *Mathematics Subject Classification.* Primary 68Q05, 52C20.

© 2000 American Mathematical Society

instance [**GJ79**]. Among the most notable is the 3CNF-SAT variation, because of the small-sized clauses and the low bound on the number of possible clauses. Notice that given m variables, there can be at most

$$\binom{m}{k} \times 2^k$$

possible clauses in a kCNF-SAT formula (choose k variables and for each one either leave intact or negate). In particular, for $k = 3$, there are

$$\binom{m}{3} \times 2^3 = \frac{8}{6} \times m(m-1)(m-2) = O(m^3)$$

possible clauses in a 3CNF-SAT formula.

The number of satisfying assignments for a given formula can be anywhere between 0 and 2^m. If it is 0, the formula is said to be *unsatisfied*, whereas if it is 2^m, the formula is a *tautology*. In all other cases, the formula is simply *satisfiable*.

2. DNA Computation and Satisfiability

The basic idea is to exploit the massive parallelism possible in DNA operations in order to emulate a non-deterministic device that solves the SAT problem in polynomial time. Such emulation can be achieved by exponential-order parallelism.

Consider a particular assignment to the boolean variables in a CNF-SAT formula. On a conventional computer it is fairly easy to check whether this particular assignment is a solution to the problem, i.e. an assignment that satisfies all the clauses in the formula. In fact, this can be done in time linear in the size of the formula. It is the huge (exponential) number of different possible assignments and thus the huge sequence of checks that makes the problem difficult on a conventional computer. If there was a way to perform this check on a DNA-based computer, all checks could be done in parallel by an exponential number of computers, completing the whole computation in polynomial time.

This work proposes a way to perform the checking procedure above on molecular substrate using 2D DNA self-assembly. By creating billions of billions of copies of the participating DNA structures (tiles, in our case), we expect that the procedure will run in parallel on all possible assignments. The assignments will be created dynamically as part of the assembly. In effect, that will make the computation time linear in the size of the formula, while pushing the exponential dimension of the problem into the large number of DNA assemblies, and thus into the space (volume) occupied by the DNA molecules. If there is a satisfying assignment, we expect that at least one of these parallel checks will discover it.

The rest of this section reviews the main proposals for biomolecular solutions to the SAT problem, briefly describes the general DNA structures and methods that are used here and delineates our work.

2.1. Related Work. Lipton was the first to propose a DNA model for satisfiability [**Lip95**]. His proposal follows Adleman's elimination method [**Adl94**], whereby the whole combinatorial space of solutions is created and subsequently the "good" ones are extracted by a series of separation steps. Lipton's method was refined in [**BDLS96**]. Later, Hagiya et al. [**HAK$^+$99**] presented an approach to evaluate and learn μ-formulas (a particular form of Boolean formulas whereby each variable occurs at most once) using a technique that has become known as *Whiplash*

PCR. This method was improved by Winfree [**Win98b**] to account for general CNF formulas. Finally, a proposal for CNF-SAT using hairpin multicrossover DNA tiles and linear assembly can be found in [**WER99**].

In Lipton's method the sequence of separate/extract and combine/union operations performed on the DNA molecules is determined by the problem instance. As a result the number of biosteps required is linear in the size of the formula. In the other two approaches, the number of required biosteps is constant. It is generally desirable to perform as less biosteps as possible due to the error rate and inaccuracies associated with such operations. "Single pot" DNA computations seem to be more promising in this sense.

The initial DNA solution in Lipton's method is not dependent on the problem instance. A feature of the other approaches and a potential practical problem is that construction of the participating DNA structures cannot really begin until a particular SAT instance is given at hand. This "instance-specific" design implies long total computation time (encoding/computation/decoding). In addition, all required man/machine resources need to be employed again and again as new instances are provided. This may also increase the likelihood of inadvertent encoding errors, which if occur will render the whole computation useless.

It is higly desirable for practical reasons to keep the number of biosteps small and the participating DNA structures (strands, tiles, etc.) as simple as possible. This will also allow experimentalists to do more experimental computations without being much concerned about costly operations and/or (re)synthesis of complex DNA structures. Obviously, there is a trade-off between these two goals.

A balanced solution to these problems, would possibly be an algorithmic design that requires minimal encoding for a given instance (some straightforward description of the input), whereas the main algorithm is coded in preconstructed "instance-independent" DNA molecules that can be created and even tested off-line in a batch fashion.

The DNA self-assembly technique employed here follows concepts and proposals from a model known as *DNA tiling computation* originally proposed by Winfree [**Win98a**] for $O(1)$ DNA computations. Basic components for DNA tiling have been prototyped and tested by Seeman and colleagues. In particular, double crossover molecules have been shown to be rigid and able to form planar lattices containing hundreds of thousands of tiles [**WLWS98**]. Winfree has shown how to solve the Hamiltonian Path problem using 2-dimensional DNA tiling [**Win98a**]. Further, LaBean, Winfree and Reif [**LWR99**] have been experimenting with parallel XOR and addition operations using DNA tiling.

2.2. 2D DNA Self-Assembly for Satisfiability. The main goal of the algorithmic design presented here is to utilize the computational power of DNA tiling to solve the SAT problem, while minimizing the dependency of the process on particular problem instances. In other words, it is an attempt to separate the algorithm (instance-independent) from the data (instance-dependent). This can be accomplished by coding the general algorithm as a library of preconstructed non-specific DNA tiles which, when combined with an appropriate encoding for a given instance (input), would perform the desired computation. A separation step afterwards could separate the successful computations from where a satisfying assignment could be drawn. Put another way, we would like an encoding that specifies the general algorithm for the problem and not one that specifies a specialized algorithm for solving a particular instance of the problem.

Our design is described at the algorithmic level. We abstract each DNA tile as a square with labels at the corners (see figure 1 below). Each label indicates a particular kind of sticky end. Two sticky ends that can complement in the Watson-Crick sense and ligate are represented by identical labels. Each tile can have from one to four labels. Non-labeled corners indicate the absence of sticky ends. Experimental work [**WYS98**] has shown that, in principle, there exist parameter conditions (temperature, pH, ionic strength, etc.) under which tile binding to a slot defined by two sticky ends can be dominated by tiles with two matching sticky ends rather than single match tiles. Although these conditions are somewhat difficult to achieve in practice, this result shows that solution parameters can be tuned such that undesired and/or corrupted assemblies are avoided. It is taken for granted here that a tile would not occupy a slot unless both labels match.

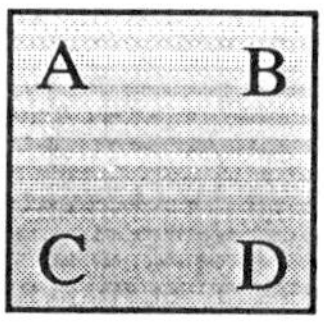

FIGURE 1. Abstracted Tile Representation.

The actual implementation details are not discussed here since they fall outside of the scope of this manuscript. However, we believe that we make no arbitrary hypotheses. In fact, our work is based on the assumptions and achievements that come with DNA tiling computation in general.

2.3. The Non-Deterministic Algorithm. As was mentioned above, our design attempts to simulate a non-deterministic algorithm for the SAT problem. Non-determinism implies that at some step(s) the algorithm makes a non-deterministic choice (as if some oracle could tell the algorithm what the right choice is). In our case, this is simulated by an exponential number of DNA assemblies, expecting to cover all possible choices. The algorithm is given below. Notice that step 3 is the nondeterministic step.

NON-DETERMINISTIC SATISFIABILITY($x_1, x_2, \ldots, x_m, C_1, C_2, \ldots, C_n$)
1. Set all clauses to be unmarked
2. **for** ($i = 1, \ldots, m$) **do**
3. Assign a value (0 or 1) to variable x_i
4. Mark all clauses $C_j, j = 1, \ldots, n$ which are satisfied by this assignment
5. **if** (all clauses are marked)
6. **then return** SUCCESS and output $\{x_i\}$
7. **else return** FAILURE

We present two slightly different designs for encoding this algorithm as a DNA tiling. The difference lies in what the tiles code. In design A tiles encode clauses; in design B tiles encode literals. After a detailed exposition of both designs a comparison is made.

3. Design A: Encoding Clauses

In this case we assume that the formula is given in 3-CNF form[1]. Since the number of possible clauses is countable in this case (see section 1), we can order all clauses and number them in some systematic way. With this mapping each clause C is represented by its corresponding number, say j. For simplicity, in what follows we number a clause C with its number in the ordering as C_j. Given a variable x_i and a clause C_j, we can easily construct a function $F(i, v, j)$ that determines whether the clause C_j is satisfied when variable x_i takes the binary value v. The function F will be actually "precoded" in the structure of the DNA tiles.

We can represent the desired computation in a table format, that facilitates the transition to the explanation of the DNA assembly. Consider the 3-CNF formula

$$(\bar{x}_1 + \bar{x}_2 + x_3)(x_1 + \bar{x}_2 + \bar{x}_3)(\bar{x}_1 + x_2 + x_3)$$

and the table given below. There are 3 variables (represented in the first column of the table) and 3 clauses (represented in the last row) with numbers 1, 4, and 3 according to some ordering. The second column represents a possible assignment. Cells marked with "*" are helper cells.

*	T	T	T	T	SS
$\mathbf{x_3}$	1	OK	OK	OK	*
$\mathbf{x_2}$	0	OK	OK	C_3	*
$\mathbf{x_1}$	1	C_1	OK	C_3	*
*	*	$\mathbf{C_1}$	$\mathbf{C_4}$	$\mathbf{C_3}$	*

Following the algorithm above, the table is filled in a bottom-up manner, one row at a time. A cell corresponding to variable x_i and clause C_j will be marked as "OK" if and only if the clause C_j is satisfied when variable x_i takes the binary value v_i indicated in the second column or the cell below is already labeled "OK". Otherwise, it is marked with the clause name "C_j". Therefore, as we move up, "C_j" is propagated up as long as the clause is not satisfied. Once it is satisfied, it turns to an "OK" label, which propagates to the top independently of the assignments of the remaining variables. In effect, the entries of the table reflect the function F.

It remains to check whether all clauses are satisfied. This is done in the first row of the table. Initially, we assume that the conjunction of all clauses is satisfied (label "T"=TRUE in the second column)[2]. The label "T" will propagate to the right as long as there are "OK"s to the right in the row below. The upper right cell is filled with "SS" (=SUCCESS) if and only if the cell to the left is "T" and the cell below is "*" (end of formula). Therefore, the formula is satisfied with this assignment if and only if the symbol "SS" appears in the table. Notice that if it was not satisfied, "T" would not had propagated to the end and "SS" would never appear, in which case we could propagate a label "F" (=FALSE) or leave the table incomplete.

The idea illustrated with the table above can be carried out almost directly by assembly of DNA tiles. The input is coded as a concatenation of tiles representing the first column and the last row of the table. This input structure is reproduced in

[1]Generalization to kCNF-SAT can be easily done, albeit increasing the number of required DNA tiles.

[2]By definition, an empty conjunction is satisfied [**DSW94**].

billions and is mixed with a DNA solution that already contains tiles from a fixed library to be described shortly. The appropriate tiles will self-assemble on this input layer. Values are assigned to the variables in a random manner. Each assembly is testing one possible assignment to the variables. The input DNA structure and the resulting tiling computation of a satisfying assignment (in fact, the one in the table above) is shown in Figure 2.

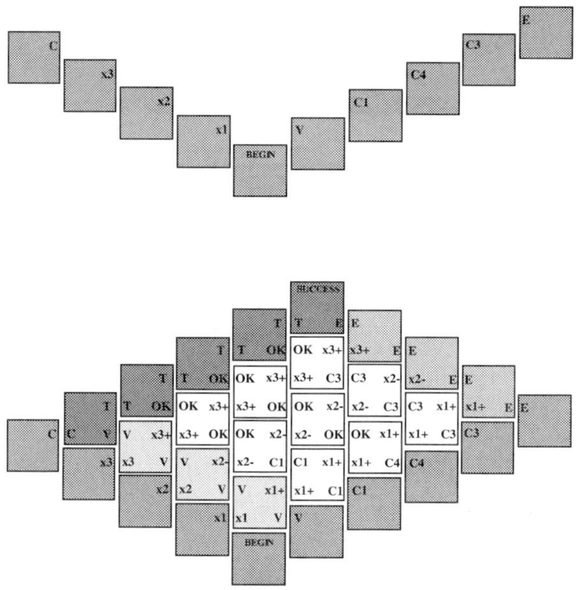

FIGURE 2. Input tiles and a successful assembly computation for design A.

For illustration purposes, the whole computation is unfolded step-by-step in Figure 3. Initially, there is only one slot where a tile can bind. This tile will be an assignment tile; the label "V" (=Value) indicates that a value is expected for the particular variable. Once a value is assigned to x_1, there are two open slots. The slot to the left "asks" for an assignment to variable x_2. The slot to the right "wants" to check whether the assignment of 1 to x_1 (shown as x_1+) satisfies clause C_1. This slot will be filled by the appropriate tile that contains the "answer". Unfortunately, x_1+ does not satisfy C_1 and thus the label C_1 is propagated up, to be checked against the remaining assignments. At the same time, x_1+ is propagated at the other side to check for the remaining clauses. At the third step, there are three slots open. One for assigning a value to x_3, one for checking C_1 against x_2-, and one for checking C_4 against x_1+. Notice that both clauses are satisfied in this case, and thus "OK" is propagated up and left. This continues until assignment of values has been completed at which time the final check begins as well (indicated by label "C" for Check). If all clauses are satisfied the assembly will continue until the "T" (=True) label meets the "E" (=End) label and the success marker will be placed on the top. If there is some unsatisfied clause, the "T" label cannot propagate and the assembly will remain incomplete and thus without the success marker. At the very end, a separation procedure that isolates the assemblies containing the success marker will provide one or more solutions to the input instance.

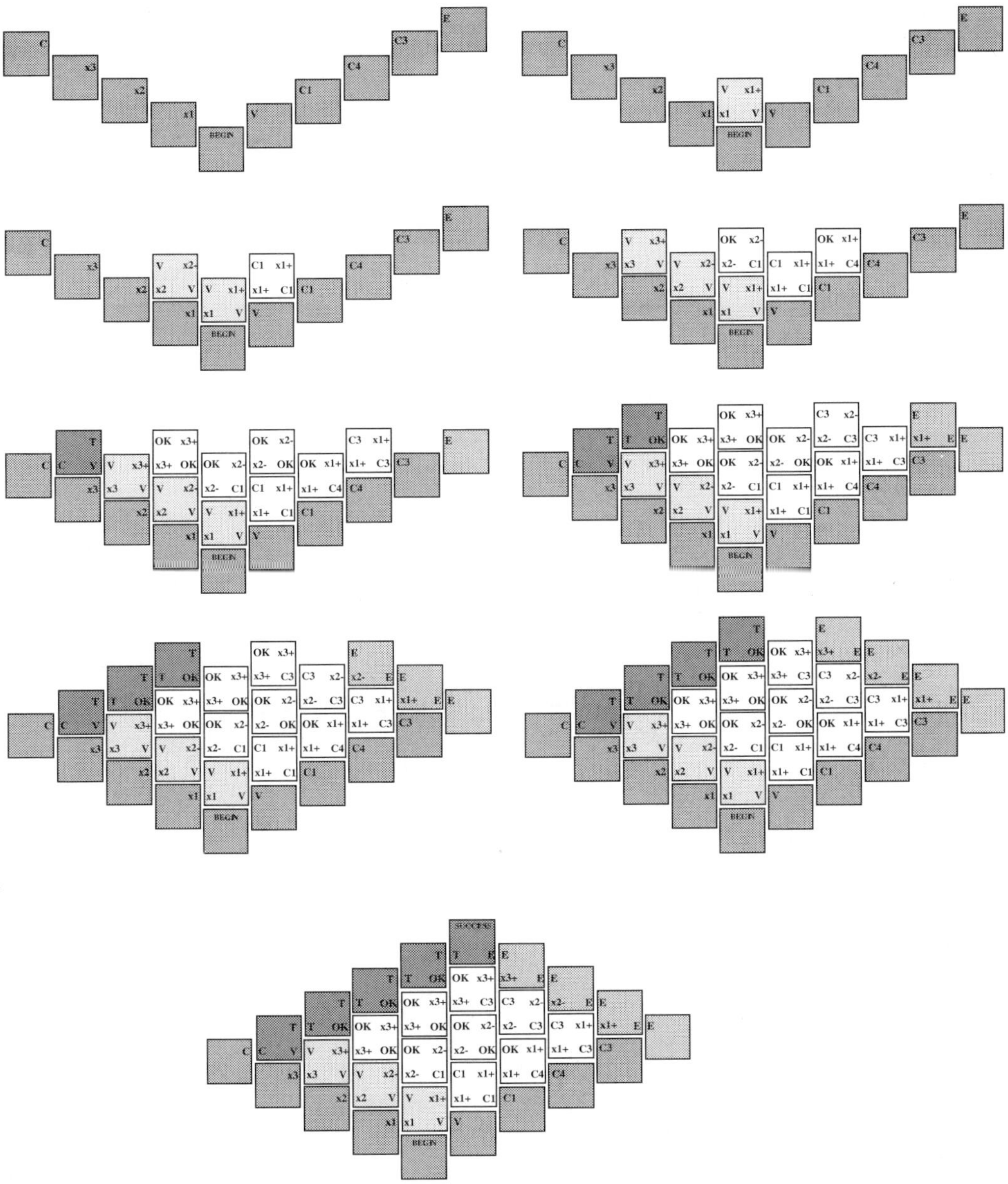

FIGURE 3. Steps of a successful DNA assembly.

3.1. Complexity of Design A. The complexity of the design is considered in terms of computation time, computation space and number of distinct tiles required. It is obvious from the examples given that the computation time $T(A)$ is equal to the depth (diagonal) of the assembly. In fact, it is

$$T(A) = (m+1) + (n+2) - 1 = m + n + 2 = \Theta(m+n) = O(m^3)$$

for m variables and n clauses. $\Theta(m+n)$ is linear in the size of the formula. $O(m^3)$ is an upper bound polynomial to the number of variables. We have used the fact that $n = O(m^3)$ for 3CNF-SAT (see section 1). The space $S(A)$ taken for each assembly is the area of the assembly.

$$S(A) = (m+2) \times (n+3) = \Theta(m \times n) = O(m^4)$$

which is upper-bounded polynomially to the number of variables. Finally, the library of fixed tiles need contain the following tiles:

- **Variables.** There must be m tiles coding m variables, where m is the maximum number of variables that can appear in a formula.

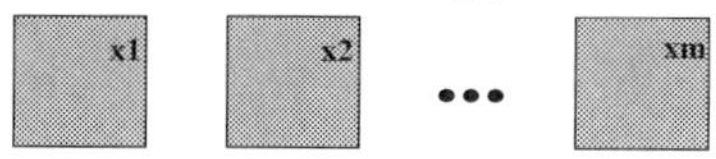

- **Clauses.** There must be $n = \frac{4}{3} \times m(m-1)(m-2) = \Theta(m^3)$ tiles coding all possible clauses.

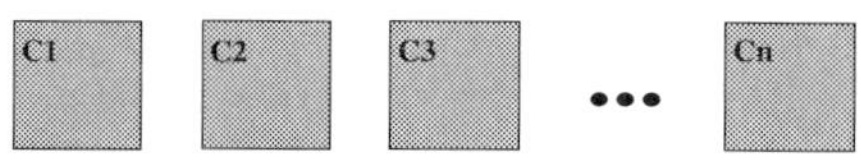

- **Input Boundaries.** There are 4 such tiles to mark the end of variable list, the end of clause list, the beginning of value assignment and the beginning of computation in the input assembly.

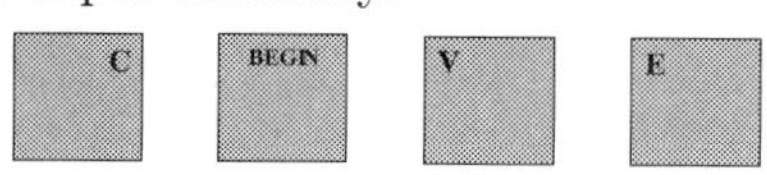

- **Assignment.** Each variable can take one of two values, so there is a total of $2m$ tiles for assigning values. A $+$ sign indicates an assignment of 1, whereas a $-$ sign indicates an assignment of 0.

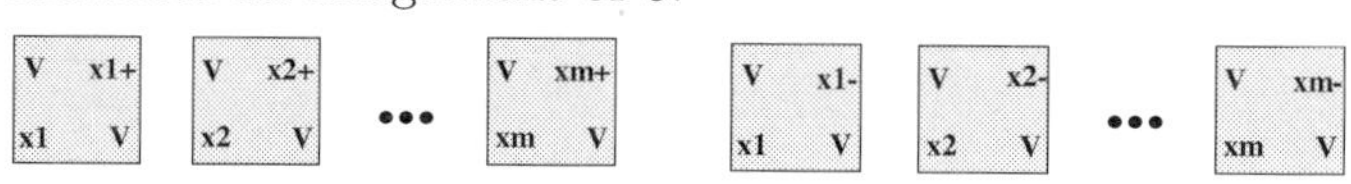

- **Computation Boundaries.** For each variable assignment there must be an ending tile, thus a total of $2m$ such tiles.

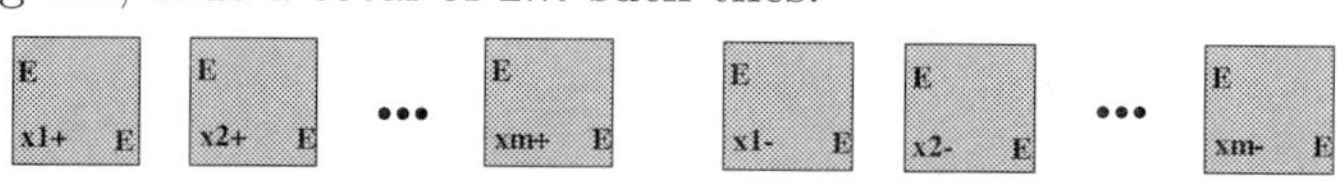

- **Computation.** For each variable assignment and for each clause there must be a tile that indicates whether the clause is satisfied or not. There are $2mn$ such tiles. Further, there must be tiles to propagate the "OK"s of the satisfied clauses to the end of the assembly. There are $2m$ of those tiles.

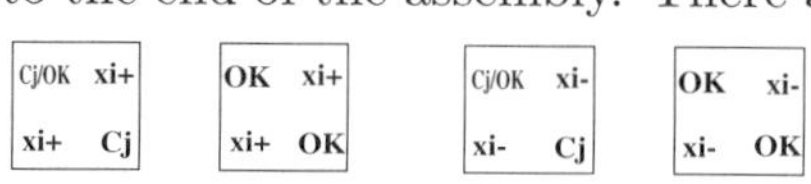

- **Final Check.** Finally, there must be some tiles to check if all clauses are satisfied and mark the result. There are 3 such tiles.

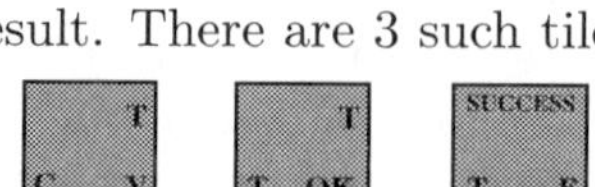

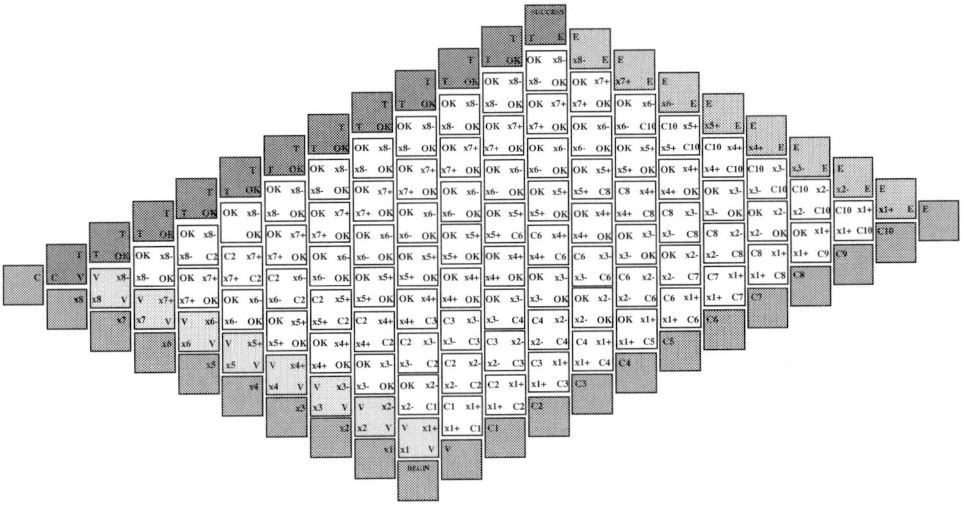

FIGURE 4. A successful DNA assembly for 3CNF-SAT instance.

Summing up all the numbers, we have the total number of tiles for design A.

$$N(A) = 7m + (2m+1)n + 7 = \Theta(m \times n) = \Theta(m^4)$$

3.2. An Example. In order to reinforce the method, Figure 4 shows another example of successful computation for a formula with 8 variables and the following 10 clauses.

$$C_1 = (\bar{x}_2 + x_3 + x_7)$$
$$C_2 = (x_3 + \bar{x}_5 + \bar{x}_8)$$
$$C_3 = (\bar{x}_1 + x_2 + x_4)$$
$$C_4 = (\bar{x}_3 + x_5 + x_6)$$
$$C_5 = (x_1 + \bar{x}_4 + x_5)$$
$$C_6 = (x_5 + x_7 + x_8)$$
$$C_7 = (\bar{x}_2 + x_4 + x_6)$$
$$C_8 = (x_3 + \bar{x}_4 + x_5)$$
$$C_9 = (x_1 + x_4 + \bar{x}_6)$$
$$C_{10} = (x_2 + \bar{x}_6 + x_8)$$

4. Design B: Encoding Literals

This design attempts to overcome the large number of tiles required by design A. In particular, it operates at a lower level, encoding literals that appear in each clause rather than the clause itself. The basic idea is the same. Consider the same 3-CNF formula

$$(\bar{x}_1 + \bar{x}_2 + x_3)(x_1 + \bar{x}_2 + \bar{x}_3)(\bar{x}_1 + x_2 + x_3)$$

and the table given below. There are 3 variables (represented in the first column of the table) and 3 clauses (represented in the last row) with all literals listed explicitly. The "s" (=SEPARATOR) is used to separate clauses. The second column represents a possible assignment as before. Cells marked with "*" are helper cells.

*	F	F	T	T	F	T	T	T	F	F	F	T	F	SS
$\mathbf{x_3}$	1	$\bar{x}_1$	OK	OK	s	OK	OK	$\bar{x}_3$	s	$\bar{x}_1$	x_2	OK	s	*
$\mathbf{x_2}$	0	$\bar{x}_1$	OK	x_3	s	OK	OK	$\bar{x}_3$	s	$\bar{x}_1$	x_2	x_3	s	*
$\mathbf{x_1}$	1	$\bar{x}_1$	$\bar{x}_2$	x_3	s	OK	$\bar{x}_2$	$\bar{x}_3$	s	$\bar{x}_1$	x_2	x_3	s	*
*	*	$\bar{\mathbf{x}}_1$	$\bar{\mathbf{x}}_2$	$\mathbf{x_3}$	s	$\mathbf{x_1}$	$\bar{\mathbf{x}}_2$	$\bar{\mathbf{x}}_3$	s	$\bar{\mathbf{x}}_1$	$\mathbf{x_2}$	$\mathbf{x_3}$	s	*

Again, following the algorithm given in section 2.3, the table is filled in a bottom-up manner, one row at a time. However, now a cell corresponding to variable x_i and literal y_j will be marked with "OK" if and only if literal y_j is TRUE when variable x_i takes the binary value v_i (indicated in the second column) or the cell below is already labeled "OK". Otherwise, it is marked with the literal name "y_j" to propagate the computation to the next row. At the end of the day, the second row of the table will contain "OK"s and/or some "y_j"s depending on what satisfies what.

The checking step is a little more involved compared to the previous one. We need to check whether each individual clause is satisfied and further whether the whole formula is satisfied. Since each clause is a disjunction, we initially assume that it is not satisfied[3]. This is denoted by the "F" (=FALSE) label in the second column. As long as the particular clause has not been satisfied (i.e., there are y_j's to the right and bottom of "F"), the label "F" propagates to the right unchanged. Once, an "OK" label is encountered to the right and bottom of an "F", it turns to a "T" (=TRUE) label, indicating that the clause has been satisfied. "T" propagates until the separator symbol "s" is met to the bottom-right. If "T" meets the separator, that implies that the current clause is satisfied and we can continue with the next one by initializing to "F" again. However, if "F" meets the separator, that means that this assignment failed to satisfy the formula and computation is halted. Finally, if the "F" label hits the "*" marker at the end of the table, it is implied that the whole formula is satisfied and therefore the success symbol "SS" marks the upper-right corner of the table. As previously, an assignment is satisfying if and only if the symbol "SS" appears at the upper-right corner of the table.

Having the concept of the table in mind, it is easy to make the transition to the DNA assembly. The basic idea is again the same as previously, but the tiles and the coding are somewhat different. Figure 5 below shows the input tile assembly and the successful computation of the table above.

Notice that by encoding literals we are not restricted to 3CNF-SAT (or to any kCNF-SAT) anymore. We can encode any SAT formula given in CNF format in a straightforward manner.

4.1. Complexity of Design B. Again, the complexity of the algorithm is considered in terms of computation time, computation space and number of distinct

[3]By definition, an empty disjunction is not satisfied [**DSW94**].

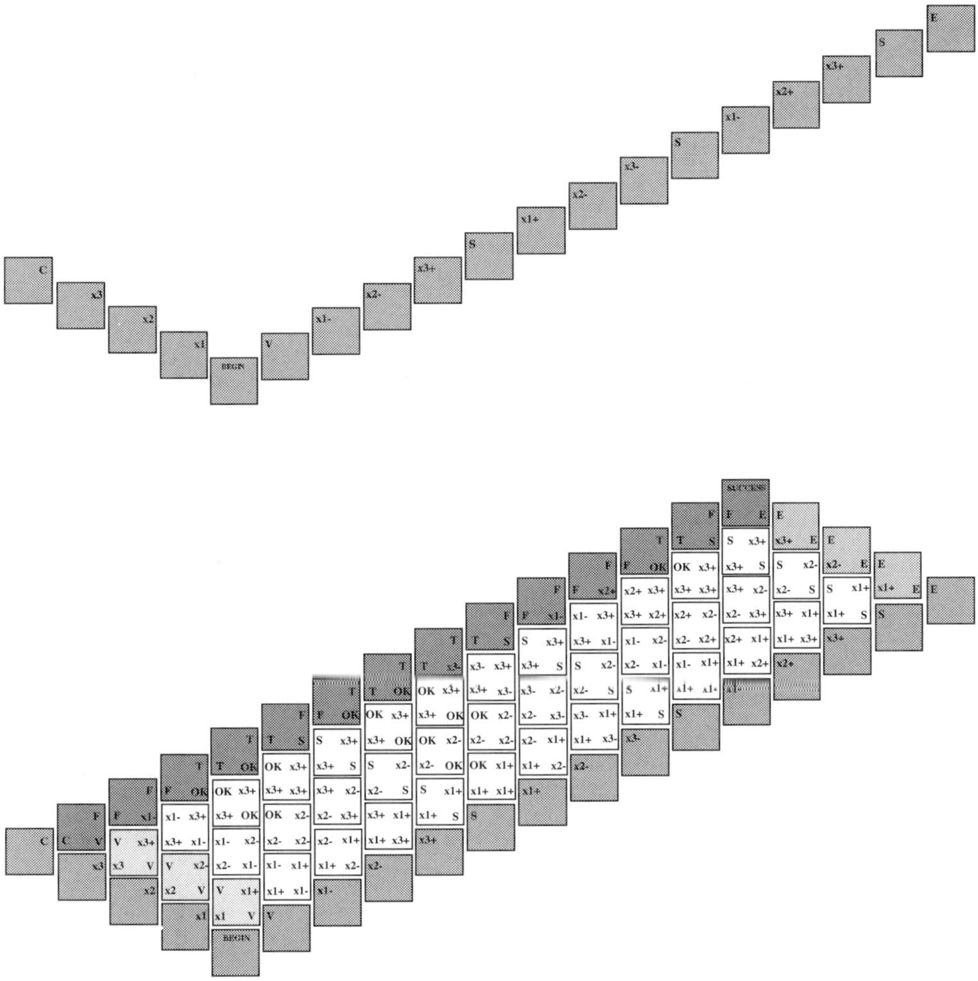

FIGURE 5. Input tiles and a successful assembly computation for design B.

tiles required. The computation time $T(B)$ is

$$\begin{aligned} T(B) &= (m+1)+(l+n+2)-1 = m+l+n+2 \\ &= \Theta(m+l+n) = O(m+mn+n) = O(mn) \end{aligned}$$

The last equality follows from the fact that in the worst case all clauses contain all the variables, that is $l = mn$, and therefore mn is an upper bound for l. If we consider 3CNF-SAT only, then $l = 3n$ and therefore

$$T(B) = (m+1)+(4n+2) = \Theta(m+n) = O(m^3)$$

which is of the same order as in the previous design. The space $S(B)$ taken for each assembly is

$$S(B) = (m+2)\times(l+n+3) = \Theta(m(l+n)) = O(m^2 n)$$

For 3CNF-SAT, $l = 3n$ and so $S(B) = \Theta(mn) = O(m^4)$ as before. The tile library is somewhat different in this case.

- **Variables.** There are m tiles coding the m variables, where m is the maximum number of variables that can appear in a formula.

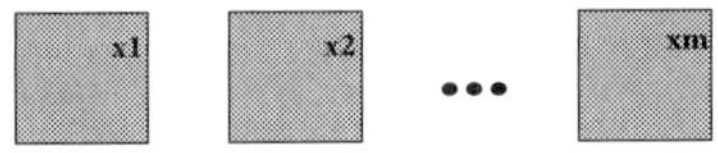

- **Literals.** There are $2m$ tiles coding all possible literals. They are used for coding the clauses of the formula.

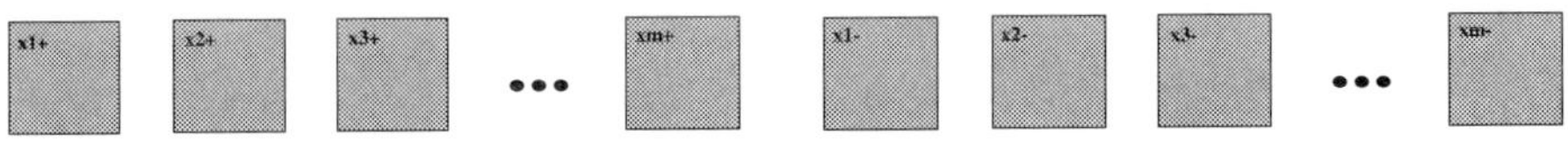

- **Input Boundaries.** There are 5 such tiles: four are the same as in design A, plus the separator.

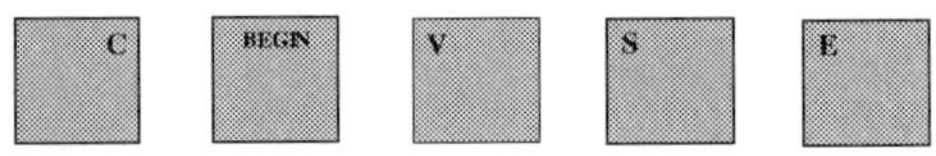

- **Assignment.** The same as before. There is a total of $2m$ such tiles.

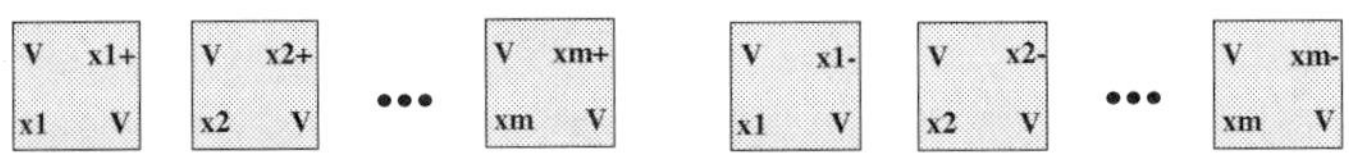

- **Computation Boundaries.** As before, for each variable assignment there must be an ending tile, thus a total of $2m$ such tiles.

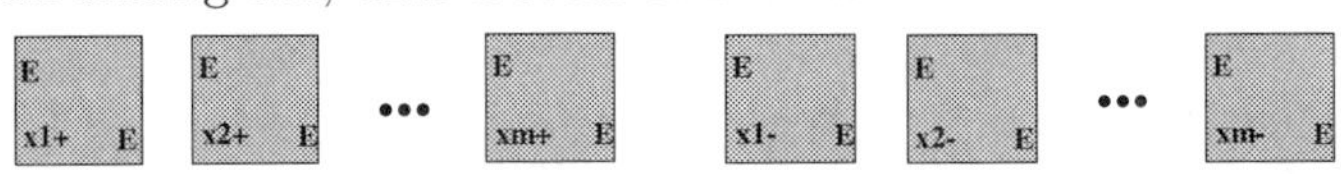

- **Computation.** For each pair of literals x_i, y_j there must be a single tile. If the literals do not refer to the same variable, or they refer to the same variable but are complementary, the result is simply propagation of the labels along the diagonals of the tile. However, if they match, x_i is propagated and the literal y_j is turned to an "OK". This requires

$$\binom{2m}{2} = m(2m-1)$$

such tiles. Further, for each variable and each possible assignment there must be tiles to propagate the "OK" labels up. There are $2m$ of those tiles.

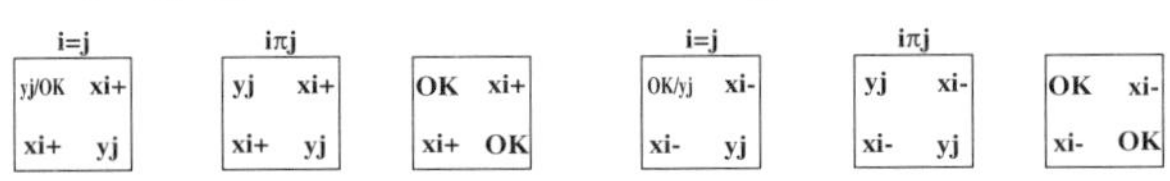

- **Final Check.** Finally, the tiles that check for satisfiability at the end are a little more involved than in the previous case. First, "F" has to propagate over literals y_j and turn into an "T" if "OK" is encountered. "T" propagates over y_j's and "OK"s. Finally, two tiles are needed for the extrema and one for resetting "T" to "F". In total, $4m + 5$ tiles.

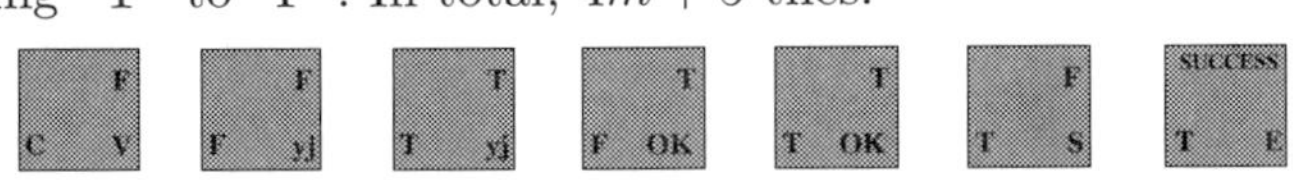

Summing up all the numbers, we have a total number of tiles:

$$N(B) = 2m^2 + 12m + 10 = \Theta(m^2)$$

which is two orders of magnitude less than the total number of tiles in design A.

5. Comparison and Discussion

The two designs implement the same algorithm in a slightly different way. It is our belief that design B is better compared to design A for two main reasons: (1) the input formula can be any CNF formula, and (2) the total number of required tile types is only $\Theta(m^2)$, where m is the maximum number of variables. In contrast, design A assumes that the formula is given in 3CNF and requires $\Theta(m^4)$ tile types. On the other hand, design A results in smaller computation time and space but asymptotically it seems that the difference disappears (see the analysis above).

There is a final detail that is crucial for the success of both algorithms. The concentrations of assignment tiles corresponding to variable x_i, i.e. tiles for x_i+ and x_i-, have to be equal so that there is equal chance of assigning either value. If this is not the case, there might be assignments that will never be explored because of this "discrimination" in assigning values.

A limitation of the algorithm, which is common for most DNA computations, comes from the fact that the exponential dimension of the problem has been pushed into the physical space (volume) occupied by the DNA molecules. This will eventually become a restrictive factor. The input size and thus the DNA volume cannot grow forever. This implies an upper bound to the size of instances that can be solved in practice. Obviously, the practicality of a DNA algorithm for satisfiability is heavily dependent on whether this upper bound is well above the upper bound for instances that can be solved on a conventional computer.

6. Future Work

It is an open question whether the design(s) will work well in practice or not, a fact that can be verified only experimentally. Encouragement comes from the recent investigations of several DNA tile structures [**WLWS98**]. In particular, TAO35 (see figure 1) is a general DNA tile that is currently being investigated for use in self-assembly computations [**LWR99**]. Moreover, it was recently demonstrated by LaBean, Winfree and Reif that input layers like the ones we use can be constructed relatively easily using a long "scaffold strand" of DNA which traverses all input tiles [**LWR99**]. The scaffold strand acts as a nucleation front for assembly of input layer tiles. The completed input layer then acts as the foundation for growth of the overall tile assembly, as described above.

Another line of research will focus on ways to enhance the algorithm will well-known heuristics for satisfiability, such as the *unit propagation* rule (if there is a clause with a single literal, force the corresponding variable to take the value that makes the clause true) and the *purification* rule (if a variable appears in the formula in exclusively negated or non-negated form, assign to this variable the value that makes all instantiations true). Actually, both of these rules can be taken into account in the current designs simply by altering the concentrations of the assignment tiles to the corresponding variables so that only the desired value is given as an option. However, this way it becomes a manual step that is perform only at the beginning. Alternatively, preprocessing of the formula could eliminate such variables. Our goal is to incorporate them in the algorithm since the need for unit propagation and/or purification might reappear during computation.

7. Conclusion

We presented an algorithmic design for solving the satisfiability problem using 2-dimensional DNA self-assembly. It seems that the peculiarity of DNA-based computation requires a reconsideration of our thinking about algorithms and computation in general. New designs or redesigns of traditional algorithms aiming to match the requirements and capabilities of DNA computation models seem to be an important step toward useful DNA-based computers.

Acknowledgements

Michail G. Lagoudakis was partially supported by the Lilian-Boudouri Foundation in Greece. This work was also supported in part by grant NSF/DARPA CCR-9725021. The authors would like to thank the reviewers for useful feedback.

References

[Adl94] Leonard M. Adleman, *Molecular computation of solutions to combinatorial problems*, Science **266** (1994), 1021–1024.

[BDLS96] Dan Boneh, Chris Dunworth, Richard J. Lipton, and Jiří Sgall, *On the computational power of DNA*, Discrete Applied Mathematics **71** (1996), 79–94.

[DSW94] Martin D. Davis, Ron Sigal, and Elaine J. Weyuker, *Computability, complexity, and languages, 2nd ed.*, Academic Press, San Diego, 1994.

[GJ79] Michael R. Garey and David S. Johnson, *Computers and intractability: A guide to the theory of NP-completeness*, Freeman, New York, 1979.

[HAK+99] Masami Hagiya, Masanori Arita, Daisuke Kiga, Kensaku Sakamoto, and Shigeyuki Yokoyama, *Towards parallel evaluation and learning of boolean μ-formulas with molecules*, DNA Based Computers III: DIMACS Workshop, June 23-25, 1997 (Providence, RI) (Harvey Rubin and David Harlan Wood, eds.), DIMACS: Series in Discrete Mathematics and Theoretical Computer Science., vol. 48, American Mathematical Society, 1999.

[Lip95] Richard J. Lipton, *DNA solutions of hard computational problems*, Science **268** (1995), 542–544.

[LWR99] Thomas H. LaBean, Erik Winfree, and John H. Reif, *Experimental progress in computation by self-assembly of DNA tilings*, Proceedings of the 5th DIMACS Meeting on DNA Based Computers, held at MIT , June 14-15, 1999 (Erik Winfree, ed.), preliminary, 1999.

[WER99] Erik Winfree, Tony Eng, and Grzegorz Rozenberg, *String tile models for DNA computing by self-assembly*, in preparation (1999).

[Win98a] Erik Winfree, *Algoritrhmic self-assembly of DNA*, Ph.D. Dissertation, California Institute of Technology (1998).

[Win98b] Erik Winfree, *Whiplash PCR for $O(1)$ computing*, Proceedings of the 4th DIMACS Meeting on DNA Based Computers, held at the University of Pennsylvania, June 16-19, 1998 (Lila Kari, Harvey Rubin, and David H. Wood, eds.), preliminary, 1998.

[WLWS98] Erik Winfree, Furong Liu, Lisa A. Wenzler, and Nadrian C. Seeman, *Design and self-assembly of two-dimensional DNA crystals*, Nature **394** (1998), 539–544.

[WYS98] Erik Winfree, Xiaoping Yang, and Nadrian C. Seeman, *Universal computation via self-assembly of DNA: Some theory and experiments*, DNA Based Computers II: DIMACS Workshop, June 10-12, 1996 (Providence, RI) (Laura F. Landweber and Eric B. Baum, eds.), vol. 44, American Mathematical Society, 1998.

(M. Lagoudakis) Department of Computer Science, Duke University, Durham, North Carolina 27708

E-mail address: `mgl@cs.duke.edu`

(T. LaBean) Department of Computer Science, Duke University, Durham, North Carolina 27708

E-mail address: `thl@cs.duke.edu`

DIMACS Series in Discrete Mathematics
and Theoretical Computer Science
Volume **54**, 2000

YAC : Yet Another Computation Model of Self-Assembly

Takashi Yokomori

ABSTRACT. This paper proposes a new model for DNA computation termed YAC based on self-assembly principle. The model has three advantages: (*i*) It has the universal computability of Turing machines. (*ii*) It requires only simple and basic molecular biological operations. Besides *annealing* and *melting* in a *one-pot* reaction, only the *detection* of a completely hybridized double stranded molecule is used. (*iii*) Therefore, a molecular biological implementation of the model seems highly feasible.

In order to make YAC computation more resource efficient, we introduce an incremental computation method.

1. Introduction

In recent intensive study of normal form theorems for the generative grammars in formal language theory, quite a few number of normal form theorems for a variety of types of grammars have been presented, ranging from regular grammars to phrase-structure grammars in Chomsky's hierarchy. Among others, some of the normal forms for phrase-structure grammars can be of crucially importance when one wants to develop new computational models. In fact, one of the normal form theorems Geffert proved in [**8**] is most interesting for our purpose, which tells us that each recursively enumerable language can be generated by a phrase-structure grammar with only 5 nonterminals $\{S, A, B, C, D\}$ (S is the initial nonterminal) and with all of the context-free rules being of the form $S \to v$ and two extra (non context-free) rules $AB \to \varepsilon$ and $CD \to \varepsilon$. In other words, each derivation of the grammar in this normal form can *almost* be represented by a derivation tree. This feature is of great importance, because this implies that *most* of the computation processes can be carried out without checking any context-sensitivity requirement.

Based on the normal form characterization for phrase-structure grammars due to Geffert, the current research presents a new model of DNA computation whose basic feature underlying its computation mechanism is classified into *self-assembly* principle. Over the other models of DNA-computing proposed and discussed so far in the literature, our models possess several advantages which seem greatly to facilitate the implementation issue in terms of the state-of-the-art technology in molecular biology.

This paper is organized as follows. As an introduction to self-assembly computation, we first discuss uncoded and coded self-assembly models of computation and

2000 *Mathematics Subject Classification.* Primary 68Q05, 68Q10, 68Q45; Secondary 68Q85, 92C40.

© 2000 American Mathematical Society

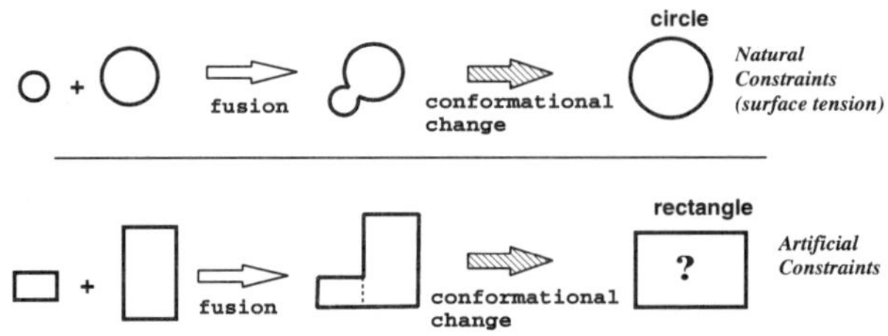

FIGURE 1. Examples of Uncoded Self-assembly

give a brief consideration into these models in Section 2. Section 3 presents preliminary existing results focusing on Geffert's normal form theorem for phrase-structure grammars. In Section 4, we propose our new DNA computation model YAC (Yet Another Computation model based on self-assembly principle). Further, Section 5 develops more resource-efficient method for implementing YAC models, describing an incremental computation schema in YAC and a computational complexity result on NP languages.

2. Self-Assembly Computations — Coded v.s. Uncoded Models

The phenomena of so-called *self-assembly* can be seen at various places, time, and occasions. For example, " · · · raindrops on a leaf autonomously merge together into one bigger drop. This is known as a typical example of *uncoded* self-assembly, where self-assembly of each component is obeyed by a simple principle of potential energy minimization. In vivo, on the other hand, protein molecules inside cells or *Bacteriophages* self-assemble its components in terms of conformational changes that is designed by *coded* instructions · · · ([**17**])".

Figure 1 illustrates an example computation process of raindrops in the uncoded self-assembly, where one basic operation *fusion* is employed under a *natural constraint* with an energy minimization. Here, a natural question arises: what kind of computation may be possible ? if arbitrary geometrical figures are taken into consideration to get *fusion* and a natural constraint is replaced by an *artificial constraint* imposed on the final process of computing desired results.

For example, suppose that a *rectangle constraint* is taken as an artificial constraint. Then, as shown in (a) of Figure 2, it is possible to carry out a computation for a polynomial factorization $\mathrm{x}^2 + 2\mathrm{xy} + \mathrm{y}^2 = (\mathrm{x} + \mathrm{y})^2$ in the framework of "uncoded self-assembly" computation. The example of (a) in Figure 2 shows a computation process where only correct solution is extracted from a pool of self-assembled components by an artificial constraint "*square*" that is supposed to be achieved by "conformational change[1]" due to some natural features on computation component materials involved (like a free energy minimization principle).

On the other hand, the same computation tasks can be performed in a simpler framework of *coded* self-assembly model, which is illustrated in (b) of Figure 2, where no concept of conformational change is necessary. Instead, coded self-assembly computation may require, in the general setting, coded components of multi-sort as is shown in (a) of Figure 2. (Note that components A and B in Figure 2 have two sides differently coded.) Thus, it seems that skillfully designed, coded components

[1]By conformational change we mean an *equivalent transformation* of an assembly, leading to a target assembly with a desired constraint.

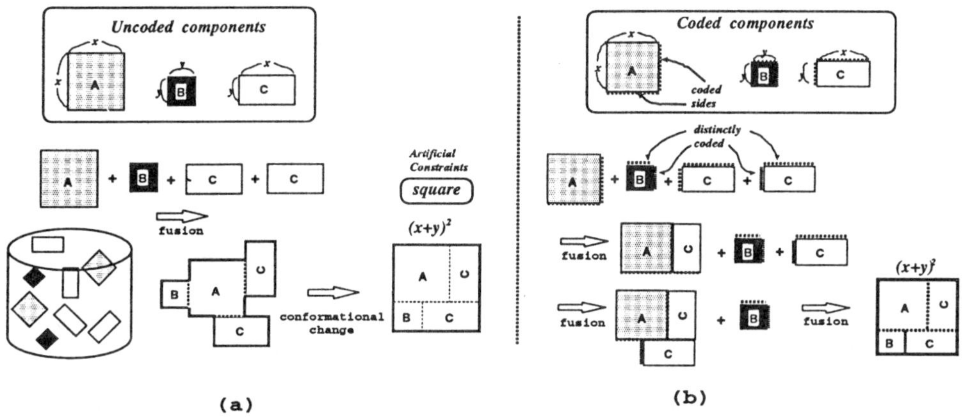

FIGURE 2. Uncoded and Coded Self-assembly Computations

make it possible to achieve a variety of fairly complicated computations, which have already been demonstrated and even proved by the work on self-assembly models by Winfree and others ([**11, 20**]). (It should be noted that no concept or knowledge of elementary arithmetic is involved in the process of self-assembly computation for factorization.) In general, there seems to be trade-off relations in computational power between the uses of "coded components" and "conformational change'.

From these observations, one may derive the following computation schema:

(Computation)=(Self-Assembly)+(Conformational Change)

where "self-assembly" depends on fusion of either uncoded or coded components, while "conformational change" is regulated by either natural or artificial constraint. Related discussion on conformation-driven computation can be found in [**4, 5, 18**]. At present, it remains open and an interesting problem to explore the computational capability of this new framework of self-assembly models. Recently by pursuing the suggested paradigm with a different interpretation of conformational change, the author introduces a new computation model termed CCC (Computation by Conformational Change), and a certain class of NP-complete problems is showed to be efficiently solved in the model CCC ([**25**]).

In this paper, we shall present a very simple type of a coded self-assembly computation model with a simple constraint but still having the Turing universal computability.

3. Preliminary Results on Grammars in Normal Form

We assume standard notions and notations from the theory of formal languages. (See, e.g., [**9**].)

Definition 1. ([**8**]) *A generative system (g-system, for short) is a 7-tuple* $\Gamma = (K, V_N, \Sigma, S, \delta, p_I, p_F)$, *where* K *is a finite set of states,* V_N *and* Σ *are mutually disjoint finite sets of nonterminal and terminal symbols, respectively,* $S (\in V_N)$ *is the initial symbol, and* p_I *and* p_F *(in* K*) are initial and final states, respectively. Further,* δ *is a finite set of transition rules of the form:* $(p, a, v, q) \in K \times V \times V^* \times K$,

TABLE 1. Definitions of h_1 and h_2 ([**8**])

$x \in \Sigma_A$	$h_1(x)$	$h_2(x)$	(remarks)
s_0	d_0	$d_0 d_1 S$	
s_1	d_1	d_2	
s_2	$d_2 q_I$	$q_F d_1$	
s_3	d_1	ε	
(q, X)	X	$q(q, X)$	for all $(q, X) \in K \times V$
(q, X, v, q')	$(q, X)q'$	v	for all $(q, X, v, q') \in \delta$

where $V = V_N \cup \Sigma$. *A binary relation* $\Rightarrow$ *on* V^* *is defined by*

$$u \Rightarrow v \overset{\text{def}}{\Longleftrightarrow} \exists (q_I, a_1, v_1, q_1), (q_1, a_2, v_2, q_2), \cdots, (q_{n-1}, a_n, v_n, q_F) \text{ such that } u = a_1 a_2 \cdots a_n \text{ and } v = v_1 v_2 \cdots v_n$$

Finally, the language of Γ *is the set* $L(\Gamma) = \{w \in \Sigma^* | S \Rightarrow^* w\}$, *where* $\Rightarrow^*$ *is the reflexive and transitive closure of* $\Rightarrow$.

Given a recursively enumerable language L over Σ, one may assume a phrase-structure grammar $G = (V_N, \Sigma, P, S)$ in a normal form generating L with the property that each rule of P is in one of the following forms: $A \to BC, AB \to CD, A \to a$, and $A \to \varepsilon$, where $A, B, C, D \in V_N$, $a \in \Sigma$. (See, e.g., [**16**])

It is easily seen that for each phrase-structure grammar G in the normal form mentioned above, there effectively exists a g-system Γ such that $L(G) = L(\Gamma)$.

Definition 2. ([**8**]) *Let* $\Sigma (= \{a_1, \cdots, a_k\})$ *be a finite alphabet. The Extended Post Correspondence (EPC, for short) over* Σ *is*

$$P = (\{(u_1, v_1), \cdots, (u_m, v_m)\}, (z_{a_1}, \cdots, z_{a_k}))$$

where $u_i, v_i, z_a \in \{0,1\}^*$ *(for each* $i = 1, \cdots, m$, *and each* $a \in \Sigma$*).*

The language of P, *denoted by* $L(P)$, *is defined as follows:*

$$L(P) = \{b_1 \cdots b_n \in \Sigma^* | \exists i_1, \cdots, i_t \in \{1, \cdots, m\} \text{ such that } v_{i_1} \cdots v_{i_t} = u_{i_1} \cdots u_{i_t} z_{b_1} \cdots z_{b_n}\}$$

Proposition 1. ([**8**]) *For each recursively enumerable language* L, *there effectively exists an EPC* P *such that* $L = L(P)$.

(Proof Sketch.) We describe the outline of the proof. For a given recursively enumerable language L over Σ, let $\Gamma = (K, V_N, \Sigma, S, \delta, p_I, p_F)$ be a g-system such that $L = L(\Gamma)$.

Now, the construction of an equivalent EPC P from Γ can be given in the following manner. First, let

$$\Sigma_A = K \times V \cup \delta \cup \{s_0, s_1, s_2, s_3\} = \{e_i | 1 \le i \le n_A\},$$
$$\Sigma_B = K \cup V \cup K \times V \cup \{d_0, d_1, d_2\} = \{a_j | 1 \le j \le n_B\}$$

(where $s_0, \cdots, s_3$ and d_0, d_1, d_2 are new symbols)

and define two homomorphisms h_1 and h_2 from Σ_A to Σ_B^* by Table 1. (Note that Σ_B properly contains $\Sigma (= \{a_1, \cdots, a_{n_L}\})$ over which L is defined, that is, it holds that $n_L < n_B$.)

Let $c : \Sigma_B \to \{0,1\}^*$ be a binary coding defined by: for each $a_i (1 \le i \le n_B)$, $c(a_i)$ is the binary notation for i. Now, we construct P as follows:

$$P = (\{(u_1, v_1), \cdots, (u_{n_A}, v_{n_A})\}, (z_{a_1}, \cdots, z_{a_{n_L}})) \quad \cdots \quad (1)$$

where for each $i = 1, \cdots, n_A$, $u_i = c(h_1(e_i)), v_i = c(h_2(e_i))$, and for each $j = 1, \cdots, n_L$, $z_{a_j} = c(a_j)$. Then, one can verify that for any $w = b_1 \cdots b_n \in \Sigma^*$, $S \Rightarrow^* w$ in Γ iff there exists $\alpha \in \Sigma_A^+$ such that $h_2(\alpha) = h_1(\alpha)w$. Since c is a bijection, this is equivalent to $c(h_2(\alpha)) = c(h_1(\alpha))c(w)$, that is, there exist $i_1, \cdots, i_t \in \{1, \cdots, n_A\}$ such that $v_{i_1} \cdots v_{i_t} = u_{i_1} \cdots u_{i_t} z_{b_1} \cdots z_{b_n}$. $\square$

Notes 1. ([8]) A careful reading of the proof above shows the following important fact. Suppose that $S = w_0 \Rightarrow w_1 \Rightarrow \cdots \Rightarrow w_r \Rightarrow w (\in \Sigma^*)$ in Γ. Then, a desired α can always be found in the form:

$$s_0(s_1(K \times V)^{n_0} s_2 H^{n_0}) \ \cdots \ (s_1(K \times V)^{n_r} s_2 H^{n_r}) s_3$$

where $n_i = |w_i|$(the length of w_i). We call a string α (in Σ_A^*) a *solution sequence* for w.

Thus, for a given $w = b_1 \cdots b_n$ in L, there exist $i_1, \cdots, i_t \in \{1, \cdots, n_A\}$ such that

$$v_{i_1} \cdots v_{i_t} = u_{i_1} \cdots u_{i_t} z_{b_1} \cdots z_{b_n} \quad \cdots \quad (2)$$

where $u_i, v_i, z_{b_j} \in \{0,1\}^*$ ($1 \leq i \leq t$ and $1 \leq j \leq n$). A string $z_w = c(w)(= z_{b_1} \cdots z_{b_n})$ is called the *binary coding of* w.

Let f and h be codings defined by: $f(0) = \mathtt{T}$, $f(1) = \mathtt{C}$ and $g(0) = \mathtt{A}$, $g(1) = \mathtt{G}$. Then, for a string x in $\{0,1\}^*$, we write $x' = f(x)$ and $x'' = g(x)$. Note that x' is in $\{\mathtt{T}, \mathtt{C}\}^*$ and x'' is in $\{\mathtt{A}, \mathtt{G}\}^*$.

From Proposition 1, Geffert derived an interesting normal form theorem for phrase-structure grammars which is rephrased here with some modifications.

Proposition 2. ([8]) *Each recursively enumerable language L over Σ can be generated by a phrase-structure grammar with $G = (\{S, \mathtt{A}, \mathtt{C}, \mathtt{G}, \mathtt{T}\}, \Sigma, R \cup \{\mathtt{TA} \to \varepsilon, \mathtt{CG} \to \varepsilon\}, S)$, where $R = R_1 \cup R_2 \cup R_3$ and*

$$R_1 = \{S \to z'^R_{a_j} S a_j \mid a_j \in \Sigma \text{ and } z_{a_j} \text{ are those of } P \text{ in } (1)\}$$

$$R_2 = \{S \to u'^R_i S v''_i \mid 1 \leq i \leq n_A \text{ and } u_i \text{ and } v_i \text{ are those of } P \text{ in } (1)\}$$

$$R_3 = \{S \to u'^R_{n_A}\} \text{ (where we may assume that } u_{n_A} = c(h_1(s_3)))$$

Notes 2. 1. For each $i = 1, 2, 3$, a rule in R_i is said to be Type-(i).
2. The proof tells that in Type-(ii) rules, v-side strings may possibly be empty, while u-side strings are never empty.
3. **Notes 1** tells that in fact there is only one Type-(iii) rule corresponding to s_3 in Table 1.
4. From **Notes 1** together with discussion given between Definition 1 and 2, whenever a Type-(ii) rule with empty v-side is used, it is immediately followed by either a Type-(ii) with non-empty v-side or the unique rule of Type-(iii).

Now, a derivation process for w in G is carried out at the three stages as follows:
[1st stage](*Generation*) Using Type-(i) rules,

$$S \Rightarrow z'^R_{b_n} S b_n \Rightarrow z'^R_{b_n} z'^R_{b_{n-1}} S b_{n-1} b_n \Rightarrow \cdots \Rightarrow z'^R_{b_n} \cdots z'^R_{b_1} S b_1 \cdots b_n = z'^R_w S w$$

[2nd stage](*Extension*) Using Type-(ii) rules and one Type-(iii) rule,

$$\Rightarrow z'^R_w u'^R_{i_t} S v''_{i_t} w \Rightarrow \cdots \Rightarrow z'^R_w u'^R_{i_t} \cdots u'^R_{i_2} u'^R_{i_1} v''_{i_1} v''_{i_2} \cdots v''_{i_t} w$$
$$(\text{where } u'^R_{i_1} = u'^R_{n_A} \text{ and } v''_{i_1} = \varepsilon)$$

[3rd stage](*Test*) Using cancellation rules $\mathtt{TA}\rightarrow \varepsilon$ and $\mathtt{CG}\rightarrow \varepsilon$,

$$\Rightarrow (u'_{i_1} \cdots u'_{i_t} z'_w)^R (v''_{i_1} \cdots v''_{i_t}) w = f(u_{i_1} \cdots u_{i_t} z_w)^R \cdot g(v_{i_1} \cdots v_{i_t}) w \Rightarrow \cdots \Rightarrow w.$$

(Note that a relation (2) was used).

Here, from the property of the phrase-structure grammar G in normal form in Proposition 2 we have the following:

[Important Observations]

1. It is not until a Type-(*iii*) rule has been applied that two extra cancellation rules ($\mathtt{TA} \rightarrow \varepsilon$ and $\mathtt{CG} \rightarrow \varepsilon$) are ready to be used.
2. Mixed use of Type-(*i*) and Type-(*ii*) rules leads to sentential forms for which no chance to get terminal strings remains.
3. Any occurrence of x ($\in \Sigma$) in any substring (except for a suffix) of a sentential form blocks a way to produce a terminal string.
4. A successful derivation process can be divided into three phases: At the first phase (*Generation*), Type-(*i*) rules are used. The second phase (*Extension*) consists of using Type-(*ii*) rules and ends up with one application of the Type-(*iii*) rule. Finally, the third phase (*Test*) erases well-formed pairings (TA or CG) using two extra rules only.
5. There is only one occurrence of either TA or CG in any sentential form to which cancellation rules can be applied.
6. *Test* phase can only be started up with the unique position where either TA exclusively or CG appears immediately after one application of the Type-(*iii*) rule.

4. YAC : Yet Another Computation of Self-Assembly

As one instance of a *coded* self-assembly computation model, we present our DNA computation model : YAC(Yet Another Computation of Self-assembly) which is based on the characterization results given in the previous section.

Suppose that given a recursively enumerable language L, let $G = (\{S, \mathtt{A}, \mathtt{C}, \mathtt{G}, \mathtt{T}\}, \Sigma, R \cup \{\mathtt{TA}\rightarrow \varepsilon,\ \mathtt{CG}\rightarrow \varepsilon\}, S)$ be the phrase-structure grammar in normal form such that $L = L(G)$ in Proposition 2.

In our model YAC, we prepare in the following manner the set of basic components to be assembled :

1. For each Type-(*ii*) rule : $S \rightarrow u'^R S v''$, construct a component with the shape illustrated in (b) of Figure 3. There are two categories of components for each Type-(*ii*) rule : one for types (α), (β) and (γ), the other for a type (δ). (Note that a type (δ) is for a rule with $v = \varepsilon$ on its v-side.) Each component in (b) is said to be **Type-(ii)**.
2. Further, for the unique Type-(*iii*) rule : $S \rightarrow u'^R$, construct a component with the shape illustrated in (c) of Figure 3, where connector is attached as an endmarker. Thus, we have exactly two components of type (α') and type (δ'). Each component in (c) is said to be **Type-(iii)**.

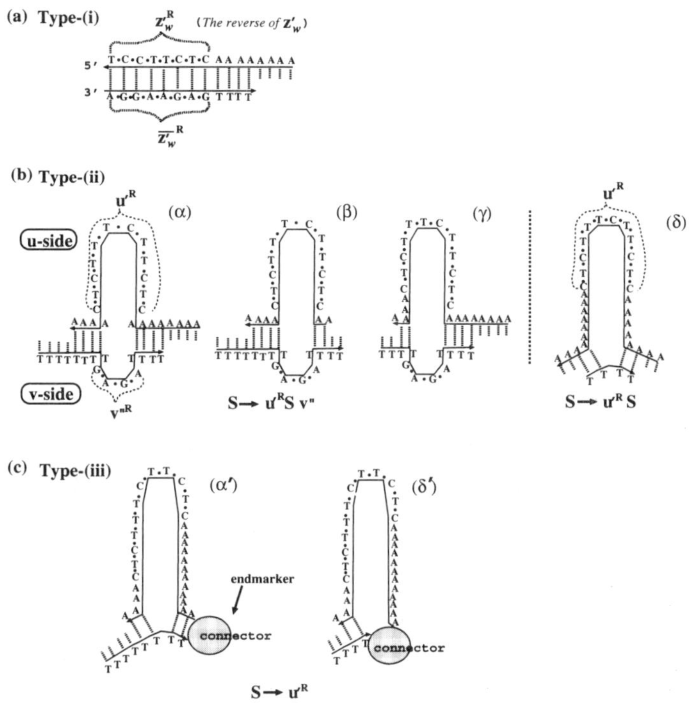

FIGURE 3. Basic Components in YAC Model :

3. Finally, for a given input string $w = b_1 \cdots b_n (\in \Sigma^*)$ to be recognized, we prepare a component in shape shown in (a) of Figure 3. Note that a string z'_w of the component in (a) is designed so that it reflects the binary coding of the input w, where $\overline{z'_w}$ denotes $z''_w (= g(z_w))$ on the analogy of complementarity of DNA molecules.

We assume that under the presence of *ligase*, all these components with sufficiently high concentration in a single pot of a test tube are provided. (See Figure 4.)

Finally, we adopt the following operation (that acts as an artificial constraint in a coded self-assembly model) :

- d-Detect : Given a test tube that may contain double stranded strings, return "*yes*" if it contains (at least) one *completely hybridized* double stranded string, and "*no*" otherwise.

Note that d-Detect is a "double stranded version" of a standard operation Detect (to detect whether or not there is a single strand string in the test tube). Figure 4 illustrates a computation schema for YAC, where a random pool is a multi-set of basic components with sufficiently high concentration that can produce all possible assembly necessary for recognizing an input string.

Besides an operation d-Detect, YAC only requires two basic operations to execute its computation process : *annealing* and *melting*. An example of a successful computation in YAC is illustrated in Figure 5.

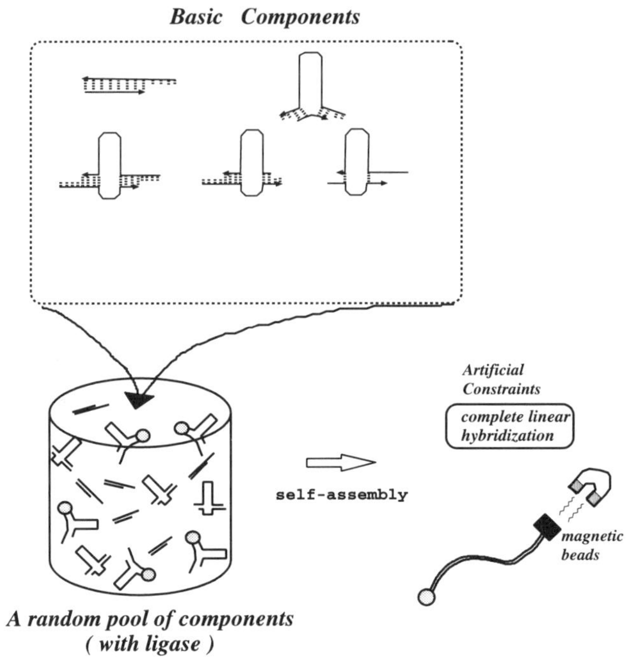

FIGURE 4. Computation Schema for YAC

[Implementation Notes]

- In implementing 1 described above, DNA complementarity is conveniently available in such a way that the symbols A and G on v-side of the rule be realized by *purine* : A, G and the symbols C and T on u-side by *pyrimidine* : C, T, respectively. This also gives us more guarantee to prevent each "loop" part on v-side (or u-side) from forming unexpected (undesirable) structures. The same feature applies to the case 2 above.
- In a practical implementation of basic components, two kind of *separaters*, consisting of 12 As and complementary 12 Ts, should be properly inserted at the "dotted position" between two symbols of each string on both u-side and v-side of each component in Figure 3, respectively, so that a complete hybridization may be successfully made. (Here we preferred the current presentation because of its simplicity in representing the basic idea underlying.)
- Implementing an operation d-Detect may contain two tasks : one is to *separate* only complexes of partially duplex DNAs (PDDs, for short) from others, the other is to *detect* if there is a double stranded molecule (coding a solution sequence) among them. (For PDDs, see (c) of Figure 5.) For the former task (what we call *PDD-separation*), we would be able to utilize the secondary structural property of non-PDDs in order to discriminate them from PDDs. More specifically, some of well-established methods for discriminating mutant DNA strands from complete double strands, such as DGGE[2] or SSCP[3], seem very promising for practical use. The latter task may be achieved by

[2]denaturing gradient gel electrophoresis
[3]single-stranded conformation polymorphism

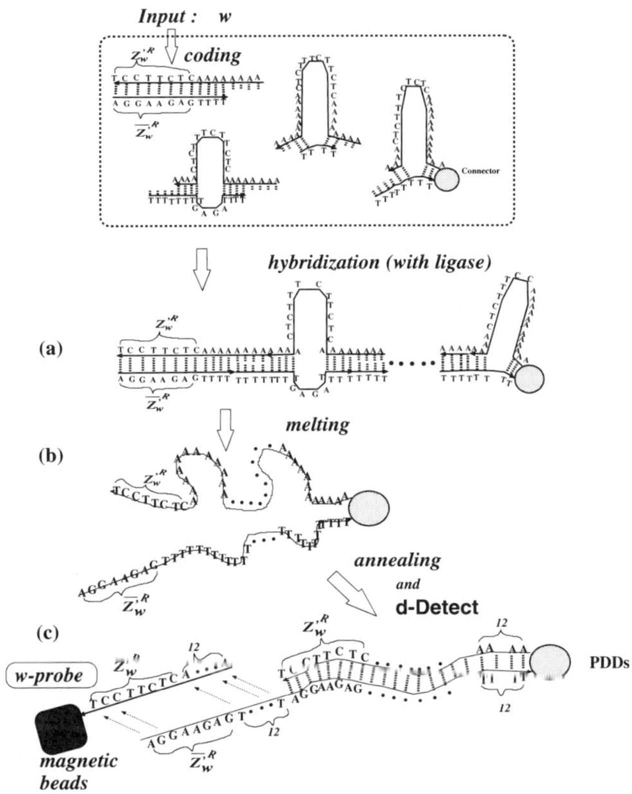

FIGURE 5. A Successful Computation Process in YAC: After hybridizing many basic components, candidate complexes for successful computations are obtained (as in (a)). Then, by melting those complexes we have DNA molecules in the shape of (b). After annealing, if the input w is in L, then a desired double stranded molecule is detected using the w-probe attached with magnetic beads in (c).

using a familiar biotechnique to probe a particular complex among many DNA molecules with the help of attached magnetic beads. (See (c) of Figures 5 or 6.)

[Universal Computability of YAC]. Under the assumption that all operations involved in the computation process of YAC are correctly performed (without any errors), it is almost straightfoward to see that the YAC model has the universal computability of Turing machines.

From Proposition 2 and its following explanations throughout the previous section, it holds that for an arbitrary input w in Σ^*,

$$
\begin{aligned}
w \in L \iff & \exists u'_{i_1} \cdots u'_{i_t}, v''_{i_1} \cdots v''_{i_t}, \text{ and } z'_w \text{ such that} \\
& S \Rightarrow^* f(u_{i_1} \cdots u_{i_t} z_w)^R \cdot g(v_{i_1} \cdots v_{i_t})w \\
& \text{where } (i_1, \cdots, i_t) \text{ satisfies the relation (2)} \\
\iff & \exists \text{ a completely hybridized double strands with } f(u_{i_1} \cdots u_{i_t} z_w) \\
& \text{on one side and } g(v_{i_1} \cdots v_{i_t}) \text{ on the other,} \\
& \text{ignoring every occurrence of separaters } \mathtt{A}^{12} \text{ and } \mathtt{T}^{12}, \\
& \text{in the final test tube of YAC computation.}
\end{aligned}
$$

In particular, **[Important Observations]** guarantee that among all of possible aggregates comprising basic components in a random pool of components only completely hybridized double strands can reflect successful computations for w in YAC. More specifically, in order to see that the above propositions are correct, we claim the following :

1. *Generation* phase of G is replaced in YAC by creating Type-(i) component with the binary coding z'_w of the input w. (See (a) of Figure 5.)
2. In simulating *Extension* phase of G by YAC, any component of type (α) in Type-(ii) is always available, and whenever a component of type (δ) must be used, instead of (α) a component of type (β) can proceed (δ), then immediately followed by a component of type (γ). (See Figure 3.) This ordering of compositions can only achieve a successful simulation of *Generation* phase containing a Type-(ii) rule with empty string on its v-side, which is assured by 4. of **Notes 2**.
3. It holds that w is in L iff there is a successful computation process in YAC at whose final stage there is a PDDs with a specific single strand : $\overline{z'}^R_w \mathtt{T}^{12}$ (see (c) of Figure 5), where $\overline{z_w}$ denotes the complement of z_w.
4. Thus, all of what remains of our task is to perform a d-Detect operation in which the w-probe (consisting of $z'^R_w \mathtt{A}^{12}$ attached with magnetic beads) is readily applied to separate only targeted double strands.

Thus, we have the following :

Proposition 3. *Any recursively enumerable language can be recognized by YAC computation model.*

A similar argument can apply to context-sensitive languages. That is, it is easily seen that any context-sensitive language can be generated by an ε-free g-system, which eventually leads us to the following result.

Proposition 4. *Any context-sensitive language can be recognized by YAC computation model in which type* (α) *of Type-*(ii) *components, type* (α') *of Type-*(iii) *components and the one Type-*(i) *component are only needed.*

5. Incremental Computation in YAC: Getting Resource Efficiency

Although the computation process of YAC described in the previous section seems to be simple enough to implement, it is easily seen that creating a random pool of components in a single pot can require a huge quantity of DNA molecules as the case may be. In order to reduce this difficulty and get more resource efficiency,

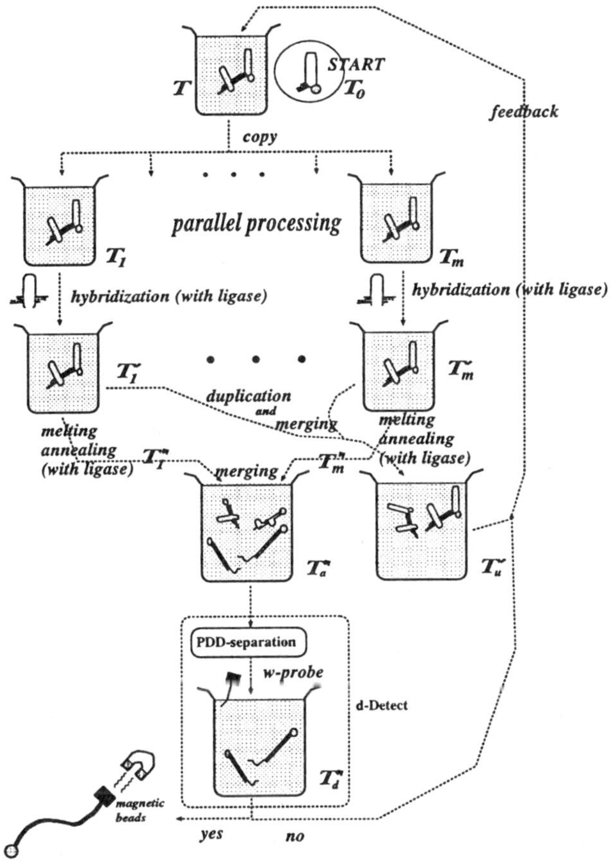

FIGURE 6. Incremental Computation in YAC

we shall now present another way of implementing the YAC computation that is much more resource-efficient than the naive one.

Instead of creating a random pool of all components, we adopt a step-by-step method to simulate via YAC a successful derivation in the normal form grammar G for L, whose basic idea has been already proposed in the literature (for example, [**12, 22**]).

Let w be an input string to be recognized with respect to L and let m be the total number of all Type-(ii) components in YAC for L.

Starting with the Type-(iii) components, we simulate a derivation process of G in a backward manner.

(0) *Initialization*: Prepare the initial test tube T_0 containing only Type-(iii) components.

(1) *Parallel simulation*: Let T be the current test tube. We make m copies $T_1, ..., T_m$ of T, and for each $i = 1, \cdots, m$, we put the i-th Type-(ii) component into T_i (See Figure 6.) Then, for each $i = 1, \cdots, m$, the i-th component is hybridized with intermediate aggregates contained in T_i, in a paralell manner, which results in producing T_i'. After duplicating each T_i', we then merge all T_i's to get a new tube $T_u'(= T_1' \cup \cdots \cup T_m')$. On the other hands, by T_i'' let us denote the test tube

obtained from T_i' by melting and annealing. Now, all T_i''s are got together into one tube T_a''. Then, after PDD separation in d-Detect we have a test tube T_d'' that contains only PDDs. Now, we make a detection to T_d'' to examine if or not it contains a completely hybridized double strand. If it turned out there is, then go to (2). Otherwise, we go back to the test tube T_u' and let it be the current tube T. Then, go to (1) again.

(2) *Halt*: Output "yes" and terminate the computation.

Note that if w is not in L, then the above procedure may not terminate forever when L is non-recursive.

[Computing NP in YAC]. We analyze the time complexity of YAC computation for the class NP. Generally, in most computational models based on self-assembly principle, it is a hard problem to introduce the definition of time complexity acceptable to everyone, because the basic operation of self-assembly takes place everywhere in parallel in a single pot, so that one can hardly define the unit time for those behaviors in a reasonable manner. Further, the fact that self-assembly is a biochemical phenomenon that strictly depends upon the concentration of the basic components makes the matter more difficult.

On the other hand, in the computation process of either YAC (in a naive implementation) or its incremental version, the length of a composed duplex (possibly coding a solution sequence) dominates the time of performing an operation d-Detect. Therefore, in order to discuss the time complexity for YAC, let us define it by the *length* of a composed duplex to be checked in the final test tube.

Now, let L be a language recognized by a nondeterministic Turing machine in time $T(n)$, where n is the input size. Then, it is shown that one can construct a phrase-structure grammar G in normal form in Propositon 2 such that $L = L(G)$ with the time complexity $T^2(n)$ ([**8**]). This implies that for a given input w of length n, the length of a solution sequence for w is bounded by $O(T^2(n))$. Thus, we have:

Proposition 5. *Any language in* NP *can be recognized by YAC computation in polynommial time.*

6. Conclusion

From a computation theoretic point of view, the basic principle of computation in the YAC model is quite similar to those of DNA-EC([**24**]) and Watson-Crick Automata ([**7**]) in that DNA computational mechanism of these models is based on the task of *equality checking* of two working tapes of DNA strands, while we claim that YAC model is much simpler than others and it gains some advantages over them. On the other hand, there are already some results on self-assembly DNA computing models. Among others Winfree's 2D self-assembly models in [**20, 21**] (W-models, for short) that may be regarded as a coded self-assembly model in our sense, are most relevent to discuss.

The advantages and distinguished features of YAC model may be summarized as follows:

1. YAC needs not to take care of controlling the state-transition process that is essential to most other computation models proposed so far, which makes YAC considerablly simpler and seems easier to implement it.
2. A normal form grammar G on which YAC depends requires only a *fixed number* of nonterminals (in fact, 5 nonterminals are sufficient) in order to simulate *any* recursively enumerable language L, while the corresponding number of states neccessary in other grammar-based models strictly depends on L.
3. Moreover, the roles of 4 nonterminals A, C, G, T out of 5 of G are very much suitable for DNAs to simulate in that the cancellation rules $\mathtt{TA} \to \varepsilon$ and $\mathtt{CG} \to \varepsilon$ can be performed by just the right task for the complementary pairs of T and A, and of C and G to hybridize, respectively.
4. In order to achieve the universal computability, YAC needs only *plane* components of simple tree-like structures in its computation process, while more complicated components (such as *double crossover molecules*) are required in W-models or others.
5. Growing up aggregates of basic components in YAC is based on a simple hybridization mechanism in a linear direction (dimension), which seems to allow YAC an easier way of implementation than any other.

Based on a new computation paradigm of a coded self-assembly with an artificial constraint, we have proposed a DNA computing model YAC having the Turing universal computability in which any recursively enumerable language can be recognized in a purely self-assembly computation manner. It should be noted that YAC employs as an artificial constraint a kind of "free energy minimization" requirement of being a completely hybridized double stranded molecule. Furthermore, the structural simplicity of YAC computation assembly units is expected to facilitate its implementation at least in its design schema. In particular, the very two groups of four DNAs (A,G: purine and C,T: pyrimidine) can be readily applied both to encoding the computational information of a language into its YAC assembly units and to checking the completeness of the computation results in the final process with its complementary property. On the other hand, it is also true that one cannot be so optimistic at the moment about the biomolecular feasibility of the YAC model. The proposed model in this paper still remains at the conceptual level, and more molecular biological investigation is needed before the model becomes truly feasible.

One of the most important open problems on YAC computation model is to analyse how many basic components in a single pot would be required for solving real world problems of practical size.

Acknowledgements

The author is grateful to an anonymous referee for many constructive comments which greatly improved both readability and consistency of the manuscript. Thanks are due to Kensaku Sakamoto for useful suggestions on biomolecular methods DGGE and SSCP.

This work is supported in part by "Research for the Future" Program No.JSPS-RFTF 96I00101 from the Japan Society for the Promotion of Science, by Grants-in-Aid for Scientific Research No.11878055 from Ministry of Education, Science,

Sports and Culture, Japan, and by Waseda University Grant for Special Research Projects No.98A-824.

References

[1] L.Adleman. Molecular Computation of Solutions to Combinatorial Problems. *Science* **266**, pp.1021-1024, 1994.

[2] L.Adleman. On Constructing A Molecular Computer. *DNA Based Computers* (Selected papers from Proc. of DIMACS Workshop on DNA Based Computers'95), DIMACS Series in Discrete Mathematics and Theoretical Computer Science, **27**, pp.1-21, 1996.

[3] D.Beaver. A Universal Molecular Computer. *DNA Based Computers* (Selected papers from Proc. of DIMACS Workshop on DNA Based Computers'95), DIMACS Series in Discrete Mathematics and Theoretical Computer Science, **27**, pp.29-36, 1996.

[4] M. Conrad. On Design Principles for a Molecular Computer. *Comm. ACM*, Vol.28, No.5, pp.464-480, 1985.

[5] M. Conrad. Molecular Computing Paradigms. *IEEE Computer*, Vol.25, No.11, pp.6-9, 1992.

[6] J.Engelfriet and G.Rozenberg. Fixed Point Languages, Equality Languages, and Representation of Recursively Enumerable Languages, *J. ACM*, **27**, pp.499-518, 1980.

[7] R.Freund, Gh.Paun, G.Rozenberg and A.Salomaa. Watson-Crick Automata. *Proc. of 3rd DIMACS Workshop on DNA based computers*, pp.305-317, 1997.

[8] V. Geffert. Normal Forms for Phrase-Structure Grammars. *RAIRO /Theoretical Informatics and Applications*, **25**, pp.473-496, 1991.

[9] J.Hopcroft and J.Ullman. *Introduction to Automata Theory, Languages and Computation.* Addison-Wesley, 1979.

[10] N. Jonoska and M. Saito. Creating 3-Dimensional Graph Structures with DNA. *Preliminary Proc. of 3rd DIMACS Workshop on DNA Based Computers*, pp.192-203, 1997.

[11] N. Jonoska, S.Karl and M. Saito. Three Dimensional DNA Structures in Computing. *Preliminary Proc. of 4th DIMACS Workshop on DNA Based Computers*, pp.189-200, 1998.

[12] N.Morimoto, M.Arita and A.Suyama. Stepwise generation of Hamiltonian path with molecules. In *Biocomputing and Emergent Computation*, pp.184-192, 1997.

[13] Gh.Paun. Regular Extended H Systems are Computationally Universal. *J. Automata, Languages and Combinatorics* (formerly *EIK*), **1**, pp.27-36, 1996.

[14] Gh.Paun., G.Rozenberg and A.Salomaa. *DNA Computing : New Computing Paradigms*, Springer-Verlag, 1998.

[15] J.H.Reif. Parallel Molecular Computation : Models and Simulations. *Proc. of Seventh Annual ACM Symposium on Parallel Algorithms and Architectures*(SPAA'95), pp.213-223, 1995.

[16] A. Salomaa. *Computation and Automata*, Academic Press, 1985.

[17] K. Saitou. Self-Assembling Automata : A Model of Conformational Self-Assembly, *Proc. of Pacific Symposium on Biocomputing'98*(Eds, R.B.Altman, et al.), pp.609-620, 1998.

[18] K. Saitou and M.J.Jakiela. On Classes of One-dimensional Self-Assembling Automata, *Complex Systems*, **10 (6)**, pp.391-416, 1996.

[19] Y.Sakakibara and C.Ferretti. Splicing on tree-like structures. *Proc. of 3rd DIMACS Workshop on DNA based computers*, pp.348-358, 1997.

[20] E.Winfree, X.Yang and N.C.Seeman. Universal Computation via Self-assembly of DNA : Some Theory and Experiments. *DNA Based Computers II* (Papers from Proc. of 2nd DIMACS Workshop on DNA Based Computers, 1996), DIMACS Series in Discrete Mathematics and Theoretical Computer Science, **44**, pp.191-213, 1999.

[21] E.Winfree. *Algorithmic Self-Assembly of DNA*. Ph.D Dissertation, submitted to California Institute of Technology, May 1998.

[22] D.H. Wood. A DNA Computing Algorithm for Directed Hamiltonian Paths. *Genetic Programming 1998* : Proc. of 3rd Annual Conference (J.R.Koza, et al. eds.), Morgan Kaufmann, pp.731-734, 1998.

[23] T.Yokomori, S.Kobayashi and C.Ferretti. On the Power of Circular Splicing Systems and DNA Computability. *Proc. of IEEE Intern. Conf. on Evol. Comput.* (ICEC'97), pp.219-224, 1997.

[24] T.Yokomori and S.Kobayashi. DNA-EC : A model of DNA computation based on Equality Checking. *Proc. of 3rd DIMACS Workshop on DNA based computers*, pp.334-347, 1997.

[25] T.Yokomori. Computation = Self-assembly + Conformational Change : Toward New Computing Paradigms. to appear in *Proc. of 4th International Conference on Developments in Language Theory (DLT'99)*, Aachen, July, 1999.

(T. Yokomori) Department of Mathematics, School of Education, Waseda University, 1-6-1 Nishi-waseda, Shinjuku-ku, Tokyo 169-8050, JAPAN

E-mail address: yokomori@mn.waseda.ac.jp

DIMACS Series in Discrete Mathematics
and Theoretical Computer Science
Volume **54**, 2000

DNA Hybridization Catalysts and Molecular Tweezers

A. J. Turberfield, B. Yurke, and A. P. Mills, Jr.

ABSTRACT. We demonstrate new methods for the control of DNA hybridization: formation of a loop with a protective strand is used to inhibit hybridization, and a DNA catalyst that opens the loop is used to catalyse hybridization. A combination of inhibition and catalysis will allow control of the bonds formed between elements of a self-assembled structure. We also demonstrate a new class of nanomachine, made of DNA and using the hybridization of DNA as a source of chemical energy to produce repeated movement.

1. Introduction

The free energy released by hybridization of two DNA strands depends sensitively on their base sequences; it is possible to find large sets of quasi-orthogonal strands within which only strands with completely complementary sequences will form stable duplexes [**1**]. This specificity may be exploited to implement massively parallel computational schemes [**2**]. Hybridization of DNA also has great potential for the engineering of nanostructures: complicated two- and three- dimensional structures have been made entirely from DNA [**3, 4, 5**] and DNA tags have been used to bind colloidal particles together [**6, 7**]. The range of potential uses for DNA structures was increased by the demonstration of DNA-templated growth of metal wires [**8**] and semiconductor nanocrystals [**9**] and of a DNA device in which conformational changes may be induced by changing the buffer solution [**10**]. In this paper we contribute to the exploration of DNA as an engineering material in two ways [**11**]: we demonstrate catalytic control of the hybridization of complementary strands, a technique which may lead to new strategies for self-assembly; and we introduce a new class of active DNA devices that use DNA both as a structural material and as a fuel.

2. Protected Self-Assembly

The specificity of DNA hybridization may be exploited to control the interactions between very large numbers of DNA-tagged nanoscale objects as shown schematically in Figure 1(a); each bond is formed by the hybridization of complementary tags. First demonstrations of DNA-controlled self-assembly have been only partially successful in creating ordered structures, however [**5, 6**]. A fundamental problem arises when it is necessary to make more than one bond to incorporate an object into the designed structure, and when more than one object is tagged with the same DNA sequence; in this very general case the final structure may not

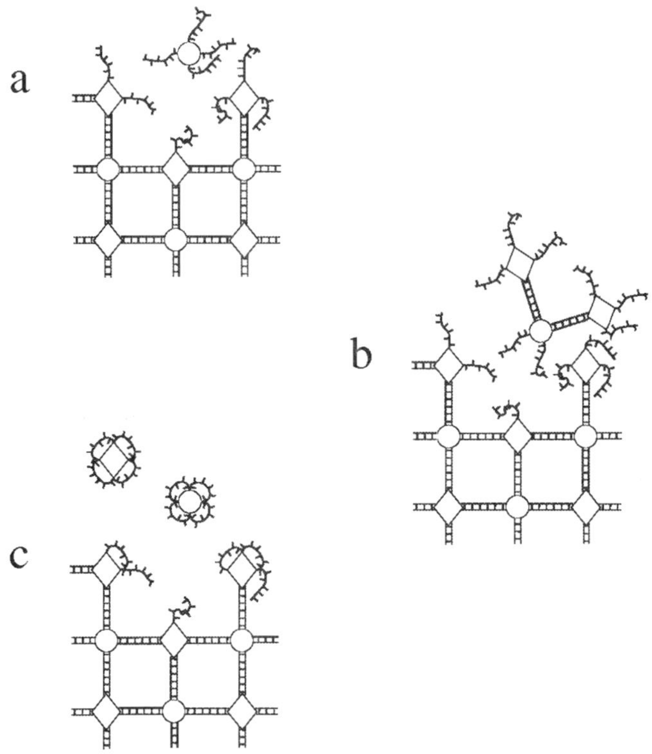

FIGURE 1. (a) Self-assembly by hybridization of DNA tags attached to colloidal particles. (b) A defect is about to be created because the tags are not unique: the circular element has already been linked to two diamonds and will form only two of the three bonds required to hold it in place correctly. (c) Protected self assembly: hybridization is inhibited by secondary loop structure and catalysed at the growing surface of the device.

be uniquely determined. Figure 1(b) illustrates this problem: the circle that has arrived at the growing surface of the structure is about to be incorporated at a site flanked by three diamonds, but a pre-existing bond between the circle and another diamond will prevent proper insertion. Slow growth at a temperature near the melting point of the duplex DNA strands linking particles will reduce the concentration of defects [**5**] but quasi-reversible growth is necessarily slow; a simulation of two-dimensional self-assembly found an inverse relationship between growth rate and the expected size of error-free structures [**12**]. We propose an alternative technique, *protected* self-assembly, which is shown schematically in Figure 1(c). This strategy has two key elements: (i) protect DNA tags from hybridization with their complements by the formation of metastable secondary structure; (ii) catalyse hybridization when the strands arrive at the designed location on the growing surface of the self-assembled structure. As we describe below, we have demonstrated both protection and catalysis by using auxiliary DNA strands both to inhibit hybridization by loop formation and to catalyse hybridization by loop opening.

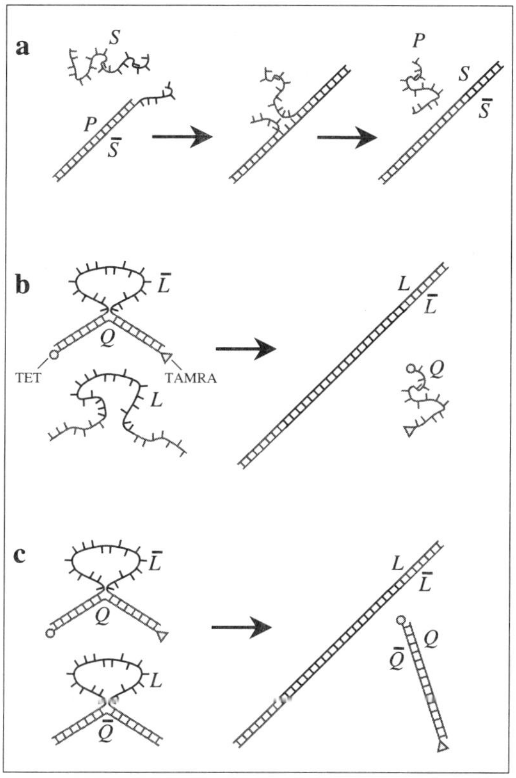

FIGURE 2. Reactions in which the hybridization of two complementary strands is inhibited by a protection strand, already hybridized to one of them.

2.1. Metastable secondary structure. We begin this section by discussing the simple strand-displacement reaction shown in Figure 2(a). Hybridization between complementary strands S and $\overline{S}$ is inhibited by a protection strand P, a truncated copy of S which has already hybridized to $\overline{S}$. Bases in the overhang section of $\overline{S}$, which are unpaired in the initial state, are hybridized in the $S\overline{S}$ duplex, reducing the free energy of the final state and driving the reaction to near-completion. Strand displacement is initiated by hybridization at the overhang and procedes by branch migration, a random walk of the junction between the competing strands S and P until strand P is completely displaced [**13**]. The nucleation of a hybridized region joining S and $\overline{S}$ is an activated process [**14**] and for oligonucleotide lengths relevant to the self-assembly applications considered here (< 100) is rate-limiting [**14, 15, 16**]. For overhang lengths greater than 8 bases the presence of strand P has little effect on the rate of reaction between strands S and $\overline{S}$ [**17**] and the rate constant for this reaction is $\sim 10^6\text{M}^{-1}\text{s}^{-1}$ [**18**] giving a time to half-completion of $\sim$1s for 1μM reactant concentrations. In this configuration the protection strand is ineffective at inhibiting hybridization between S and $\overline{S}$.

Our strategy for inhibiting hybridization between complementary strands is shown in Figure 2(b) and (c). The protection strand Q is designed to hybridize with strand $\overline{L}$ to form a loop. Table 1 gives the base sequences for the system that we have studied; Q consists of 30 bases and hybridizes with 15-base sections at either end of 70-base strand $\overline{L}$. The reaction of the $Q\overline{L}$ complex with the complementary strand L, shown in Figure 2(b) is driven by the free energy released by hybridization of the 40 unpaired bases in the belly of the loop. Although we expect that hybridization between L and $\overline{L}$ will be readily initiated at unpaired bases near the centre of $\overline{L}$ its progression is inhibited by the protection strand in two ways. Hybridization of two strands of DNA requires that one be twisted around the other to form a double helix, so for the reaction between L and $\overline{L}$ to proceed the ends of L must be threaded through the aperture of the loop. Although single-stranded DNA is relatively flexible (it has a persistence length only ~3 bases at 1M salt concentration [**19, 20**]) the loop opening is small (< 40 Å) and this process is hindered. The perimeter of the loop is less than half the persistence length of double-stranded DNA [**21, 22**] so hybridization of more than approximately half of the bases in the loop creates strain that can only be relieved by opening the loop. Both of these factors mean that in this configuration the protection strand is effective in inhibiting the reaction between L and $\overline{L}$ even thought the initial and final states are designed to differ by 40 base pairs or ~70kcal/mol [**23**][a].

We use dye labels TET (5' tetrachloro fluorescein phosphoramidite) and TAMRA (carboxy tetramethyl rhodamine) on the ends of strand Q to follow the reaction shown in Figure 2(b). TET has an emission spectrum which overlaps the absorption spectrum of TAMRA: resonant Förster energy transfer from TET to TAMRA quenches fluorescence from TET with an efficiency that depends on the inverse sixth power of the separation between the dyes [**25**]. The ends of Q are further apart in the $Q\overline{L}$ complex than in the random coil configuration of displaced Q, so the TET fluorescence is quenched as the reaction proceeds. Figure 3(a) shows the time-dependence of the fluorescence intensity after adding L to the $Q\overline{L}$ complex with 1μM reactant concentrations The second-order rate constant for this reaction[bc] is $220(80)\mathrm{M}^{-1}\mathrm{s}^{-1}$ corresponding to a time to half-completion of 4.5×10^3s.

Figure 2(c) shows an even more effective protection strategy; both strands L and $\overline{L}$ are protected (complex $\overline{Q}L$ is complementary to complex $Q\overline{L}$). Strands L

---

[a]We estimate the free energy difference between initial and final states using 'Mfold' secondary structure prediction software [**23, 24**]. The predicted corrections to the free energies of the initial and final states due to unintended secondary structure are approximately equal so the estimated energy difference between these states is equal to the difference between the designed numbers of base pairs (40) multiplied by the mean energy per base pair (-1.8 kcal mol^{-1} at 20°C [**23**])

[b]Standard reaction conditions for all experiments described in this paper were T = 20°C in SPSC buffer: 50mM Na_2HPO_4 and 1M NaCl at pH 6.5.

[c]TET fluorescence was excited with the 514.5nm line of an argon ion laser and selected by an interference filter with bandpass 10nm centred at 540nm. Strands L and $\overline{L}$ were titrated against Q using dye quenching as an indicator [**26**]: TET emission increases as $\overline{L}$ is added to Q until the $Q-\overline{L}$ reaction is complete; emission decreases again as $\overline{L}$ is displaced from the $Q\overline{L}$ complex by addition of L (the reactants were annealed for 4 minutes at 80°C after each addition of L). We estimate an error of 10% in determining relative concentrations. Absolute concentrations were obtained from uv absorption measurements by the commercial supplier of the oligonucleotides; from our measurements of relative concentrations we estimate that the error in absolute concentrations is of order 30%. Reactants (L and the $Q\overline{L}$ complex) were mixed to give equal initial concentrations c = 1μM. The reaction was followed for 10^5s, after which the reactants were annealed for 4 minutes at 80°C to check that the reaction was complete and to determine the fluorescence intensity at the end point I_∞. The second-order rate constant for this reaction k_2 was obtained by fitting the fluorescence intensity decay curve to the function $I(t) = (I_o - I_\infty) / (1 + k_2 ct) + I_\infty$ which is appropriate for a second-order reaction between reactants of equal concentration.

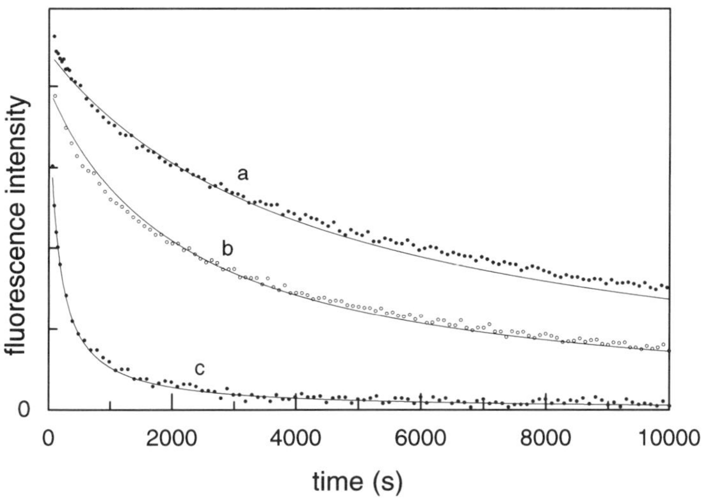

FIGURE 3. Time-dependence of TET fluorescence during reactions in which Q is displaced from a loop complex (initial reactant concentration 1μM). The ends of strand Q are labelled with dyes TET and TAMRA; when Q is displaced from the loop complex it assumes a random coil configuration and the TET fluorescence is quenched by resonant Förster energy transfer from TET to TAMRA. The residual quenched intensity from free Q has been subtracted, so the the data plotted is directly proportional to the concentration of the loop complex. Solid lines are fits assuming second-order kinetics. a) $Q\overline{L}+L \rightarrow Q+L\overline{L}$; b) $Q\overline{L'}+L \rightarrow Q+L\overline{L'}$ catalysed by 0.06μM M; c) $Q\overline{L'}+L \rightarrow Q+L\overline{L'}$ catalysed by 1μM M

and $\overline{L}$ now cannot twist around each other without opening at least one of the loops. We have established a lower limit to the time to half-completion for reaction between the complexes at 5μM concentration of 6×10^4s. In this reaction the fluorescence intensity increases with time because the ends of strand Q are further apart when it is hybridized with $\overline{Q}$ than in the $Q\overline{L}$ complex. The time-dependence of the fluorescence during this reaction is not well represented by either first- or second-order decay kinetics.

We have established above that it is possible to increase the reaction time for hybridization of complementary strands of DNA by five orders of magnitude by use of secondary loop structures as shown in Figure 2(b) and (c). To complete the demonstration of the elements of our strategy for protected self-assembly we must now show that it is possible to catalyse hybridization of the protected strands at the growing surface of a self-assembled structure.

2.2. Catalysis by DNA. Figure 4 shows how we have implemented catalytic control of the reaction shown in Figure 2(b). Strand $\overline{L}'$ has been extended by a 12-base overhang section which gives catalyst M a toehold. Strand M is complementary to a section of $\overline{L}'$ starting with 6 bases at the overhang section, continuing with a 15-base section which competes with strand Q for binding to $\overline{L}'$, and ending with a 10-base section that hybridizes to the first 10 bases of the belly of the loop

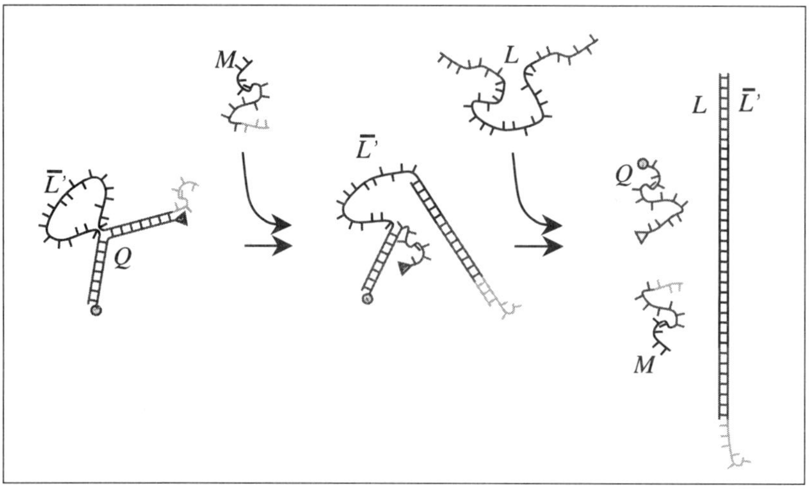

FIGURE 4. Catalysis of the reaction of a DNA loop with a complementary strand of DNA.

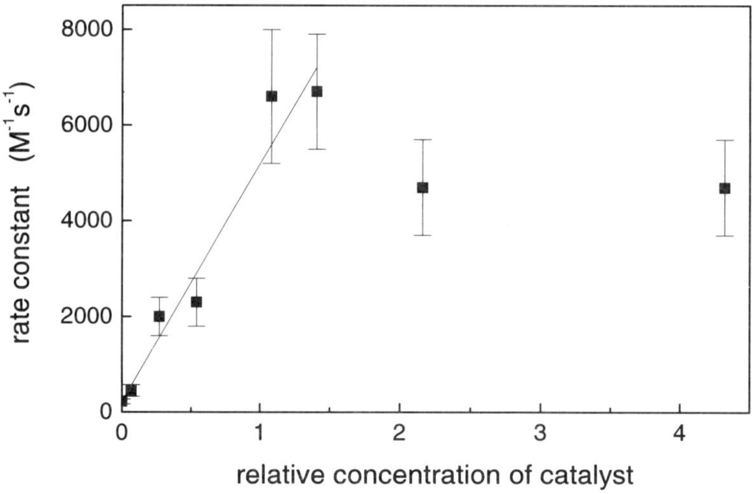

FIGURE 5. Reaction rate constant as a function of the concentration of DNA catalyst M relative to that of the other reactants ($[Q] = [L] = [\overline{L}'] = 1\mu\text{M}$).

formed in the $Q\overline{L}'$ complex. Hybridization of M to the overhang section of $\overline{L}'$ is unimpeded by the presence of Q; M then displaces half of Q from $\overline{L}'$ and thereby opens the loop. Once the loop is open, hybridization of strand L to $\overline{L}'$ is no longer impeded; L displaces M (regenerating the catalyst) and completes the displacement of Q.

Figure 3(b) and (c) are fluorescence decay curves for reactions catalysed by two different concentrations of the catalyst M. Figure 5 shows the dependence of the reaction rate constant[d] on the concentration of the catalyst M for $[Q] = [L] = [\overline{L}'] = 1\mu M$. At low concentrations of M the increase in the reaction rate is approximately proportional to $[M]$. The maximum reaction rate, 30× faster than the uncatalysed rate, occurs for a stoichiometric quantity ($[M]\sim 1\mu M$); this corresponds to a reaction time to half-completion of 150s. At higher catalyst concentrations the reaction slows significantly by ~30%; this suggests that excess catalyst may act as a competitive inhibitor. Binding of excess M to short complementary sequences away from the intended position on $\overline{L}'$ is possible and would impede hybridization with L, but the maximum free energy of hybridization (estimated to be ~5 kcal mol^{-1} using 'Mfold' secondary structure prediction software [**24, 23**]) is insufficent to form a stable complex.

2.3. Implementation of Protected Self-Assembly. We have shown above that we are able to use a DNA strand to catalyse the hybridization of strands whose interaction is otherwise inhibited by designed secondary loop structure. We have therefore demonstrated the two key types of control required to implement protected self-assembly: we may both inhibit and reactivate the hybridization of complementary strands of DNA.

Our strategy for the implementation of protected self-assembly is to ensure that all DNA tags attached to elements of the structure are initially protected from hybridization by the formation of secondary loops. After growth has been initiated, either by thermal activation or by seeding with a small number of catalyst strands, each hybridization reaction that links a new element to the growing structure will release the sequence required to catalyze the formation of the next bond from a protected position in a loop. Research on the use of this catalytic control of DNA by DNA to produce an ordered self-assembled structure is in progress.

3. Molecular Tweezers

The first active DNA device relied on a B-Z phase transition (from a right- to a left-handed double helix) of a short DNA duplex; the transition was induced by a change in the composition of the buffer solution [**10**]. We introduce a new type of active device in which the hybridization of DNA is used as a source of chemical energy to produce motion.

The structure of the machine in its open state is shown schematically in Figure 6(a). The machine has two stiff arms formed by hybridization of 18-base sections of strands B and C with complementary sequences at the ends of strand A. A four-base single-stranded region at the centre of A, between strands B and C, acts as a hinge; the machine looks like a pair of tweezers. Strands B and C are 42 bases long; the remaining 24-base portions of B and C, which are used to bind the fuel strand which closes the tweezers, are unhybridized in the open state.

Figure 6(b) shows how the tweezers are cycled between the open and closed states using fuel and removal strands F and $\overline{F}$. The 56-base fuel strand F consists of consecutive 24-base sections which are complementary to the dangling ends of B and C, with an additional 8-base overhang section. When F hybridizes with the

[d]Rate constants were derived from fluorescence intensity decay curves as described above. Error bars include uncertainties in relative concentrations and due to the curve fitting procedure; they do not take into account the systematic error due to uncertainty in the absolute concentration of the reactants which has no effect on the relative values of the rate constants. The decrease of the rate constant at high catalyst concentrations is significant.

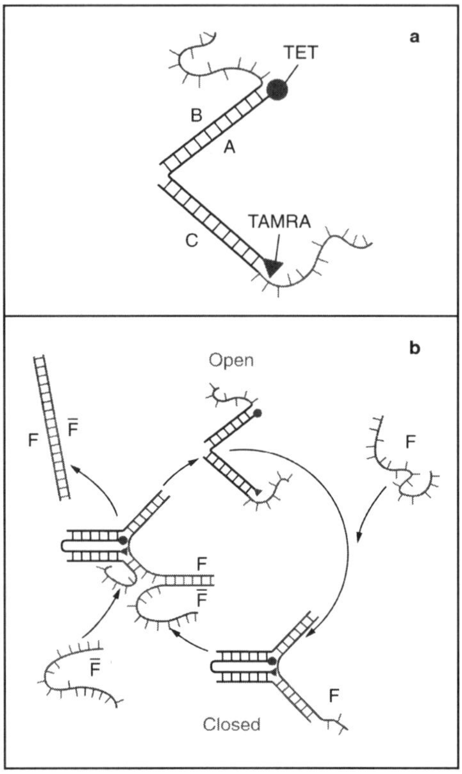

FIGURE 6. (a) A schematic diagram of the molecular tweezers, which are formed by hybridization of three strands of DNA. (b) Cycling the molecular tweezers. Fuel strand F binds to the ends of strands B and C to pull the tweezers closed. Removal strand $\overline{F}$ displaces F from the tweezers to open them again.

unhybridized ends of strands B and C it pulls the ends of the tweezers together. We estimate the force exerted from the average free energy of hybridization of a complementary base pair (-1.8 kcal mol^{-1} at 20°C [**23**]) and the separation between base pairs in single-stranded DNA (0.43nm [**27**]); the ratio gives an average closing force ~15 pN which is consistent with the force required for mechanical separation of double-stranded DNA [**28**]. This force, generated by a device made entirely from DNA, is at the upper end of the range of measured forces exerted by single-molecule kinesin [**29, 30, 31**] and myosin [**32, 33**] motors. Removal strand $\overline{F}$ is the complement of F: it hybridizes with F, starting at the overhang, and removes it from the machine. A double-stranded waste product is formed, and the tweezers are returned to the open state.

The base sequences chosen for the three strands are given in Table 1. The tweezer structure is formed by mixing stoichiometric quantities of strands A, B and C at a concentration of 1μM in SPSC buffer at 20°C. Strand A, like strand Q, is labelled at the 5' and 3' ends with dyes TET and TAMRA. Fluorescence quenching is used as an indicator [**26**] to titrate strands B, C, F and $\overline{F}$ against A. As half of A is straightened from a random coil by hybridization with B (or C) the mean

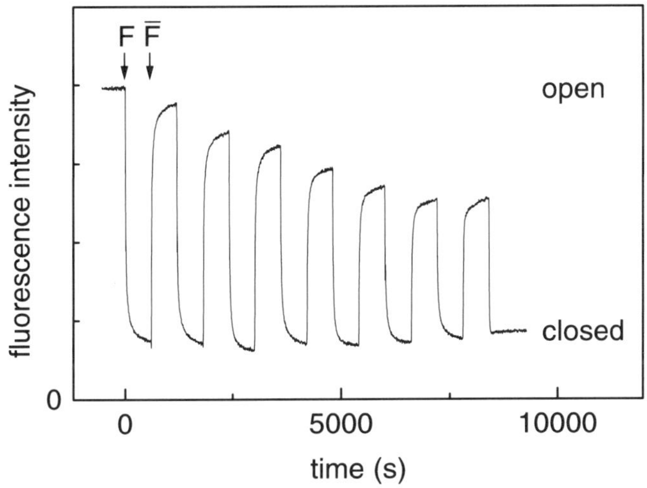

FIGURE 7. Cycling the molecular machine by successive addition of fuel strand F and removal strand $\overline{F}$. Dye labels on the ends of the tweezer arms are closer together when the tweezers are pulled closed by the F, resulting in fluorescence quenching.

separation between the dye molecules on A increases leading to a five-fold increase in the fluorescence intensity. The cumulative effect of hybridization with both B and C is an seven-fold increase in fluorescence intensity in the open state. The fluorescence intensity drops by a factor of six when the tweezers are closed by the addition of fuel strand F, and recovers on removal of F by a stoichiometric amount of removal strand $\overline{F}$. Figure 7 shows the fluorescence intensity variation as the tweezers are cycled seven times between the open and closed states by successive additions of strands F and $\overline{F}$. The switching time for the machine is approximately 20s.

The recovery of a high fluorescence intensity after each cycle of closing and opening indicates that the dye-labelled ends of A are relatively far apart. We have used polyacrylamide gel electrophoresis to check that strand A is still intact after three such cycles: we deduce that the initial configuration, in which the two halves of A are straightened by hybridization with B and C, is recovered after each cycle. The low fluorescence intensity in the closed state tells us only that the dyes are relatively close together. To test whether the duplex structures formed by hydrogen bonding between A, B and A, C survive the force exerted by hybridization with the fuel strand we have used an additional strand $\overline{A}$, the complement of A, to test the integrity of the machine by displacing B and C from A. In the absence of other strands $\overline{A}$ hybridizes with A (producing an eight-fold increase in fluorescence intensity) with a time to half-completion of 17s for $[\overline{A}]=[A] = 1\mu$M. When a stoichiometric quantity of $\overline{A}$ is introduced to pre-assembled tweezers in the open state the time to half-completion for the reaction is increased to 4.8×10^3 s by the need for thermally-activated displacement of parts of B and C from A before hybridization with $\overline{A}$ can begin. The interaction of $\overline{A}$ with tweezers held in the closed state is even slower, by an order of magnitude, than in the open state (time to half-completion 5×10^4 s; the closed structure inhibits the winding and unwinding of strands that accompanies strand exchange. The high stability of the tweezers, even in the closed state where the hinge region is likely to be strained, indicates that the A, B and A, C duplexes are not severely disrupted when the tweezers are closed. We have also checked that when one of the tweezer components B, C is left out changes in fluorescence intensity due to successive additions of F and $\overline{F}$ are less than 8% of the change achieved with the complete structure, and that there is no observable interaction between unhybridized strand A and either F or $\overline{F}$. We conclude that unanticipated interactions between strands are weak and that, when the machine is fully assembled, the observed changes in fluorescence intensity are caused by the machine operating as designed.

4. Conclusions

We have demonstrated that hybridization between complementary strands of DNA may be inhibited by the deliberate formation of loop structures. Hybridization may be triggered by the presence of a catalyst strand of DNA which opens the loop by a strand-displacement reaction. Inhibition and catalysis of hybridization are key elements of a strategy for protected self-assembly: DNA bonding between elements of a self-assembled structure may be controlled by catalyst strands present at the growing surface of the structure.

We have also demonstrated a simple nanomachine constructed of DNA that uses DNA as a fuel. Our tweezer structure allows the separation between two dye molecules to be changed repeatably on a nanometre length scale; a similar structure could be used to investigate the interaction between chemically active components attached to the ends of the tweezers or, if combined with a DNA cage [**3**], to alternately hide and reveal a target molecule. More sophisticated devices based on this principle may include, for example, levers made of the rigid double-crossover structures demonstrated by Seeman and co-workers [**34**] to increase the force exerted, rotary motors in which the rotor is pulled sequentially to points spaced around a stator, and free-running machines using metastable DNA loops as fuel. Because the binding between fuel and machine is sequence-specific the DNA strands that act as fuel may also serve as information carriers to coordinate components of a complex machine or to carry signals between machines. The use of DNA fuel allows precise control of movements on a nanometre length scale without prodding by a scanning microscope tip; it makes possible coordinated motion of between 2 and 10^{18} (or more) devices which may incorporate elements chosen for their biochemical, optical or electrical properties. Controlled motion provides exciting new degrees of freedom in the design of nanostructure devices.

References

1. Deaton, R., Garzon, M., Murphy, R. C., Rose, J. A., Franceschetti, D. R. & Stevens, S. E., Jr. *Reliablility and Efficiency of a DNA-Based Computation.* Phys. Rev. Lett. **80** (1998), 417–420
2. Adleman, L. M. *Molecular Computation of Solutions to Combinatorial Problems.* Science **266** (1994), 1021-1024; Lipton, R. J. *DNA Solution of Hard Computational Problems.* Science **268** (1995), 542-545; Ouyang, Q., Kaplan, P. D., Liu, S. & Libchaber, A. *DNA Solution of the Maximal Clique Problem.* Science **278** (1997), 446–449
3. Chen, J. & Seeman, N.C. *Synthesis from DNA of a molecule with the connectivity of a cube.* Nature **350** (1991), 631–633
4. Mao, C., Sun, W. & Seeman, N. C. *Assembly of Borromean rings from DNA.* Nature **386** (1997), 137–138

5. Winfree, E., Liu, F., Wenzler, L. A. & Seeman, N. C. *Design and self-assembly of two-dimensional DNA crystals.* Nature **394** (1998), 539–544
6. Mirkin, C. A., Letsinger, R. L., Mucic, R. C. & Storhoff, J. J. *A DNA-based method for rationally assembling nanoparticles into macroscopic materials.* Nature **382** (1996), 607–609
7. Alivisatos, A. P., Johnsson, K. P., Peng, X., Wilson, T. E., Loweth, C. J., Bruchez, M. P., Jr. & Schultz, P. G. *Organization of 'nanocrystal molecules' using DNA.* Nature **382** (1996), 609–611
8. Braun, E., Eichen, Y., Sivan, U. & Ben-Yoseph, G. *DNA-templated assembly and electrode attachment of a conducting silver wire.* Nature **391** (1998), 775–778
9. Coffer, J. L., Bigham, S. R., Li, X., Pinizzotto, R. F., Rho, Y. G., Pirtle, R. M. & Pirtle, I. L. *Dictation of the shape of mesoscale semiconductor nanoparticle assemblies by plasmid DNA.* Appl. Phys. Lett. **69** (1996), 3851–3853
10. Mao, C., Sun, W., Shen, Z. & Seeman, N. C. *A nanomechanical device based on the B-Z transition of DNA.* Nature **397** (1999), 144–146
11. Turberfield, A. J., Yurke, B. & Mills, A. P., Jr. *Coded Self-Assembly of DNA Nanostructures.* Bull. Am. Phys. Soc. **44** (1999), 1711
12. Winfree, E. *Simulations of Computing by Self-Assembly* in Proc. 4th International Meeting on DNA Based Computers, Kari, L., Rubin, H. & Wood, D. H., Eds. (to be published in DIMACS Series in Discrete Mathematics and Theoretical Computer Science, American Mathematical Society, Providence, R.I.)
13. Lee, C. S., Davis, R. W. & Davidson, N. *A Physical Study by Electron Microscopy of the Terminally Repetitious, Circularly Permuted DNA from the Coliphage Particles of Escherichia coli 15.* J. Mol. Biol. **48**, (1970) 1–22
14. Green, C. & Tibbetts, C. *Reassociation rate limited displacement of DNA strands by branch migration.* Nucleic Acids Research **9** (1981), 1905–1918
15. Wetmur, J.G. & Davidson, N. *Kinetics of Renaturation of DNA.* J. Mol. Biol. **31** (1968), 349–370
16. Radding, C. M., Beattie, K. L., Holloman, W. K. & Wiegand, R. C. *Uptake of Homologous Single-stranded Fragments by Superhelical DNA: IV Branch Migration.* J. Mol. Biol. **116** (1977), 825–839
17. Yurke, B and Mills, A. P. Jr., *Toehold-enhanced DNA strand exchange: Fuel for nanomachines*, to be published.
18. Morrison, L. E. & Stols, L. M. *Sensitive Fluorescence-Based Thermodynamic and Kinetic Measurements of DNA Hybridization in Solution.* Biochemistry **32** (1993), 3095–3104
19. Tinland, B., Pluen, A., Sturm, J. & Weill, G. *Persistence Length of Single-Stranded DNA.* Macromolecules **30** (1997), 5763–5765
20. Smith, S. B., Yujia, C. & Bustamante, C. *Overstretching B-DNA: The Elastic Response of Individual Double-Stranded and Single-Stranded DNA Molecules.* Science **271** (1996), 795–799
21. Manning, G. S. *A Procedure for Extracting Persistence Lengths from Light-Scattering Data on Intermediate Molecular Weight DNA.* Biopolymers **20** (1981), 1751–1755
22. Smith, S. B., Finzi, L. & Bustamante, C. *Direct Mechanical Measurements of the Elasticity of Single DNA Molecules by Using Magnetic Beads.* Science **258** (1992), 1122–1126
23. SantaLucia, J., Jr., *A unified view of polymer, dumbbell, and oligonucleotide DNA nearest-neighbor thermodynamics.* Proc. Natl. Acad. Sci. USA **95** (1998), 1460-1465
24. Zuker, M., Mathews, D.H. & Turner, D.H. *Algorithms and Thermodynamics for RNA Secondary Structure Prediction: A Practical Guide* in *RNA Biochemistry and Biotechnology* (Barciszewski, J. & Clark, B.F.C., Eds.) NATO ASI Series, Kluwer Academic Publishers, 1999; Mathews, D.H., Sabina, J., Zuker, M. & Turner, D.H. *Expanded Sequence Dependence of Thermodynamic Parameters Provides Robust Prediction of RNA Secondary Structure.* J. Mol. Biol. **288** (1999), 911-940
25. Stryer, L. & Haugland, R. P. *Energy Transfer: A Spectroscopic Ruler.* Proc. Natl. Acad. Sci. USA **58** (1967), 719–726
26. Heller, M. J. & Morrison, L. E. in *Rapid Detection and Identification of Infectious Agents* (Kingsbury, D. T. & Falkow, S., Eds.) Academic Press, New York, 1985, pp. 245–256
27. Record, M. T., Jr., Anderson, C. F. & Lohman, T. M. *Thermodynamic analysis of ion effects on the binding and conformational equilibria of proteins and nucleic acids: the roles of ion association or release, screening, and ion effects on water activity.* Quart. Rev. Biophys. **11** (1978), 103–178

28. Bockelmann, U., Essevaz-Roulet, B. & Heslot, F. *Molecular Stick-Slip Motion Revealed by Opening DNA with Piconewton Forces.* Phys. Rev. Lett. **79** (1997), 4489–4492
29. Svoboda, K., Schmidt, C. F., Schnapp, B. J. & Block, S. M. *Direct observation of kinesin stepping by optical trapping interferometry.* Nature **365** (1993), 721–727
30. Kuo, S. C. & Sheetz, M. P. *Force of Single Kinesin Molecules Measured with Optical Tweezers.* Science **260** (1993), 232–234
31. Hunt, A. J., Gittes, F. & Howard, J. *The Force Exerted by a Single Kinesin Molecule Against a Viscous Load.* Biophysical J. **67** (1994), 766–781
32. Finer, J. T., Simmons, R. M. & Spudich, J. A. *Single myosin molecule mechanics: piconewton forces and nanometre steps.* Nature **368** (1994), 113–119
33. Ishijima, A., Harada, Y., Kojima, H., Funatsu, T., Higuchi, H. & Yanagida, T. *Single-molecule analysis of the actomyosin motor using nano-manipulation.* Biochem. Biophys. Res. Commun. **199** (1994), 1057–1063
34. Fu, T.-J. & Seeman, N. C. *DNA double-crossover molecules.* Biochemistry **32** (1993), 3211–3220

TABLE 1. Oligonucleotide Sequences

Name	Sequence[e]
Q	5' AATGAGGGTGACTTCGACCGTAAACGACGA 3'
L	5' AATGAGGGTGACTTCAACCTAGCGAGCGAACGTGCCAATTCTGATCTACTGTGTGGACCGTA AACGACGA 3'
$\overline{L}$	5' TCGTCGTTTACGGTCCACACAGTAGATCAGAATTGGCACGTTCGCTCGCTAGGTTGAAGTCA CCCTCATT 3'
$\overline{L}'$	5' GCATCTCTGGAATCGTCGTTTACGGTCCACACAGTAGATCAGAATTGGCACGTTCGCTCGCT AGGTTGAAGTCACCCTCATT 3'
M	5' CTACTGTGTGGACCGTAAACGACGATTCCAG 3'
A	5' TGCCTTGTAAGAGCGACCATCAACCTGGAATGCTTCGGAT 3'
B	5' GGTCGCTCTTACAAGGCACTGGTAACAATCACGGTCTATGCG 3'
C	5' GGAGTCCTACTGTCTGAACTAACGATCCGAAGCATTCCAGGT 3'
F	5' CGCATAGACCGTGATTGTTACCAGCGTTAGTTCAGACAGTAGGACTCCTGC TACGA 3'

[e]Oligonucleotides were supplied by Integrated DNA Technolgies, Inc; strands A and Q were HPLC purified, other strands were PAGE purified. Strands A and Q were dye labelled 5' TET, 3' TAMRA (NHS ester hand tagged) by the manufacturer.

(A. Turberfield) Bell Laboratories, Lucent Technologies, 600 Mountain Avenue, Murray Hill NJ 07974, and, University of Oxford, Department of Physics, Clarendon Laboratory, Parks Road, Oxford OX1 3PU, U.K.

E-mail address: a.turberfield@physics.oxford.ac.uk

(B. Yurke) Bell Laboratories, Lucent Technologies, 600 Mountain Avenue, Murray Hill NJ 07974

E-mail address: yurke@lucent.com

(A. Mills) Bell Laboratories, Lucent Technologies, 600 Mountain Avenue, Murray Hill NJ 07974

E-mail address: apm@lucent.com

DIMACS Series in Discrete Mathematics
and Theoretical Computer Science
Volume **54**, 2000

Designing and Selecting Components for Nucleic Acid Computers

Michael P. Robertson, Jay Hesselberth, J. Colin Cox, and Andrew D. Ellington

ABSTRACT. Just as the design and function of modern electronic computers is dependent on the nature of their components, the design and function of nucleic acid computers will likely be dependent on the types of molecules used in their construction. We show that ribozymes can function as switches and as molecular logic gates, and propose that nucleic acid catalysts might therefore be used to generate circuitry for biological computation. Nucleic acids can also self-assemble into stochastic, organized architectures that may prove to be the equivalent of circuit boards for nucleic acid components.

1. Introduction

Most nucleic acid computations that have been carried out or conceived of to date rely on 'hybridization logic,' the embedding of information or algorithms in the Watson-Crick base-pairs of double-stranded DNA or RNA. Of course, the most famed example of this is Adleman's solution of a limited Hamiltonian path problem (HPP) [**1**]. A graph containing seven nodes connected by six edges was encoded via one set of oligonucleotides that represented the nodes and a partially overlapping, complementary set of oligonucleotides that represented the edges connecting these nodes. The node and path oligonucleotides were hybridized, ligated, amplified, and sieved to correctly reveal the existence of a Hamiltonian path. Other notable, executed computations based on hybridization include binary integer addition [**2**], the solution of a maximal clique problem [**3**], word design strategies [**4**] and satisfiability problems other than the HPP [**5**].

We have previously argued [**6, 7**] that computations based on hybridization logic are inherently limited in utility, complexity, and fidelity. We instead advocate that, just as the development of physical components (vacuum tubes, transistors) that could efficiently support electronic switching preceded the development of electronic computers, that the development of biological components that can efficiently support biological switching should precede the development of biological computers. Just as it would likely have been difficult to imagine the blueprints for ENIAC in a world where there were no vacuum tubes, biologists, physicists, and computer scientists may be overreaching themselves in trying to design functional DNA computers at the current time.

© 2000 American Mathematical Society

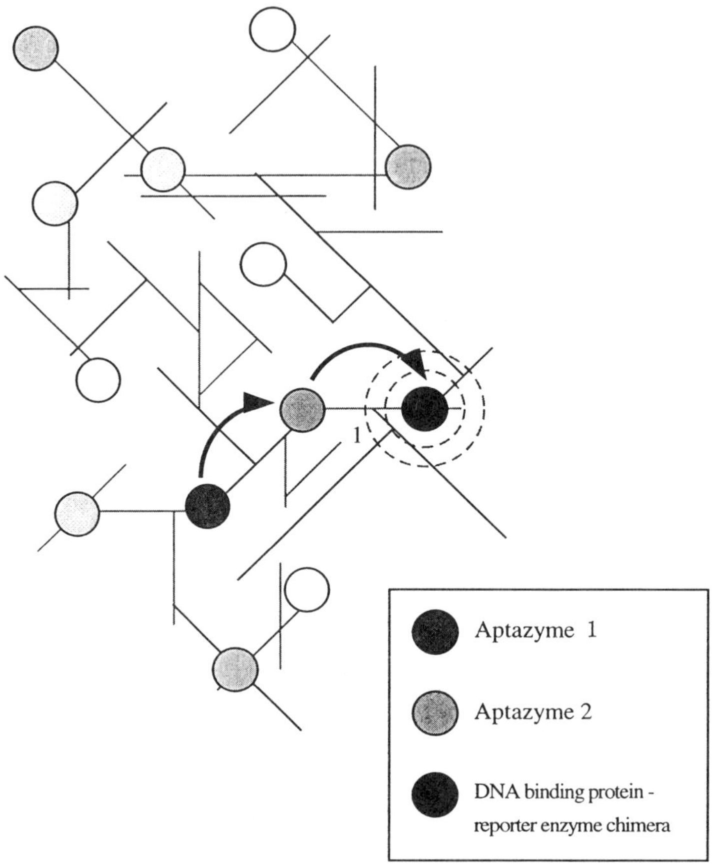

FIGURE 1. Schematic for a DNA computer. The lines represent a DNA matrix, while the shaded circles represent various logical or readout components that have been randomly localized in the matrix. When several components that can operate on one another are co-localized, a logical statement is evaluated. The exact nature of each component is explained in the text.

2. Overview of Machine Design

The following schematic is not meant to solve a particular problem, does not execute a particular algorithm, and does not represent a functional computer. Rather, it is hoped that by designing biological analogs of some of the components of a silicon computer that researchers may be stimulated to use these biological analogs to develop algorithms appropriate for biological computation or circuitry that can more easily interface with biology.

It would seem that there are three types of components required for a functional biological computer (Figure 1): some form of input / output device, switches or gates, and a matrix upon which other components can be positioned and information can be stored (if only as a position-dependent change in state). In general, it should be relatively simple to develop minimal input / output devices for biological computation. Nucleic acids and enzymes can specifically recognize substrates and can also be engineered to return visual or electrochemical signals. However, more complex input / output devices that can 'read' and translate a variety of different signals into a uniform code, and then 'write' based on this code a similar variety of signals are much more difficult to imagine. In other words, it is currently difficult to design a functional biological computer that would be able to recognize numerous different word strings, act on those word strings, and return diverse answers that could themselves be easily read. Because of this, we will assume that biological computers will be restricted to simple inputs and simple outputs, at least for the near term. For example, the types of components we will describe might be suitable for an extremely simple computation device that sensed the relative levels of glucagon and insulin, and returned as an output the release of the appropriate hormones or the catabolism of sugar.

While nucleic acids and enzymes can be used in isolation to read and write signals, there are currently no readily engineered molecular analogues of logic gates or ordered matrices in which to embed these gates. This is because most biological computation occurs at the level of the cell, and we cannot yet engineer with precision the signaling pathways and architecture of the cell. Nonetheless, it may be possible to contrive molecular gates and matrices that do function analogously to those in silicon devices. To this end, we present nascent results on the development of ribozyme logic gates and nucleic acid matrices.

3. Ribozyme Logic Gates

Lizzardi, Breaker and their co-workers have generated allosteric ribozymes (also known as 'aptazymes') that can sense nucleic acid or other inputs and generate catalytic outputs, such as the cleavage of a particular nucleic acid strand [**8, 9**]. These aptazymes are similar in conception to electronic transducers, such as transistors (Figure 2). Because of this, it may be possible to construct simple logic gates based on nucleic acids. It should be noted that while allosteric protein enzymes might also be similarly engineered, it is likely that the task would be far more formidable: while the simple base-pairing rules that govern nucleic acid structure and function can be altered at will, the more complex amino acid interactions that govern protein structure do not admit to a similarly malleable code. For example, while the substrate preferences of the hammerhead ribozyme or Group I self-splicing intron are pliable and programmable, the substrate preferences of, say, a protein kinase or phosphatase cannot be similarly programmed.

We have previously identified an allosteric ribozyme ligase (L1) that can sense nucleic acids or other inputs and catalyze the adjoining of two separate sequences [**10**] (Figure 3). These ligases can be thought of as performing an operation that is the negation of the one catalyzed by the hammerhead ribozyme. The allosteric properties of the L1 ligase are quite remarkable: it is activated 10,000-fold by its cognate nucleic acid primer, and the specificity of the ribozyme for the primer can be reprogrammed by simply altering the primer binding site. Moreover, just as Breaker and his co-workers were able to 'addict' the hammerhead ribozyme to non-nucleic acid effectors, we have been similarly able to addict the L1 ligase to

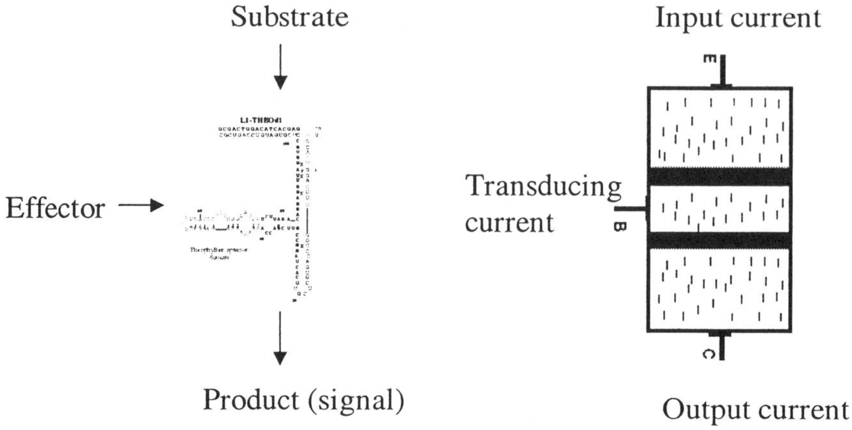

FIGURE 2. Comparison of allosteric ribozymes as signal transducers and transistors as electronic transducers.

non-nucleic acid effectors. When one of the stem-loops of the L1 ligase is replaced with aptamers that can bind small organic molecules, the activity of the L1 ligase is frequently found to be dependent on the presence of the appropriate small organic molecules. For example, when an anti-ATP aptamer is appended to the L1 ligase, the ligase can be activated 830-fold by ATP [**10**] (see also Figure 4). When an anti-theophylline aptamer is appended to the L1 ligase, the ligase can be activated 1,600-fold by theophylline. Structural studies have revealed that when these aptamers bind their respective small molecules that there is an induced fit conformational change; we believe that the mechanism of allosteric activation involves transduction of this induced fit to the helical junction between the aptamer and the ribozyme. Stabilization of the helical junction in turn stabilizes the ribozyme active site. In support of this model, we have found that both designed and selected sequence changes in the helical junction region can greatly improve the allosteric responsiveness of the ribozymes.

Based on these experiments, we reasoned that it might be possible to transduce multiple input signals in parallel and to create a ribozyme logic gate. To this end, we replaced both helices of the L1 ligase with aptamers: the ATP-dependent aptazyme we had previously created was augmented with an anti-theophylline aptamer on the adjacent helical stem. Astoundingly, the ribozyme was now dependent on both organic inputs for maximal catalytic activity (Figure 4). This ribozyme may be the smallest known engineered nanomachine, a molecular 'and' gate that can integrate the presence of two signals (ATP and theophylline) and return a catalytic output (sequence ligation). Moreover, it is in fact a logic gate with three inputs, since the oligonucleotide primer that was originally necessary for full L1 ligase activity is required for activity in this construct as well. We are currently determining the sensitivity and switching parameters of binary and trinary input ribozyme logic gates.

To the extent that the L1 ligase can operate in trans, integrated ribozyme circuitry capable of performing basal assessments of an environment and returning

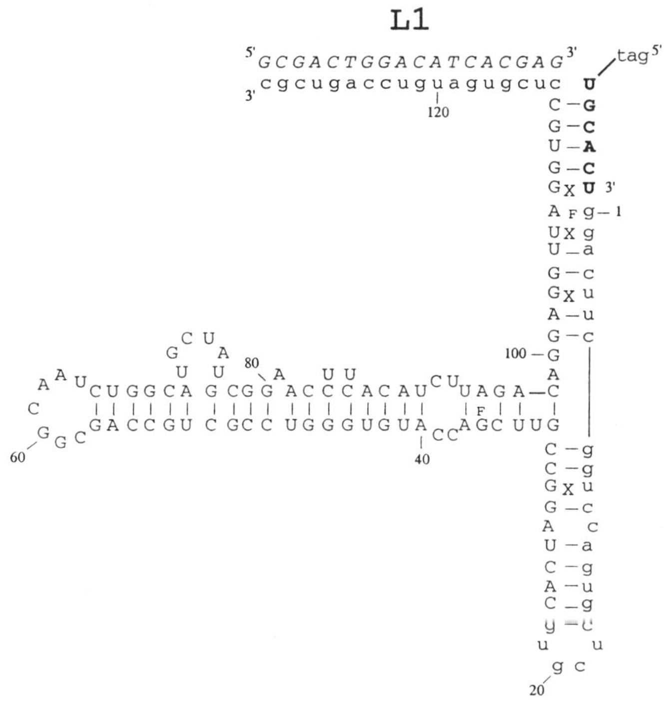

FIGURE 3. The allosteric L1 ligase. The ribozyme aligns and joins a substrate tag to its $5'$ end (denoted by the number 1). This reaction can only occur if an oligonucleotide primer (italicized) is present. In the absence of the primer, the ligation reaction is 10,000 times slower.

appropriate responses can be imagined. For example, the L1 ligase 'and' gate described above might put together a hammerhead ribozyme that would in turn respond to a third effector and return a fluorescence polarization signal by cleaving a larger fluorescent substrate into smaller pieces. The logical statement that would be evaluated by this circuit is $[(A \wedge B \wedge C) \wedge D]$. The relation between hardware and formalism is shown in more detail Figure 5.

4. Nucleic Acid Matrices

For the past several decades, Ned Seeman and his co-workers have single-handedly developed the science of constructing nucleic acid matrices. Their efforts have culminated in the incredible synthesis of uniform two-dimensional DNA arrays. It is possible that such arrays may eventually be used for biological computation [**11, 12**].

However, one potential difficulty with the matrices proposed by Seeman and his co-workers is that that they are sequentially constructed via ligation and in

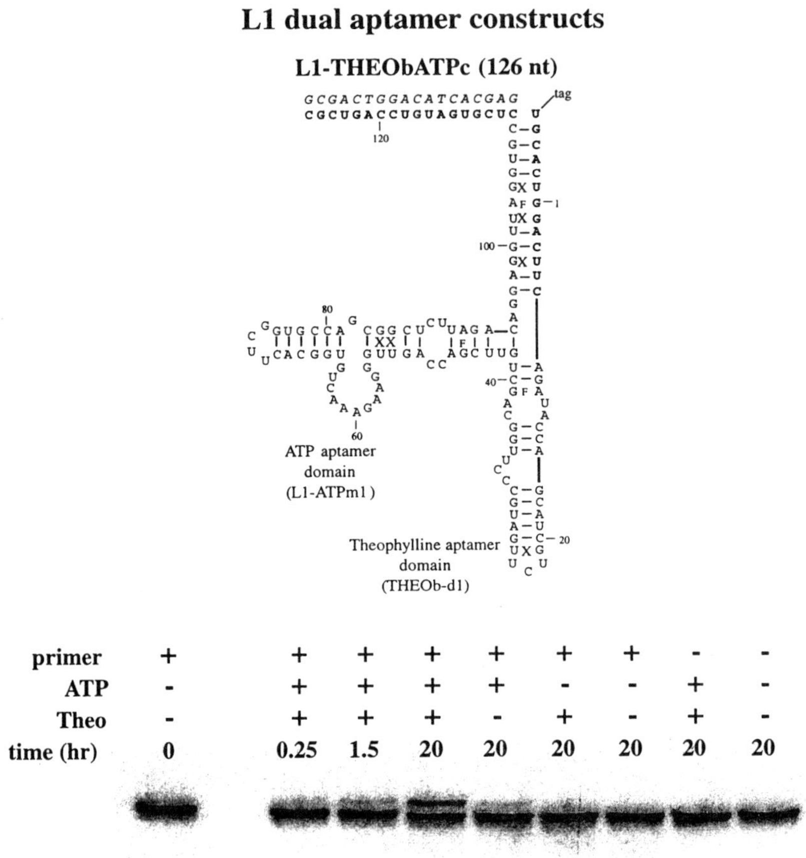

FIGURE 4. A ribozyme logic gate. Anti-adenosine (left) and anti-theophylline (lower) aptamers were appended to the L1 ligase. In consequence, its activity became dependent on the small molecule effectors (gel, inset). The gel shows a radiolabeled L1 incubated with a cold substrate tag. Ligation produces a larger radiolabeled product. Reactions were carried out in the presence of saturating (1 mM) small molecule effectors.

consequence may be difficult to assemble in quantity or on large surfaces. For this reason, we have begun experiments with an alternative type of self-assembling DNA array. To gain these advantages, however, required a Faustian bargain in that we have traded away the symmetry and uniformity of the arrays championed by Seeman.

Our arrays are built from a simple, self-assembling molecule, the G wire [**13**]. Eric Henderson and his co-workers originally showed that a DNA decamer, namely (dGGGGTTGGGG), could self-assemble to form relatively long (100 - 1000 nm) linear quadruplex structures, and Protozanova and MacGregor [**14, 15, 16**] have examined similar structures with frayed interstices. We have now designed a G wire

For the logical statement (A^B^C)^D, let:

A = ATP
B = Theophylline
C = Oligonucleotide
D = Flavin
Fluorescent signal = True

The L1 ligase (dashed line) will sense the presence of A^B^C and return the hammerhead ribozyme, which will in turn sense D and return the fluorescent signal 'True.'

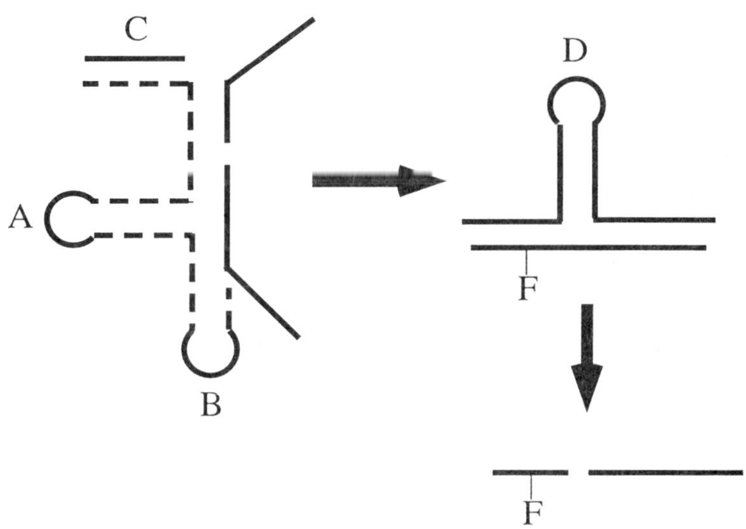

FIGURE 5. A fictional schematic for a series of ribozymes that would evaluate a logical statement. The adaptation of logical formalisms to ribozyme hardware may not be precise, especially since the L1 ligase 'produces' an operator that assists in the evaluation of the statement.

'branch' that facilitates the assembly of two-dimensional matrices (Figure 6). By mixing differing quantities of linear and branched G wire monomers we can form lines, simple right angular structures (L's and T's and X's), loose matrices with large pores, and tight matrices with small pores (Figure 7). While these various structures propagate over many micrometers as seen by an atomic force microscope, they are not regular in the sense of being locally symmetric. Nonetheless, we believe that they may be useful for positioning ribozyme logic gates and other circuit elements, as vaguely shown in Figure 1. In particular, even though the assembly of the G wire matrices is stochastic, it is likely that on average different

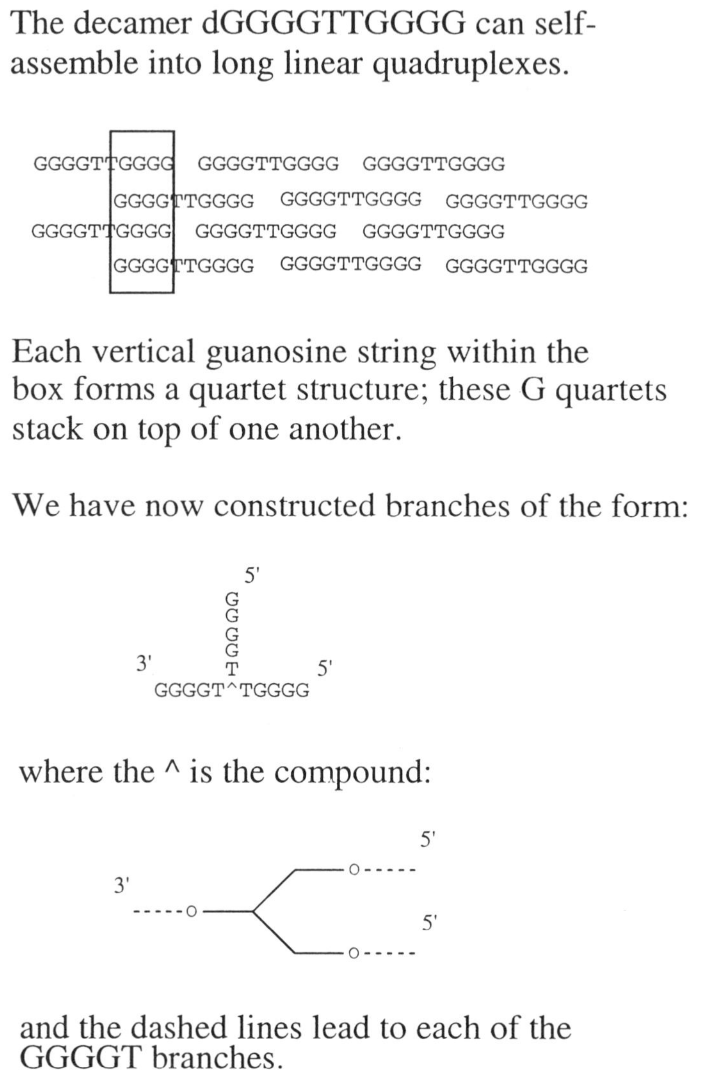

FIGURE 6. DNA matrices based on G wires. Linear wires can be formed from a simple decamer; branched structures of varying degrees of complexity could potentially be formed by mixing the linear monomers with branch-forming monomers.

circuit elements would lie adjacent to one another in such a way that reliable (albeit simple) computations similar to those described above could be performed and the results 'stored' on the matrix. From this vantage, both the ordered arrays of Seeman and our own semi-ordered arrays should have equivalent utility.

The same principles that facilitate the design and assembly of G wire matrices should similarly allow these matrices to be peppered with input / output devices

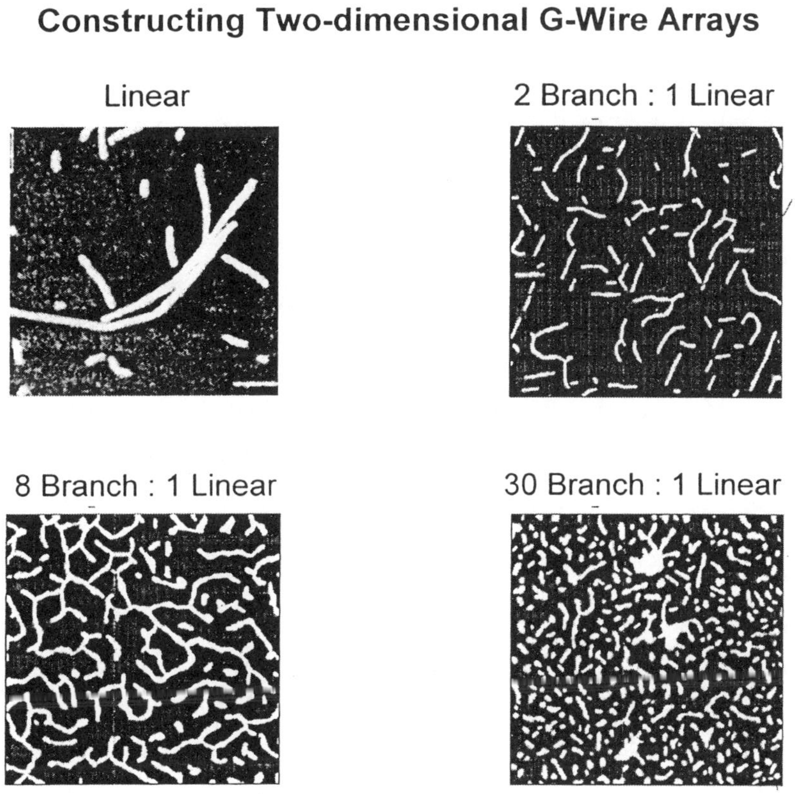

FIGURE 7. AFM images of branched G-wires. Atomic force microscopy has been used to assess formation of stochastic two-dimensional arrays made of DNA. Tet 1.5 (5′ GGGGTTGGGG 3′) was the basic building block of the arrays. Branching was randomly initiated using a bifurcated version of Tet 1.5, as described in Figure 6. The branch-forming oliognucleotide was synthesized via a symmetric, branching phosphoramidite (Clontech). The upper left panel is a figure which has appeared previously [**13**]. AFM images of branched G-wires are by Ellington and Henderson (unpublished)

and logic gates. For example, the simple G wire extension motif (dGGGGT) can be coupled to nucleic acids bearing aptamers, aptazymes, or protein binding sites. These functional elements would then be intermixed with linear and branched G wire monomers and the self-assembled matrices would stochastically join functional elements or place them in proximity to one another (Figure 1). Further, the range of G wire architectures could be augmented by the inclusion of other structural elements, such as dual branching monomers, multi-helix junctions, or even the crossover structures characteristic of Seeman's ordered arrays. In this way, it may be possible to build self-assembling three-dimensional circuits.

5. Conclusion and Prospects

While the design of biological computers is still in its infancy, and while we believe that the principles and components described above may offer the best chance for functional devices, it is still extremely unlikely that any biological computer (short of man himself) will challenge silicon-based devices for supremacy. However, the goal of biological computation should not be to challenge silicon but rather to interface with silicon [**17**]. Thus, the components described above should be viewed as a potential route towards building better 'alligator clips' that will ultimately connect the world of metabolites and cells with the world of transistors and computers. In this respect, the self-assembling matrices and circuits we have made or envision should enjoy one final advantage: while they may not be supreme calculators in and of themselves, they may prove to be interesting masks for photoresists, and thus may allow nanoscale electronic devices with biological components to be patterned and constructed. We are currently working towards proofs of such hybrid devices.

References

[1] L. Adleman. Molecular Computation of Solutions to Combinatorial Problems. *Science*, 266: 1021–1024, 1994.

[2] F. Guarnieri, M. Fliss, C. Bancroft. Making DNA Add. *Science* 273: 220–223, 1996.

[3] Q. Ouyang, P. D. Kaplan, S. Liu, A. Libchaber. DNA solution of the maximal clique problem. *Science*, 278: 446–449, 1997.

[4] A. G. Frutos, Q. Liu, A. J. Thiel, A. M. Sanner, A. E. Condon, L. M. Smith, R. M. Corn. Demonstration of a word design strategy for DNA computing on surfaces. *Nucleic Acids Res.* 25: 4748–4757, 1997.

[5] A. R. Cukras, D. Faulhammer, R. J. Lipton, L. F. Landweber. Chess Games: A Model for RNA Based Computation. in press, 1998.

[6] K. D. James, A. R. Boles, D. Henckel, A. D. Ellington. The fidelity of template-directed oligonucleotide ligation and its relevance for DNA computation. *Nucleic Acids Res.* 26: 5203–5211, 1998.

[7] J. C. Cox, D. S. Cohen, A. D. Ellington. The Complexities of DNA Computation. *Trends Biotechnol.* 17: 151–154, 1999.

[8] J. Tang, R. R. Breaker. Rational Design of Allosteric Ribozymes. *Chem. Biol.* 4: 453–459, 1997.

[9] H. Porta, P. M. Lizardi. An allosteric hammerhead ribozyme. *Biotechnology* 13: 161–164, 1995.

[10] M. P. Robertson, A. D. Ellington. In vitro selection of an allosteric ribozyme that transduces analytes to amplicons. *Nat. Biotechnol.* 17: 62–66, 1999.

[11] N. C. Seeman. DNA nanotechnology: novel DNA constructions. *Annu. Rev. Biophys. Biomol. Struct.* 27: 225–248, 1998.

[12] E. Winfree, F. Liu, L. A. Wenzler, N. C. Seeman. Design and self-assembly of two-dimensional DNA crystals. *Nature* 394: 539–544, 1998.

[13] T. C. Marsh, J. Vesenka, E. Henderson. A new DNA nanostructure, the G-wire, imaged by scanning probe microscopy. *Nucleic Acids Res.* 23: 696–700, 1995.

[14] E. Protozanova, R. B. Macgregor, Jr. Frayed wires: a thermally stable form of DNA with two distinct structural domains. *Biochemistry* 35: 16638–16645, 1996.

[15] E. Protozanova, R. B. Macgregor, Jr. Analysis of the electrophoretic migration of DNA frayed wires. *Biophys. Chem.* 75: 249–257, 1998.

[16] E. Protozanova, R. B. Macgregor, Jr. Circular dichroism of ¿DNA frayed wires. *Biophys. J.* 75: 982–989, 1998.

[17] R. F. Service. Neurons and silicon get intimate. *Science* 284: 578-579, 1999.

(M. P. Robertson) Institute for Cellular and Molecular Biology, University of Texas at Austin, Austin, TX 78712
E-mail address: mprobertson@mail.utexas.edu

(J. Hesselberth) Department of Chemistry and Biochemistry, University of Texas at Austin, Austin, TX 78712
E-mail address: jhessel@ccwf.cc.utexas.edu

(J. C. Cox) Institute for Cellular and Molecular Biology, University of Texas at Austin, Austin, TX 78712
E-mail address: j.colin.cox@mail.utexas.edu

(A. D. Ellington) Department of Chemistry and Biochemistry, University of Texas at Austin, Austin, TX 78712
E-mail address: andy.ellington@mail.utexas.edu

DIMACS Series in Discrete Mathematics
and Theoretical Computer Science
Volume **54**, 2000

Forbidding and Enforcing

Andrzej Ehrenfeucht, Hendrik Jan Hoogeboom, Grzegorz Rozenberg,
and Nikè van Vugt

ABSTRACT. DNA molecules and various operations on them can be conveniently expressed as strings and operations on strings. Hence, many models of DNA computation have been formulated within formal language theory.

We propose a novel kind of model, which is based on two types of boundary conditions : forbidding and enforcing. *Forbidding conditions* say that a 'conflicting' group of components may not be present in a system, *enforcing conditions* say that if a certain group of molecules is present in the system, then some other molecules will eventually be present in the system.

Such forbidding-enforcing systems are 'tolerant' in describing results of (molecular) computations : one system describes the whole family of outcomes all of which obey the forbidding and enforcing constraints of the system, specifying a possibly infinite family of languages. This should be contrasted with standard formal language theory (grammars and automata) where a grammar specifies one language of all words that can be generated.

We illustrate the use of forbidding-enforcing systems for the description of the structure of DNA molecules, the description of splicing systems, and the description of the satisfiability problem.

Next to standard issues such as normal forms, we investigate two central issues : finiteness and the structure of computation.

1. Introduction

The research on DNA computing has lead to many novel models of computation, see for example [**HPP97**], [**Amo97**], [**Win98**]. Since DNA molecules can be expressed in a very natural way as strings (or double strings), and various operations on DNA can be expressed then as operations on strings, many of these models have been formulated within formal language theory. In a typical model of this sort (see, e.g., splicing systems), after the initial set of molecules (strings) and the operations on them (productions) are specified, one gets a rewriting system which defines a single language. This language represents the set of all molecules that can be obtained from the initial ones using the given operations.

We propose here a different kind of model, which is based on boundary conditions. We propose a description of molecular systems based on two types of boundary conditions : forbidding and enforcing. *Forbidding conditions* say that if a certain group of components is present in the system, then the system will 'die' (e.g., it will lose its functionality) – hence such a combination of components is

2000 *Mathematics Subject Classification.* Primary 68Q05, 68Q45.

© 2000 American Mathematical Society

forbidden. *Enforcing conditions* say that if a certain group of molecules is present in the system, then (as the result of a molecular reaction) some other molecules will eventually be present in the system. Hence the evolution of a system described by forbidding conditions $\mathcal{F}$ and enforcing conditions $\mathcal{E}$ will proceed according to the reactions described by $\mathcal{E}$ but restricted in such a way that none of the forbidding conditions from $\mathcal{F}$ will be created. Such forbidding-enforcing systems, *fe systems* for short, are much more 'tolerant' in describing results of (molecular) computations : one fe system describes the whole family of outcomes all of which obey the forbidding and enforcing constraints of the system. Thus in case when we model the molecules by strings (as we do in this paper), one fe system specifies a possibly infinite family of languages. A language belongs to this family if and only if it is consistent with the forbidding conditions and it satisfies the enforcing conditions – nothing else is required from the language. As a matter of fact, fe systems follow the rule "everything that is not forbidden is allowed", while standard formal language theory (grammars and automata) follows the dual rule "everything that is not allowed is forbidden". Hence fe systems lead in our opinion to models of computation novel from both the (modeling of) molecular systems and computation theory point of view.

This paper is organized as follows. First we introduce and investigate forbidding sets and enforcing sets, and then we combine the two and introduce forbidding-enforcing systems, which constitute our model of molecular systems. Next to standard issues such as, e.g., normal forms, we investigate two central issues.
(1) *Finiteness.* This issue is always important from the 'real world' point of view. We investigate many facets of finiteness, and we also discuss the very different role that finiteness plays in our model when compared to standard grammatical models.
(2) *Structure of computation.* Although fe systems are defined by boundary conditions, they turn out to be intrinsically computational. Each fe system, when transformed into a suitable normal form, can be *completely* represented by a tree which contains the information about the family of languages defined by the system, and also about all possible evolving computations of the system.

Finally we want to point out that we have not included the proofs of our results because the size of the paper would then become too big for this volume.

2. Preliminaries

We denote the set of (positive) natural numbers by $\mathbb{N}$, and the empty set by $\varnothing$.

In this paper we consider *non-empty* words only. For the ease of notation, we will fix one alphabet Σ, i.e., every word, language or family of languages is over Σ, unless clear otherwise. The set of all non-empty words over Σ is denoted by Σ^+ (occasionally we will also use the notation Σ^* which denotes the set of all words over Σ, including the empty word).

The length of a word w is denoted by $|w|$. For a language K and an integer $n \geq 1$, $K|_{\leq n} = \{w \in K \mid |w| \leq n\}$. A sequence of languages $K_1, K_2, \ldots$ is called *ascending* if $K_1 \subseteq K_2 \subseteq \ldots$. A sequence of languages $K_1, K_2, \ldots$ *converges* to a language K, denoted $\langle K_n \rangle_{n \in \mathbb{N}} \to K$, if, for each $\ell \geq 1$, there is an $m \geq 1$ such that $K_n|_{\leq \ell} = K|_{\leq \ell}$ for every $n \geq m$.

A word $x \in \Sigma^+$ is a *subword* of a word $y \in \Sigma^+$, denoted $x \ \underline{\text{sub}}\ y$, if $y = uxv$ for some $u, v \in \Sigma^*$. The set of subwords of a word x is denoted by $\underline{\text{sub}}(x)$. The set of subwords of a language K, denoted $\underline{\text{sub}}\,(K)$, equals $\bigcup_{x\in K} \underline{\text{sub}}\,(x)$. A language K is called *subword free* if, for all $x, y \in K$, $x\ \underline{\text{sub}}\ y$ implies $x = y$.

3. Forbidding sets

In this section we formalize the first sort of boundary conditions, viz., forbidding conditions. To this aim we introduce the notion of a forbidding set.

3.1. Definitions and basic properties.

DEFINITION 1. A *forbidding set* is a (possibly infinite) family of finite languages; these finite languages are called *forbidders*.

A language K is said to be *consistent* with a forbidder F, denoted $K\ \underline{\text{con}}\ F$, if $F \not\subseteq \underline{\text{sub}}\,(K)$. A language K is consistent with a forbidding set $\mathcal{F}$, denoted $K\ \underline{\text{con}}\ \mathcal{F}$, if K is consistent with every forbidder in $\mathcal{F}$. □

EXAMPLE 2. Consider the forbidding set $\mathcal{F} = \{\{ab, ba\}, \{aa, bb\}\}$ and alphabet $\{a, b\}$. Then it is easily seen that $K\ \underline{\text{con}}\ \mathcal{F}$ iff $K \subseteq K_i$ for some $i \in \{1, 2, 3, 4\}$, where $K_1 = ab^* \cup b^+$, $K_2 = a^*b \cup a^+$, $K_3 = ba^* \cup a^+$, $K_4 = b^*a \cup b^+$. □

For a forbidding set $\mathcal{F}$ we define the family of ($\mathcal{F}$-)consistent languages $\mathcal{L}(\mathcal{F}) = \{K \mid K\ \underline{\text{con}}\ \mathcal{F}\}$, and the family of finite consistent languages $\mathcal{L}_{\text{fin}}(\mathcal{F}) = \{K \mid K$ is finite and $K \in \mathcal{L}(\mathcal{F})\}$.

THEOREM 3. *Let $\mathcal{F}$ be a forbidding set and K a language.*
(1) $K\ \underline{con}\ \mathcal{F}$ implies $\underline{sub}\,(K)\ \underline{con}\ \mathcal{F}$
(2) If $K' \subseteq K$ and $K\ \underline{con}\ \mathcal{F}$, then $K'\ \underline{con}\ \mathcal{F}$
(3) If $K_1, K_2, \ldots$ is an ascending sequence of languages with $K_i\ \underline{con}\ \mathcal{F}$ for all $i \geq 1$, then $(\bigcup_{i\geq 1} K_i)\ \underline{con}\ \mathcal{F}$

In the third property above, we infer the consistency of the language K from a converging sequence of fragments of K. This property follows from our assumption that forbidders are finite sets.

3.2. A useful minimal normal form. In this subsection we consider three technically useful normal forms for forbidding sets. We use the following obvious notion of equivalence : two forbidding sets $\mathcal{F}_1$ and $\mathcal{F}_2$ are *equivalent* if $\mathcal{L}(\mathcal{F}_1) = \mathcal{L}(\mathcal{F}_2)$.

Let $\mathcal{F}$ be a forbidding set, and let $F_1, F_2 \in \mathcal{F}$ with $F_1 \neq F_2$. We say that F_1 and F_2 are *subword incomparable* if neither $\underline{\text{sub}}\,(F_1) \subseteq \underline{\text{sub}}\,(F_2)$, nor $\underline{\text{sub}}\,(F_2) \subseteq \underline{\text{sub}}\,(F_1)$, and we call $\mathcal{F}$ subword incomparable if each pair of distinct forbidders $F_1, F_2 \in \mathcal{F}$ is subword incomparable. This is a normal form : for every forbidding set $\mathcal{F}$ there exists an equivalent subword incomparable forbidding set $\mathcal{F}'$. Note that $\mathcal{F}'$ is not always unique : for instance, constructing a subword incomparable equivalent of $\{\{a, ab\}, \{b, ab\}\}$ can yield both $\{\{a, ab\}\}$ and $\{\{b, ab\}\}$, since $\underline{\text{sub}}\,(\{a, ab\}) = \underline{\text{sub}}\,(\{b, ab\})$.

We call a forbidding set *subword free* if all its forbidders are subword free. For every forbidding set there exists an equivalent subword free forbidding set.

A forbidding set $\mathcal{F}$ is in *minimal normal form* if $\mathcal{F}$ is subword free and subword incomparable.

THEOREM 4. *For every forbidding set $\mathcal{F}$ there exists an equivalent forbidding set $\mathcal{F}'$ in minimal normal form. $\mathcal{F}'$ is unique.*

EXAMPLE 5. The (fortunately finite) equivalent in minimal normal form of the infinite forbidding set $\mathcal{F} = \{\{a^i, b^i, a^ib^i\} \mid i \geq 1\}$ is $\mathcal{F}' = \{\{ab\}\}$, since $\{a^i, b^i, a^ib^i\} \not\subseteq \underline{\text{sub}}(K)$ for all $i \geq 1$ iff $\{a, b, ab\} \not\subseteq \underline{\text{sub}}(K)$ iff $\{ab\} \not\subseteq \underline{\text{sub}}(K)$. □

A forbidding set in minimal normal form is indeed minimal, or 'redundancy free', in the sense that removing one of its forbidders or even one element from one forbidder yields a forbidding set that is not equivalent to the original one.

3.3. Finiteness. We turn now to the issue of finiteness for forbidding sets. First we observe that we cannot restrict ourselves to finite forbidding sets. On the other hand, the finite consistent languages characterize all consistent languages of a forbidding set.

For some specific infinite forbidding sets it is possible to find an equivalent *finite* forbidding set : $\mathcal{F}_2 = \{\{ab\}\}$ clearly is a finite 'summary' of $\mathcal{F}_1 = \{\{ab\}, \{a^2b^2\}, \{a^3b^3\}, \ldots\}$. Unfortunately this is not always the case, which is seen as follows. For a finite forbidding set $\mathcal{F}$, let $\ell = \max\{|w| \mid w \in \bigcup_{F \in \mathcal{F}} F\}$. Now, if for two words x and y it is the case that $\underline{\text{sub}}(x)|_{\leq \ell} = \underline{\text{sub}}(y)|_{\leq \ell}$, then $\mathcal{F}$ cannot distinguish between x and y, which means that $\{x\} \in \mathcal{L}(\mathcal{F})$ if and only if $\{y\} \in \mathcal{L}(\mathcal{F})$.

Now consider $\mathcal{F}_{\text{even}} = \{\{ab^2a\}, \{ab^4a\}, \{ab^6a\}, \ldots\}$. For each ℓ, the words $x = ab^\ell a$ and $y = ab^{\ell+1}a$ differ only on subwords of length greater than ℓ. Clearly $\{x\} \in \mathcal{L}(\mathcal{F}_{\text{even}})$ if and only if $\{y\} \notin \mathcal{L}(\mathcal{F}_{\text{even}})$. Hence $\mathcal{L}(\mathcal{F}_{\text{even}}) \neq \mathcal{L}(\mathcal{F})$ for all finite $\mathcal{F}$.

Theorem 3 (2) and (3) together yield the following result, which states that consistence of certain specific finite parts of a language K is necessary and sufficient to ensure the consistence of K itself.

LEMMA 6. *K $\underline{con}$ $\mathcal{F}$ if and only if $K|_{\leq i}$ $\underline{con}$ $\mathcal{F}$ for every $i \geq 1$.*

As a matter of fact, finite languages are central elements of every family of consistent languages, as demonstrated by the following result.

THEOREM 7. *For all forbidding sets $\mathcal{F}_1$ and $\mathcal{F}_2$ the following holds : $\mathcal{L}(\mathcal{F}_1) = \mathcal{L}(\mathcal{F}_2)$ if and only if $\mathcal{L}_{\text{fin}}(\mathcal{F}_1) = \mathcal{L}_{\text{fin}}(\mathcal{F}_2)$.*

4. Enforcing sets

In this section we formalize the second sort of boundary conditions, viz., enforcing conditions. To this aim we introduce the notion of enforcing sets.

4.1. Definitions and basic properties.

DEFINITION 8. An *enforcing set* is a (possibly infinite) family of ordered pairs (X, Y), where X and Y are finite languages with $Y \neq \varnothing$; such a pair (X, Y) is called an *enforcer*.

A language K *satisfies* an enforcer (X, Y), denoted K $\underline{\text{sat}}$ (X, Y), if $X \subseteq K$ implies $Y \cap K \neq \varnothing$. A language K satisfies an enforcing set $\mathcal{E}$, denoted K $\underline{\text{sat}}$ $\mathcal{E}$, if K satisfies every enforcer in $\mathcal{E}$. □

EXAMPLE 9. The family $\mathcal{E} = \{(\{u, v\}, \{uv, vu\}) \mid u, v \in \Sigma^+\}$ is an enforcing set. If $K \subseteq \Sigma^+$ satisfies $\mathcal{E}$, then K is closed under 'weak catenation' : for any words $u, v \in K$ at least one of the words uv, vu is in K. Note that there are infinitely many languages satisfying $\mathcal{E}$, each resulting from a different 'implementation' of the weak catenation. □

Note the essential use in the above example of the non-determinism of enforcers. Given an enforcer (X, Y) and a language K, if the set X of *premises* is included in K, then at least one element of the set Y of *consequences* will be included in K. Thus allowing Y to include more than one element accounts for the non-determinism. In the world of molecular reactions the result of molecules from X reacting together may depend quite essentially on all kinds of conditions under which the reaction takes place (e.g., temperature, pH, ...).

EXAMPLE 10. Let $\mathcal{E}$ be the enforcing set $\{(\varnothing, \{b^n\}) \mid n \text{ is even }\}$. We refer to each enforcer $(\varnothing, \{b^n\})$ as a 'brute' enforcer, because the empty set is included in every language. Thus if K satisfies $\mathcal{E}$, then K *must* contain the language $\{b^n \mid n \text{ is even }\}$. □

For an enforcing set $\mathcal{E}$, $\mathcal{L}(\mathcal{E}) = \{K \mid K \,\underline{\text{sat}}\, \mathcal{E}\}$ is the family of ($\mathcal{E}$-)satisfying languages. Similarly, the family of finite satisfying languages is defined by $\mathcal{L}_{\text{fin}}(\mathcal{E}) = \{K \mid K \text{ is finite and } K \in \mathcal{L}(\mathcal{E})\}$.

THEOREM 11. *Let $\mathcal{E}$ be an enforcing set. If $K_1, K_2, \ldots$ is an ascending sequence of languages with K_i sat $\mathcal{E}$ for all $i \geq 1$, then $(\bigcup_{i\geq 1} K_i)$ sat $\mathcal{E}$.*

4.2. Computing is evolving through enforcing. Let $\mathcal{E}$ be an enforcing set. An enforcer $(X, Y) \in \mathcal{E}$ is *applicable to* a language K (*K-applicable*) if $X \subseteq K$. If (X, Y) is K-applicable, but $Y \cap K = \varnothing$, then (X, Y) is a *K-violator.*

Let K_0 be a language, $\mathcal{E}$ an enforcing set, and assume that it is not true that $K_0 \,\underline{\text{sat}}\, \mathcal{E}$, i.e., there are $(X, Y) \in \mathcal{E}$ with $X \subseteq K_0$ but $Y \cap K_0 = \varnothing$. Now add, for each of these K_0-violators (X, Y), at least one element of Y to K_0, and denote the (possibly infinite) superset of K_0 constructed non-deterministically in this way by K_1. Then clearly none of the K_0-violators is a K_1-violator, but some enforcers that were not applicable to K_0 may become now K_1-violators. If so, then repeat the construction described above, and so on.

This iterative 'repair procedure' is illustrated in Figure 1.

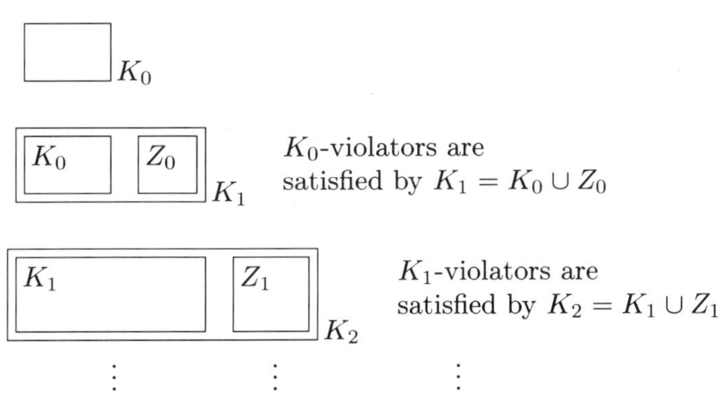

FIGURE 1.

The underlying idea of this 'evolving procedure' is formalized as follows.

DEFINITION 12. For an enforcing set $\mathcal{E}$ and languages K, K' with $K \subseteq K'$ we say that K' is an *$\mathcal{E}$-extension* of K, written $K \vdash_{\mathcal{E}} K'$, if $X \subseteq K$ implies $K' \cap Y \neq \varnothing$, for each $(X, Y) \in \mathcal{E}$. □

The $\mathcal{E}$-extension relation expresses the basic computation step induced by $\mathcal{E}$: a molecular system that has to satisfy $\mathcal{E}$ evolves according to $\vdash_{\mathcal{E}}$. It is also our basic notion for studying computations in forbidding-enforcing systems.

The following theorem says that the iterative repair procedure above yields the desired result.

THEOREM 13. *Let $\mathcal{E}$ be an enforcing set and let $K_1, K_2, \ldots$ be an infinite ascending sequence of languages. If $K_i \vdash_{\mathcal{E}} K_{i+1}$ for each $i \geq 1$, then $(\bigcup_{i\geq 1} K_i)$ $\underline{sat}$ $\mathcal{E}$.*

It is instructive to see that the previous result does not hold for finite ascending sequences : take for instance $\mathcal{E} = \{(\{a\}, \{b\}), (\{a, b\}, \{c\})\}$, and let $K_1 = \{a\}$ while $K_2 = \{a, b\}$. Then $K_1 \subseteq K_2$ and $K_1 \vdash_{\mathcal{E}} K_2$ whereas $K_1 \cup K_2$ does not satisfy $\mathcal{E}$.

This certainly agrees with our intuition : the molecular reactions go on all the time, providing that the needed components (molecules) are available. Thus such reactions may lead out of a finite language.

4.3. Finitary normal form. In this subsection we consider an important normal form for enforcing sets. We begin by distinguishing two finiteness properties of enforcing sets.

We say that two enforcing sets $\mathcal{E}_1$ and $\mathcal{E}_2$ are equivalent if $\mathcal{L}(\mathcal{E}_1) = \mathcal{L}(\mathcal{E}_2)$. For a finite language Z we define $\mathcal{E}(Z) = \{(X, Y) \in \mathcal{E} \mid X = Z\}$.

DEFINITION 14.
(1) An enforcing set $\mathcal{E}$ is *finitary* if, for each finite language Z, $\mathcal{E}(Z)$ is finite.
(2) An enforcing set $\mathcal{E}$ is *weakly finitary* if, for each finite language K_1 there exists a finite language K_2 such that $K_1 \vdash_{\mathcal{E}} K_2$. □

If an enforcing set $\mathcal{E}$ is finitary, it means that each premise set Z can have only a finite number of different consequence sets included in $\mathcal{E}$. This is a *syntactic* feature of $\mathcal{E}$. On the other hand the property of being weakly finitary is more of a *semantic* property – it says something about the effect that the enforcing specified by $\mathcal{E}$ has on finite languages. This effect is required to be 'continuous' : each finite language *can* always evolve according to $\mathcal{E}$ to a finite language. Thus one can start with a finite language and evolve it according to $\mathcal{E}$ in a smooth way without 'exploding in one step' into an infinite set.

The basic relationship between finitary and weakly finitary enforcing sets is given by the following result.

THEOREM 15.
(1) Every finitary enforcing set is weakly finitary.
(2) There exist weakly finitary enforcing sets that are not finitary.

Our next result says that languages K that satisfy finitary enforcing sets $\mathcal{E}$ play for such sets the role of the universe (Σ^+), meaning that $\mathcal{E}$ becomes weakly finitary for all finite languages within such K.

THEOREM 16. *Let $\mathcal{E}$ be a finitary enforcing set, and let K be a language such that K $\underline{sat}$ $\mathcal{E}$. For every finite language $L \subseteq K$, there exists a finite language $L' \subseteq K$ such that $L \vdash_{\mathcal{E}} L'$.*

Note that the above result does not hold if we require that $\mathcal{E}$ is weakly finitary rather than finitary. To see this consider the weakly finitary enforcing set $\mathcal{E} = \{(\varnothing, \{a, b^n\}) \mid n \geq 1\}$. Let $K = \{b^n \mid n \geq 1\}$, then obviously K $\underline{sat}$ $\mathcal{E}$. Now let

$L \subseteq K$ be finite and assume that $L \vdash_{\mathcal{E}} L'$ for a finite language L'. Let $m = \underline{\max}\ \{n \mid b^n \in L'\}$ and consider the enforcer $E = (\varnothing, \{a, b^{m+1}\})$. Obviously $L'\ \underline{\text{sat}}\ E$ does not hold, contradicting $L \vdash_{\mathcal{E}} L'$. Hence L' cannot be finite.

The following theorem is one of the main results of the forbidding-enforcing theory.

THEOREM 17. *For every enforcing set there exists an equivalent finitary enforcing set.*

4.4. Finiteness. In our approach we allow infinite enforcing sets. Unfortunately one cannot restrict oneself to finite enforcing sets only.

LEMMA 18. *If $\mathcal{E}$ is a finite enforcing set, then there is an $n \geq 1$ such that, for each $K \in \mathcal{L}(\mathcal{E})$ and each L, if $K|_{\leq n} \subseteq L \subseteq K$, then $L \in \mathcal{L}(\mathcal{E})$.*

Consequently, if $\mathcal{E}$ is finite, then $\mathcal{L}(\mathcal{E})$ contains a finite language, viz., $K|_{\leq n}$. Thus, e.g., for the enforcing set $\mathcal{E}$ in Example 10 there is no equivalent finite enforcing set, since every language in $\mathcal{L}(\mathcal{E})$ contains the infinite set $\{b^n \mid n \text{ is even}\}$. This gives the following result.

LEMMA 19. *There are infinite enforcing sets for which there is no equivalent finite enforcing set.*

We define $\mathcal{E}|_{\leq n}$ to be $\{(X, Y) \in \mathcal{E} \mid |w| \leq n \text{ for all } w \in X \cup Y\}$. Analogous to the situation with forbidding sets, satisfaction of an enforcing set $\mathcal{E}$ by a language K is guaranteed by the satisfaction of certain specific finite parts of $\mathcal{E}$ by certain specific finite parts of K.

LEMMA 20. *$K\ \underline{\text{sat}}\ \mathcal{E}$ if and only if $K|_{\leq i}\ \underline{\text{sat}}\ \mathcal{E}|_{\leq i}$ for every $i \geq 1$.*

Unlike for forbidding sets, finite languages are not particularly important for families of satisfying languages. This can be shown using the fact that we can construct any language K as a singleton family, by enforcing each of its elements. Similarly, we can enforce that every element of the complement of K be present whenever one of its elements is present. Formally, let K be a language over Σ, and let $\langle w_i \rangle_{i \in \mathbb{N}}$ be an arbitrary but fixed ordering of the words of K. Similarly, let $\langle v_i \rangle_{i \in \mathbb{N}}$ be an arbitrary but fixed ordering of the elements of $\Sigma^+ - K$. Now consider the enforcing set $\mathcal{E}_K = \{(\varnothing, \{w_1\})\} \cup \{(\{w_i\}, \{w_{i+1}\}) \mid i \geq 1\} \cup \{(\{v_i\}, \{v_1\}), (\{v_i\}, \{v_{i+1}\}) \mid i \geq 1\}$. Then $\mathcal{L}(\mathcal{E}_K) = \{K, \Sigma^+\}$, and clearly for every finite language K we have $\mathcal{L}_{\text{fin}}(\mathcal{E}_K) = \{K\}$, whereas for infinite K we have $\mathcal{L}_{\text{fin}}(\mathcal{E}_K) = \varnothing$. Hence for any two different infinite languages K and K' we have $\mathcal{L}_{\text{fin}}(\mathcal{E}_K) = \mathcal{L}_{\text{fin}}(\mathcal{E}_{K'}) = \varnothing$, whereas $\mathcal{L}(\mathcal{E}_K) \neq \mathcal{L}(\mathcal{E}_{K'})$.

5. Forbidding-enforcing systems

We will investigate now systems that combine forbidding and enforcing.

DEFINITION 21. A *forbidding-enforcing system* (fe system for short) is a construct $\Gamma = (\mathcal{F}, \mathcal{E})$, where $\mathcal{F}$ is a forbidding set and $\mathcal{E}$ is an enforcing set. □

The corresponding *forbidding-enforcing family* (fe family), denoted $\mathcal{L}(\mathcal{F}, \mathcal{E})$, consists of all languages that are both $\mathcal{F}$-consistent and $\mathcal{E}$-satisfying. Hence $\mathcal{L}(\mathcal{F}, \mathcal{E}) = \mathcal{L}(\mathcal{F}) \cap \mathcal{L}(\mathcal{E})$.

EXAMPLE 22. Let $\Sigma = \{a, b\}$ and let $\Gamma = (\mathcal{F}, \mathcal{E})$ be the fe system obtained by combining the forbidding set $\mathcal{F} = \{\{aa, bb\}, \{ab, ba\}\}$ from Example 2 and the enforcing set $\mathcal{E} = \{(\varnothing, \{b^n\}) \mid n \text{ is even}\}$ from Example 10. Then a language $K \subseteq \Sigma^+$ is in $\mathcal{L}(\mathcal{F}, \mathcal{E})$ if and only if $K = K' \cup \{b^n \mid n \text{ is even}\}$ where either $K' \subseteq ab^*$ or $K' \subseteq b^*a$. □

We carry over the 'finitary' qualification of enforcing sets to fe systems in an obvious way : an fe system $\Gamma = (\mathcal{F}, \mathcal{E})$ is *finitary* if and only if $\mathcal{E}$ is finitary.

5.1. Examples.

DNA molecules – their structure and processing. DNA molecules, partially single or double stranded, can be coded over a suitable alphabet of base pairs. For the matching base pairs we may use the symbols $\binom{A}{T}, \binom{T}{A}, \binom{C}{G}, \binom{G}{C}$. For the single stranded pieces we can use $\binom{A}{.}, \binom{T}{.}, \binom{C}{.}, \binom{G}{.}$ (upper strand) and $\binom{.}{A}, \binom{.}{T}, \binom{.}{C}, \binom{.}{G}$ (lower strand). For this example we consider languages over the alphabet Σ_{base} consisting of these twelve symbols.

We will give some natural requirements on a formal language representing the set of linear DNA molecules. These requirements can be formulated in our forbidding-enforcing framework.

First, of course, no proper molecule can have an unmatched base in the upper strand next to an unmatched base in the lower strand. This leads to forbidders $\{\ \binom{\sigma}{.}\binom{.}{\tau}\ \}$ and $\{\ \binom{.}{\sigma}\binom{\tau}{.}\ \}$ for each $\sigma, \tau \in \{A, T, C, G\}$.

Additionally, the molecule described by the string $\binom{A}{T}\binom{C}{G}\binom{C}{.}\binom{C}{G}\binom{T}{.}$ is also described by the inverted string $\binom{.}{T}\binom{G}{C}\binom{.}{C}\binom{G}{C}\binom{T}{A}$. We denote the operation of inversion, which is the composition of mirror image and replacing each symbol $\binom{\tau}{\sigma}$ in Σ_{base} by $\binom{\sigma}{\tau}$, by $\underline{\text{inv}}$. In this way we get an infinite number of enforcers, $(\{x\}, \{\underline{\text{inv}}\,(x)\})$ with x over the alphabet Σ_{base}.

The above set of forbidders and the set of enforcers together yield a fe system that admits only *correct* and *all* correct descriptions of linear DNA molecules.

The effect of cutting by restriction enzymes can easily be translated into enforcing rules. E.g., for the restriction enzyme TaqI (see, e.g., [**NEB**]) we have the enforcer $(\{x\binom{T}{A}\binom{C}{G}\binom{G}{C}\binom{A}{T}y\}, \{x\binom{T}{A}\binom{.}{G}\binom{.}{C}\})$ for each pair $x, y \in \Sigma^*_{base}$. Note that we have enforced only one of the two halves, the other follows by palindromicity and inversion.

Recombination is then modelled by reversing the rules. E.g., ligating two pieces with overhang GC (one resulting from cutting with TaqI and the other a sticky end produced by the restriction enzyme NarI (see, e.g., [**NEB**])) can be enforced by

$$\left(\{x\begin{pmatrix}T\\A\end{pmatrix}\begin{pmatrix}.\\G\end{pmatrix}\begin{pmatrix}.\\C\end{pmatrix}, \begin{pmatrix}C\\.\end{pmatrix}\begin{pmatrix}G\\.\end{pmatrix}\begin{pmatrix}C\\G\end{pmatrix}\begin{pmatrix}C\\G\end{pmatrix}y\}, \{x\begin{pmatrix}T\\A\end{pmatrix}\begin{pmatrix}C\\G\end{pmatrix}\begin{pmatrix}G\\C\end{pmatrix}\begin{pmatrix}C\\G\end{pmatrix}\begin{pmatrix}C\\G\end{pmatrix}y\}\right).$$

Splicing systems. In splicing systems [**HPP97**] the above operations are abstracted to the use of splicing rules. The effect of such a splicing rule (u_1, v_1, u_2, v_2), which says that two words $x_1u_1v_1y_1$ and $x_2u_2v_2y_2$ can be spliced to form the word $x_1u_1v_2y_2$ can be described by an enforcing set as follows : $\{(\{x_1u_1v_1y_1, x_2u_2v_2y_2\}, \{x_1u_1v_2y_2\}) \mid x_1, x_2, y_1, y_2 \in \Sigma^*\}$.

More attractively, and more directly, one may have the rules of the form $(\{x, y, r\}, \{w\})$ where r is the restriction enzyme (after all it is a molecule) and x and y are spliced into w according to r. This seems to be attractive because, as various molecules are created during the evolution of such a system, new (restriction) enzymes may become available and so their effects will also be produced – in this way we can deal with dynamically changing sets of rules.

A splicing rule may be specified in its usual string representation $u_1\#v_1\$u_2\#v_2$, whereas an enzyme may be given by its amino acid encoding, hence by a word over an alphabet of 20 symbols.

Satisfiability problem. We explain now a representation of the satisfiability problem by fe systems using the following example. Let $\Psi = (\neg x_1 \vee x_3 \vee x_6) \wedge (\neg x_1 \vee x_2 \vee \neg x_6)$ be a Boolean formula in 3-conjunctive normal form (see, e.g., [**GJ79**]). It consists of two clauses, each of which has to be satisfied.

The two truth assignments to the variable x_i can be encoded as the strings $\mathtt{x}\underline{\mathrm{bin}}\,(i)\mathtt{f}$ for 'x_i is false' and $\mathtt{x}\underline{\mathrm{bin}}\,(i)\mathtt{t}$ for 'x_i is true', where $\underline{\mathrm{bin}}\,(i) \in \{\mathtt{0},\mathtt{1}\}^*$ is a suitable binary encoding of i. Any language coding a truth assignment must have exactly one of the assignments for each variable. This can be achieved by having the brute enforcers $(\varnothing, \{\mathtt{x}v\mathtt{f}, \mathtt{x}v\mathtt{t}\})$ for each $v \in \{\mathtt{0},\mathtt{1}\}^*$ (this constitutes the enforcing set $\mathcal{E}_U$), and by having the forbidder $\{\mathtt{x}v\mathtt{f}, \mathtt{x}v\mathtt{t}\}$ for each $v \in \{\mathtt{0},\mathtt{1}\}^*$ (this constitutes the forbidding set $\mathcal{F}_U$).

Besides these universally valid restrictions, the formula Ψ itself places additional restrictions on the truth assignment. For instance, the clause $\neg x_1 \vee x_3 \vee x_6$ demands to assign true either to $\neg x_1$, or to x_3, or to x_6. Equivalently, it demands not to assign false to all of $\neg x_1$, x_3, and x_6 at the same time, i.e., it forbids the words $\mathtt{x1t}$, $\mathtt{x11f}$, and $\mathtt{x110f}$ to occur at the same time. Hence, the first clause is represented by the forbidder $\{\mathtt{x1t}, \mathtt{x11f}, \mathtt{x110f}\}$ and the second clause by the forbidder $\{\mathtt{x1t}, \mathtt{x10f}, \mathtt{x110t}\}$.

In this way, representing each clause by a forbidder, we get the forbidding set $\mathcal{F}_\Psi$. Now the 'universal' forbidders and enforcers together with the forbidders defined by the formula Ψ yield the fe system $\Gamma_\Psi = (\mathcal{F}_U \cup \mathcal{F}_\Psi, \mathcal{E}_U)$. This Γ_Ψ provides a succinct representation for the satisfiability of Ψ: Ψ is satisfiable if and only if $\mathcal{L}(\Gamma_\Psi)$ is non-empty.

5.2. The structure of computation in forbidding-enforcing systems. We move now to consider the structure of computations in fe systems, and in particular we claim that for the systems in finitary normal form there is an elegant representation, in the form of a tree, of all the computations in a given fe system.

We use here the standard notion of a tree. The trees will be rooted, node-labelled, they may be infinite but always finitely branching. This means that each node has only a finite number of children (however, we do not assume that there is a common bound on the number of children for each node). The label of a node v in a tree τ is denoted by $\underline{\mathrm{lab}}_\tau(v)$. We call a path in a tree a *full* path if it starts at the root and either ends at a leaf or is infinite.

DEFINITION 23. Let $\Gamma = (\mathcal{F}, \mathcal{E})$ be an fe system, and let τ be a tree. Then τ is a Γ-*tree* if

(1) each node label is an element of $\mathcal{L}_{\mathrm{fin}}(\mathcal{F})$,

(2) if a node v_2 is a descendant of a node v_1, then $\underline{\mathrm{lab}}_\tau(v_1) \vdash_{\mathcal{E}} \underline{\mathrm{lab}}_\tau(v_2)$. □

Hence the influence of the forbidding set is expressed by the sort of node labels that are admitted, while the influence of the enforcing set is expressed through the condition on the sort of languages that can follow each other on a single path – this is illustrated in Figure 2.

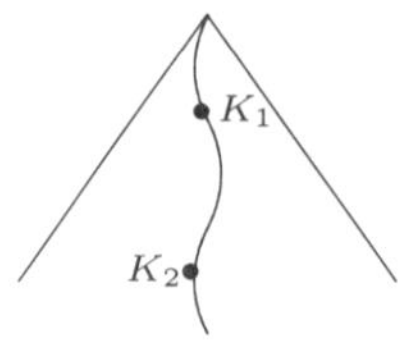

FIGURE 2. $K_1, K_2 \in \mathcal{L}_{\text{fin}}(\mathcal{F})$, $K_1 \subseteq K_2$, and $K_1 \vdash_{\mathcal{E}} K_2$

We now consider a Γ-tree where all the languages from $\mathcal{L}(\Gamma)$, finite and infinite, are represented.

DEFINITION 24. Let $\Gamma = (\mathcal{F}, \mathcal{E})$ be an fe system. A Γ-tree τ is a *complete* Γ-tree if
(1) if $K \subseteq \Sigma^+$ is finite and $K \in \mathcal{L}(\Gamma)$, then K is a node label of τ,
(2) if $K \subseteq \Sigma^+$ is infinite and $K \in \mathcal{L}(\Gamma)$, then there exists an infinite path π in τ such that $K = \bigcup_{v \in \pi} \underline{\text{lab}}_{\tau}(v)$. □

We know already that finitary fe systems constitute a normal form for forbidding-enforcing systems meaning that as far as the specifications of fe families are concerned one can restrict oneself to finitary fe systems. However, the real attractiveness of finitary fe systems stems from the following result.

THEOREM 25. *For each finitary fe system Γ there exists a complete Γ-tree.*

This means that every finitary fe system Γ can be 'completely' represented by a complete Γ-tree τ, that represents both *all languages* defined by Γ and *all computations* taking place within Γ :
(1) all finite languages in $\mathcal{L}(\Gamma)$ occur as node labels in τ,
(2) by taking for each infinite path the union of all languages along this path we get all infinite languages in $\mathcal{L}(\Gamma)$,
(3) by following all full paths in τ we get all evolving computations of Γ.

Since finitary fe systems form a normal form for fe systems, one can represent all languages defined by an arbitrary fe system Γ by a tree, viz., the Γ-tree of an equivalent finitary fe system. What will not carry over is the structure of computations in the original fe system.

THEOREM 26. *For each fe system Γ there exists a finitely branching tree τ with nodes labelled by finite languages such that for each language K,*
$K \in \mathcal{L}(\Gamma)$ iff there exists a full path π in τ such that $K = \bigcup_{v \in \pi} \underline{lab}_{\tau}(v)$.

5.3. The role of finite languages. As we have indicated already fe systems are novel also from formal language theoretic point of view in the sense that a single fe system defines a possibly infinite family of languages (rather than a single language – which is standard in formal language theory). However, there is another very important, and more 'semantical', difference with traditional formal language theory. In classical formal language theory finite languages are 'irrelevant' in the following sense : given any standard sort of grammar G (e.g. context-free grammar, context-sensitive grammar, ET0L system, ...) and a finite language F, the languages $L(G) - F$ and $L(G) \cup F$ can be defined by the same sort of grammar. Actually in all these grammars each finite language can be defined in a trivial (meaningless) way : every element of the finite language can be generated in one step from the axiom of the grammar. The situation for fe systems is drastically different. An fe family is determined by certain finite parts of languages, as formalized by the following result.

THEOREM 27. *Let $\mathcal{K}$ be an fe family, and K a language. If, for all $n \geq 1$, there is an $L \in \mathcal{K}$ with $K|_{\leq n} = L|_{\leq n}$, then $K \in \mathcal{K}$.*

COROLLARY 28. *Let $\mathcal{K}$ be an fe family, and $\Delta \subseteq \Sigma$. If all finite languages over Δ are in $\mathcal{K}$, then all languages over Δ are in $\mathcal{K}$.*

COROLLARY 29. *Let $\mathcal{K}_1$ and $\mathcal{K}_2$ be fe families. If $\{K|_{\leq n} \mid K \in \mathcal{K}_1\} = \{K|_{\leq n} \mid K \in \mathcal{K}_2\}$ for all $n \geq 1$, then $\mathcal{K}_1 = \mathcal{K}_2$.*

5.4. Some topology. Since investigation of computations in fe systems often leads to convergence problems of (evolving) sequences of languages, one comes in a natural way to some basic topological notions. The following theorem, a topological reformulation of Theorem 27, is a generalization of Theorem 3 (3) and Theorem 11, in the sense that we do not require an ascending sequence of languages, the union of which equals a certain language K, but rather a sequence of languages that converges to K.

THEOREM 30. *Let $\Gamma = (\mathcal{F}, \mathcal{E})$ be an fe system and let K be a language. If $K_1, K_2, \ldots$ is a sequence of languages with $\langle K_n \rangle_{n \in \mathbb{N}} \to K$ and $K_i \in \mathcal{L}(\mathcal{F}, \mathcal{E})$ for all $i \in \mathbb{N}$, then $K \in \mathcal{L}(\mathcal{F}, \mathcal{E})$.*

This is exactly the definition of closed sets (in topological sense) : thus the language sets $\mathcal{L}(\mathcal{F}, \mathcal{E})$ are closed sets. In fact, it can be shown that the sets of languages for which a tree as in Theorem 26 exists, are exactly the closed sets (of languages).

6. Discussion

In this paper we have introduced a model of computation which we believe is novel from both molecular systems and computation theory point of view. We believe that forbidding-enforcing is an interesting paradigm to be investigated. We see (at least) three lines of research continuing this paper :
(1) Investigating the use of fe systems for the description of various types of computing, both dry and wet. Examples given in Section 5.1 form only a beginning of a step in this direction.
(2) Developing the theory of fe systems. In particular we plan here to explore Theorem 25, and to investigate basic decision problems and complexity issues. There are also many combinatorial and topological questions that, in view of the current paper, should be pursued.
(3) There is nothing in the basic idea of forbidding-enforcing systems that restricts them to strings only. It is well known that graphs are useful for describing the basic structure of molecules. Moreover, using forbidden subgraphs to specify classes of graphs (e.g., planar graphs) is quite well understood. We plan to investigate fe systems over graphs.

References

[Amo97] Martyn Amos. *DNA Computation.* PhD thesis, Department of Computer Science, University of Warwick, UK, September 1997.

[GJ79] Michael R. Garey and David S. Johnson. *Computers and Intractability, A Guide to the Theory of NP-Completeness.* Bell Telephone Laboratories, 1979.

[HPP97] Thomas Head, Gheorghe Păun, and Dennis Pixton. Language theory and molecular genetics : Generative mechanisms suggested by DNA recombination. In Grzegorz Rozenberg and Arto Salomaa, editors, *Handbook of Formal Languages*, volume 2, chapter 7, pages 295–360. Springer-Verlag, 1997.

[NEB] New England BioLabs Catalog 1998/1999.

[Win98] Erik Winfree. *Algorithmic Self-Assembly of DNA*. PhD thesis, California Institute of Technology, Pasadena, California, May 19 1998.

(A. Ehrenfeucht) Department of Computer Science, University of Colorado at Boulder, Boulder, CO 80309, USA

(H. J. Hoogeboom) LIACS, Universiteit Leiden, Niels Bohrweg 1, 2333 CA Leiden, The Netherlands

E-mail address: `hoogeboo@wi.leidenuniv.nl`

(G. Rozenberg) LIACS, Universiteit Leiden, Niels Bohrweg 1, 2333 CA Leiden, The Netherlands, and, Department of Computer Science, University of Colorado at Boulder, Boulder, CO 80309, USA

E-mail address: `rozenber@wi.leidenuniv.nl`

(N. van Vugt) LIACS, Universiteit Leiden, Niels Bohrweg 1, 2333 CA Leiden, The Netherlands

E-mail address: `nvvugt@wi.leidenuniv.nl`

DIMACS Series in Discrete Mathematics
and Theoretical Computer Science
Volume **54**, 2000

Computational Power of Gene Rearrangement

Lila Kari and Laura F. Landweber

ABSTRACT. In [**8**] we proposed a model to describe the homologous recombinations that take place during massive gene rearrangements in hypotrichous ciliates. Here we develop the model by introducing the dependency of homologous recombinations on the presence of certain contexts. We then prove that such a model has the computational power of a Turing machine. This indicates that, in principle, some unicellular organisms may have the capacity to perform any computation carried out by an electronic computer.

1. Introduction and notation

The process we model is gene rearrangement in ciliates, unicellular eukaryotes (nucleated cells) that possess two types of nuclei: an active *macronucleus* (soma) and a functionally inert *micronucleus* (germline) which contributes only to sexual reproduction. The somatically active macronucleus forms from the germline micronucleus after sexual reproduction, during the course of development. The genomic copies of some protein-coding genes in the micronucleus of hypotrichous ciliates are obscured by the presence of intervening non-protein-coding DNA sequence elements (internally eliminated sequences, or *IES*s). These must be removed before the assembly of a functional copy of the gene in the somatic macronucleus. Furthermore, the protein-coding DNA segments (macronuclear destined sequences, or *MDS*s) in species of *Oxytricha* and *Stylonychia* are sometimes present in a permuted order relative to their final position in the macronuclear copy. (See [**8**] for a review.)

The developing ciliate macronuclear "computer" (Figure 1) apparently relies on the information contained in short direct repeat sequences to act as minimal guides in a series of homologous recombination events. These guide-sequences act in principle as splints, and the process of recombination results in linking the protein-encoding segments (MDSs) that belong next to each other in the final protein coding sequence. As such, the unscrambling of these protein-coding genes accomplishes an impressive feat of cellular computation. Other structural components of the ciliate chromatin presumably play a significant role, but the exact details of the mechanism remain elusive [**8**].

Before introducing the formal model, we summarize our notation. An alphabet Σ is a finite, nonempty set. A sequence of letters from Σ is called a string (word) over Σ and in our interpretation corresponds to a linear strand. The words are denoted by lowercase letters such as u, v, α_i, x_{ij}. The length of a word w is denoted by $|w|$ and represents the total number of occurrences of letters in the word. A

2000 *Mathematics Subject Classification.* Primary 68Q05, 03D10, 68Q42, 68Q45, 68Q99.

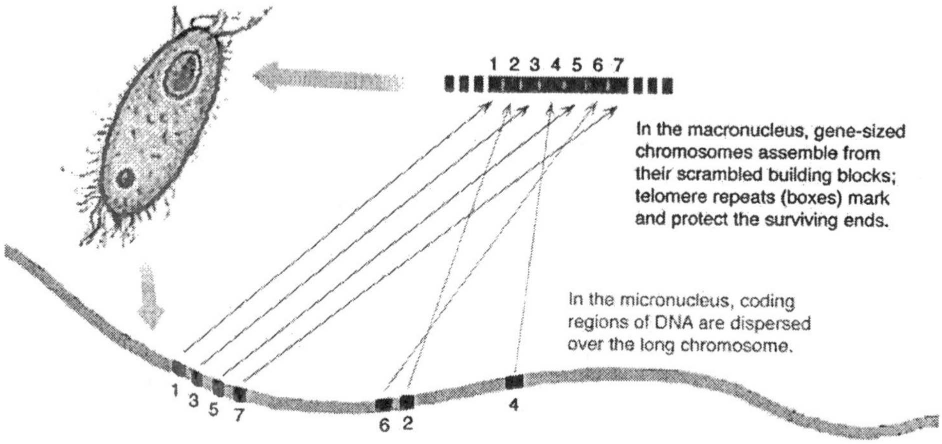

FIGURE 1. **Overview of gene unscrambling.** Dispersed coding MDSs 1-7 reassemble during macronuclear development to form the functional gene copy (top), complete with telomere addition to mark and protect both ends of the gene. (From [**8**].)

word with 0 letters in it is called an empty word and is denoted by λ. The set of all possible words consisting of letters from Σ is denoted by Σ^*, and the set of all nonempty words by Σ^+. We also define circular words over Σ by declaring two words to be equivalent if and only if (iff) one is a cyclic permutation of the other. In other words, w is equivalent to w' iff they can be decomposed as $w = uv$ and $w' = vu$, respectively. Such a circular word $\bullet w$ refers to any of the circular permutations of the letters in w. Denote by $\Sigma^\bullet$ the set of all circular words over Σ.

A rewriting system $TM = (S, \Sigma \cup \{\#\}, P)$ is called a *Turing machine*, [**12**], iff:

(i) S and $\Sigma \cup \{\#\}$ (with $\# \notin \Sigma$ and $\Sigma \neq \emptyset$) are two disjoint alphabets referred to as the *state* and the *tape* alphabets.

(ii) Elements s_0 and s_f of S, and B of Σ are the *initial* and *final* state, and the *blank symbol*, respectively. Also a subset T of Σ is specified and referred to as the *terminal* alphabet. It is assumed that T is not empty.

(iii) The productions (rewriting rules) of P are of the forms

(1) $s_i a \longrightarrow s_j b$ (overprint)
(2) $s_i ac \longrightarrow as_j c$ (move right)
(3) $s_i a\# \longrightarrow as_j B\#$ (move right and extend workspace)
(4) $cs_i a \longrightarrow s_j ca$ (move left)
(5) $\#s_i a \longrightarrow \#s_j Ba$ (move left and extend workspace)
(6) $s_f\, a \longrightarrow s_f$
(7) $a\, s_f \longrightarrow s_f$

where s_i and s_j are states in S, $s_i \neq s_f$, $s_j \neq s_f$, and a, b, c are in Σ. For each pair (s_i, a), where s_i and a are in the appropriate ranges, P either contains no productions (2) and (3) (resp.(4) and (5)) or else contains both (3) and (2) for every c (resp.contains both (5) and (4) for every c). There is no pair (s_i, a) such that the word $s_i a$ is a subword of the left side in two productions of the forms (1), (3), (5).

A configuration of the TM is of the form $\#w_1 s_i w_2\#$, where $w_1 w_2$ represents the contents of the tape, #s are the boundary markers, and the position of the state symbol s_i indicates the position of the read/write head on the tape: if s_i is positioned at the left of a letter a, this indicates that the read/write head is placed over the cell containing a. The TM changes from one configuration to another according to its rules. For example, if the current configuration is $\#ws_i aw'\#$ and the TM has the rule $s_i a \longrightarrow s_j b$, this means that the read/write head positioned over the letter a will write b over it, and change its state from s_i to s_j. The next configuration in the derivation will be thus $\#ws_j bw'\#$.

The Turing machine TM *halts* with a word w iff there exists a derivation that, when started with the read/write head positioned at the beginning of w eventually reaches the final state, i.e. if $\#s_0 w\#$ derives $\#s_f\#$ by succesive applications of the rewriting rules (1) - (7). The language $L(TM)$ *accepted* by TM consists of all words over the terminal alphabet T for which the TM halts. Note that TM is *deterministic*: at each step of the rewriting process, the application of at most one production is possible.

2. Computational power of gene rerrangement

In this section we define the notion of a *guided recombination system* that models the process taking place during gene rearrangement, and prove that such systems have the computational power of a Turing machine, the most widely used theoretical model of electronic computers.

The following strand operations generalize the intra- and intermolecular recombinations defined in [**8**] and illustrated in Figure 2 by assuming that homologous recombination is influenced by the presence of certain contexts, i.e., either the presence of an IES or an MDS flanking a junction sequence. The observed dependence on the old macronuclear sequence for correct IES removal in *Paramecium* suggests that this is the case ([**9**]). This restriction captures the fact that the guide sequences do not contain all the information for accurate splicing during gene unscrambling.

Using an approach developed in [**7**] we use contexts to restrict the use of recombinations. A *splicing scheme*, [**3**], [**4**] is a pair $(\Sigma, \sim)$ where Σ is the alphabet and $\sim$, the pairing relation of the scheme, is a binary relation between triplets of nonempty words satisfying the following condition: If $(p, x, q) \sim (p', y, q')$ then $x = y$.

In the splicing scheme $(\Sigma, \sim)$ pairs $(p, x, q) \sim (p', x, q')$ now define the contexts necessary for a recombination between the repeats x. Then we define *contextual intramolecular recombination* as

$$\{uxwxv\} \Longrightarrow \{uxv, \bullet wx\}, \text{ where } u = u'p, w = qw' = w''p', v = q'v'.$$

This constrains intramolecular recombination within $uxwxv$ to occur only if the restrictions of the splicing scheme concerning x are fulfilled, i.e., the first occurrence of x is preceded by p and followed by q and its second occurrence is preceded by p' and followed by q'.

Similarly, if $(p, x, q) \sim (p', x, q')$, then we define *contextual intermolecular recombination* as

$$\{uxv, \bullet wx\} \Longrightarrow \{uxwxv\} \text{ where } u = u'p, v = qv', w = w'p' = q'w''.$$

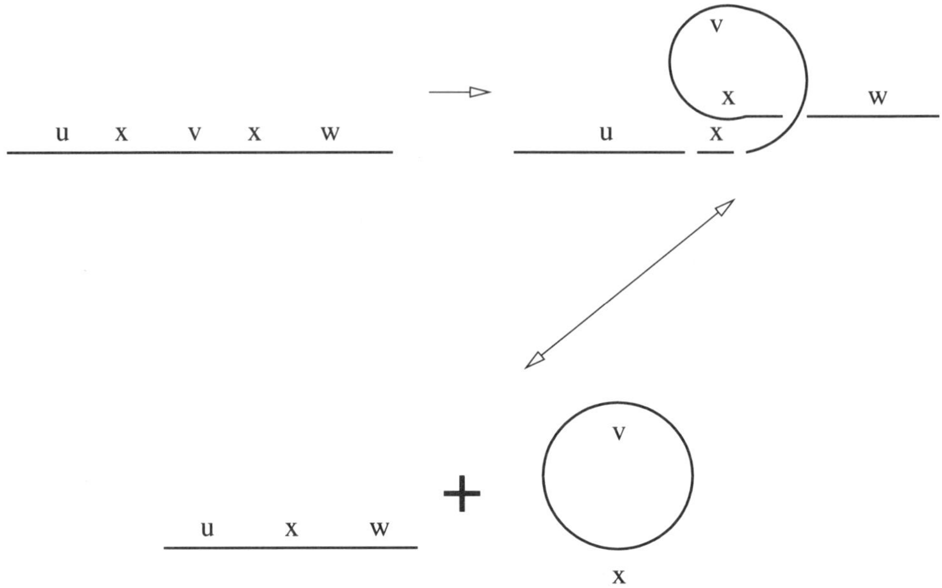

FIGURE 2. Intra- and intermolecular recombinations using repeats x. During intramolecular recombination, after x finds its second occurrence in $uxvxw$, the molecule undergoes a strand exchange in x that leads to the formation of two new molecules: a linear DNA molecule uxw and a circular one $\bullet vx$. The reverse operation is intermolecular recombination.

Informally, intermolecular recombination between the linear strand uxv and the circular strand $\bullet wx$ may take place only if the occurrence of x in the linear strand is flanked by p and q and its occurrence in the circular strand is flanked by p' and q'. Note that sequences p, x, q, p', q' are nonempty, and that both contextual intra- and intermolecular recombinations are reversible by introducing pairs $(p, x, q') \sim (p', x, q)$ in $\sim$.

The above operations resemble the "splicing operation" introduced by Head in [**3**] and "circular splicing" ([**4**], [**13**], [**11**]). [**10**], [**1**] and subsequently [**14**] showed that these models have the computational power of a universal Turing machine. (See [**5**] for a review.)

The operations defined in [**8**] are particular cases of guided recombination, where all the contexts are empty, i.e, $(\lambda, x, \lambda) \sim (\lambda, x, \lambda)$ for all $x \in \Sigma^+$. This corresponds to the case where recombination may occur between every repeat sequence, regardless of the contexts. These unguided (context-free) recombinations are computationally not very powerful: we have proved that they can only generate regular languages.

If we use the classical notion of a set, we can assume that the strings entering a recombination are available for multiple operations. Similarly, there would be no restriction on the number of copies of each strand produced by recombination. However, we can also assume some strings are only available in a limited number of copies. Mathematically this translates into using *multisets*, where one keeps track of the number of copies of a string at each moment. In the style of [**2**], if $\mathbf{N}$ is the set of natural numbers, a multiset of Σ^* is a mapping $M : \Sigma^* \longrightarrow \mathbf{N} \cup \{\infty\}$, where, for a word $w \in \Sigma^*$, $M(w)$ represents the number of occurrences of w. Here, $M(w) = \infty$ means that there are unboundedly many copies of the string w. The set $\text{supp}(M) = \{w \in \Sigma^* | \ M(w) \neq 0\}$, the *support of* M, consists of the strings that are present at least once in the multiset M.

We now define a *guided recombination system* that captures the series of dispersed homologous recombination events that take place during scrambled gene rearrangements in ciliates.

Definition *A guided recombination system is a triple* $R = (\Sigma, \sim, A)$ *where* $(\Sigma, \sim)$ *is a splicing scheme, and* $A \in \Sigma^+$ *is a linear string called the axiom.*

A guided recombination system R defines a *derivation relation* that produces a new multiset from a given multiset of linear and circular strands, as follows. Starting from a "collection" (multiset) of strings with a certain number of available copies of each string, the next multiset is *derived from* the first one by an intra- or inter-molecular recombination between existing strings. The strands participating in the recombination are "consumed" (their multiplicity decreases by 1) whereas the products of the recombination are added to the multiset (their multiplicity increases by 1).

For two multisets S and S' in $\Sigma^* \cup \Sigma^\bullet$, we say that S derives S' and we write $S \Longrightarrow_R S'$, iff one of the following two cases hold:
(1) there exist $\alpha \in \text{supp}(S)$, $\beta, \bullet\gamma \in \text{supp}(S')$ such that
– $\{\alpha\} \Longrightarrow \{\beta, \bullet\gamma\}$ according to an intramolecular recombination step in R,
– $S'(\alpha) = S(\alpha) - 1$, $S'(\beta) = S(\beta) + 1$, $S'(\bullet\gamma) = S(\bullet\gamma) + 1$;
(2) there exist $\alpha', \bullet\beta' \in \text{supp}(S)$, $\gamma' \in \text{supp}(S')$ such that
– $\{\alpha', \bullet\beta'\} \Longrightarrow \{\gamma'\}$ according to an intermolecular recombination step in R,
– $S'(\alpha') = S(\alpha') - 1$, $S'(\bullet\beta') = S(\bullet\beta') - 1$, $S'(\gamma') = S(\gamma') + 1$.

Those strands which, by repeated recombinations with initial and intermediate strands eventually produce the axiom, form the language of the guided recombination system. Formally,

$$L_a^k(R) = \{w \in \Sigma^* | \ \{w\} \Longrightarrow_R^* S \text{ and } A \in \text{ supp}(S)\},$$

where the the multiplicity of w equals k. Note that $L_a^k(R) \subseteq L_a^{k+1}(R)$ for any $k \geq 1$.

Theorem. *Let L be a language over T^* accepted by a Turing machine $TM = (S, \Sigma \cup \{\#\}, P)$ as above. Then there exist an alphabet Σ', a sequence $\pi \in \Sigma'^*$, depending on L, and a recombination system R such that a word w over T^* is in L if and only if $\#^6 s_0 w \#^6 \pi$ belongs to $L_a^k(R)$ for some $k \geq 1$.*

Proof. Consider that the rules of P are ordered in an arbitrary fashion and numbered. Thus, if TM has m rules, a rule is of the form $i : u_i \longrightarrow v_i$ where $1 \leq i \leq m$.

We construct a guided recombination system $R = (\Sigma', \sim, A)$ and a sequence $\pi \in \Sigma'^*$ with the required properties. The alphabet is $\Sigma' = S \cup \Sigma \cup \{\#\} \cup \{\$_i | \ 0 \leq i \leq m+1\}$. The axiom, i.e., the target string to be achieved at the end of the computation, consists of the final state of the TM bounded by markers:

$$A = \#^{n+2} s_f \ \#^{n+2} \$_0 \$_1 \ldots \$_m \$_{m+1},$$

where n is the maximum length of the left-side or right-side words of any of the rules of the Turing machine.

The sequence π consists of the catenation of the right-hand sides of the TM rules bounded by markers, as follows:

$$\pi = \$_0\ \$_1 e_1 v_1 f_1 \$_1\ \$_2 e_2 v_2 f_2 \$_2 \ldots \$_m e_m v_m f_m \$_m\ \$_{m+1},$$

where $i:\ u_i \longrightarrow v_i$, $1 \le i \le m+1$ are the rules of TM and $e_i, v_i \in \Sigma \cup \{\#\}$.

If a word $w \in T^*$ is accepted by the TM, a computation starts then from a strand of the form $\#^{n+2} s_0 w \#^{n+2} \pi$, where we will refer to the subsequence starting with $\$_0$ as the "program", and to the subsequence at the left of $\$_0$ as the "data".

We construct the relation $\sim$ so that

(i) The right-hand sides of rules of TM can be excised from the program as circular strands which then interact with the data.

(ii) When the left-hand side of a TM rule appears in the data, the application of the rule can be simulated by the insertion of the circular strand encoding the right-hand side, followed by the deletion of the left hand side.

To accomplish *(i)*, for each rule $i : u \longrightarrow v$ of the TM, we introduce in $\sim$ the pairs

$$(C) \qquad (\$_{i-1}, \$_i, evf) \sim (evf, \$_i, \$_{i+1}),$$

for all $e, f \in \Sigma \cup \{\#\}$.

To accomplish *(ii)* for each rule $i : u \longrightarrow v$ of the TM, add to the relation $\sim$ the pairs

$$(A) \qquad (ceu, f, d) \sim (\$_i ev, f, \$_i ev),$$

$$(B) \qquad (c, e, uf\$_i) \sim (uf\$_i, e, vfd),$$

for all $c \in \{\#\}^*\Sigma^*$, $d \in \Sigma^*\{\#\}^*$, $|c| = |d| = n$, $e, f \in \Sigma \cup \{\#\}$.

Following the above construction of the alphabet Σ', sequence π and recombination system R, for any $x, y \in \Sigma'$ we can simulate a derivation step of the TM as follows:

$$\{xceufdy\$_0 \ldots \$_{i-1}\$_i evf\$_i\$_{i+1} \ldots \$_{m+1}\} \Longrightarrow_R$$

$$\{xceufdy\$_0 \ldots \$_{i-1}\$_i\$_{i+1} \ldots \$_{m+1}, \bullet\$_i evf\} \Longrightarrow_R$$

$$\{xceuf\$_i evfdy\$_0 \ldots \$_{i-1}\$_i\$_{i+1} \ldots \$_{m+1}\} \Longrightarrow_R$$

$$\{xcevfdy\$_0 \ldots \$_{i-1}\$_i\$_{i+1} \ldots \$_{m+1},\ \bullet\$_i euf\}.$$

The first step is an intramolecular recombination using contexts (C) around the repeat $\$_i$ to excise $\bullet\$_i evf$. Note that if the current strand does not contain a subword $\$_i evf\$_i$, this can be obtained from another copy of the original linear strand, which is initially present in k copies. The second step is an intermolecular recombination using contexts (A) around the repeat f, to insert $\$_i evf$ after $ceuf$. The third step is an intramolecular recombination using contexts (B) around the direct repeat e to delete $\$_i euf$ from the linear strand. Thus, the "legal" insertion/deletion succession that simulates one TM derivation step claims that any u in the data, that is surrounded by at least $n+1$ letters on both sides may be replaced by v. This explains why in our choice of axiom we needed $n+1$ extra symbols $\#$ to provide the contexts allowing recombinations to simulate all TM rules, including (3) and (5).

From the fact that a TM derivation step can be simulated by recombination steps we deduce that, if the TM accepts a word w, then we can start a derivation in R from

$$\#^{n+2}s_0w\#^{n+2}\pi = \#^{n+2}s_0w\#^{n+2}\$_0\$_1\ldots\$_ie_iv_if_i\$_i\ldots\$_m\$_{m+1}$$

and reach the axiom by only using recombinations according to R. This means that our word is accepted by R, that is, it belongs to $L_a^k(R)$ for some k. Note that if some rules of the TM have not been used in the derivation then they can be excised in the end, and that k should be large enough so that we do not exhaust the set of rewriting rules.

For the converse implication, it suffices to prove that starting from the strand $\#^{n+2}s_0w\#^{n+2}\pi$, no other recombinations except those that excise rules of TM from the program and those that simulate steps of the TM in the data are possible in R.

In the beginning of the derivation we start with no circular strands and k copies of the linear strand

$$\#^{n+2}s_0w\#^{n+2}\$_0\ldots\$_ie_iv_if_i\$_i\ldots\$_{m+1}, w\in T^*,$$

where $i : u_i \longrightarrow v_i$ are TM rules, $e_i, f_i \in \Sigma\cup\{\#\}$, $1\leq i\leq m$.

Assume now that the current multiset contains linear strands of the form $\delta_0\pi$, where $\delta_0\in\Sigma'^*$ contains only one state symbol and no $\$_i$ symbols and

$$\pi = \$_0r_1r_2\ldots r_m\$_{m+1},$$

with r_i either encoding the right-hand side of a rule or being the remnant of a rule, i.e., $r_i\in\{\$_ie_iv_if_i\$_i\}\cup\{\$_i\}$, $1\leq i\leq m$. Moreover, assume that the circular strands present in the multiset are of the form $\bullet\$_ie_iv_if_i$, with e_i, v_i, f_i as before.

Then:

(i) We cannot use (A) or (B) to insert or delete in the program because that would require the presence of strands $ceufd$ or $\$_iev f\$_iev$ (if we want to use (A)) or $ceuf\$_i$ or $uf\$_ievfd$ (if we want to use (B)). However none of these strands can appear in the program. Indeed, the 1st, 3rd, and 4th word all contain subwords over $\Sigma\cup\{\#\}$ of length at least $n+3$, and this is more than the length of the longest subword over $\Sigma\cup\{\#\}$ present in the program. The 2nd word cannot appear in the program because no marker $\$_i$ appears alone in p, as p contains always at least two consecutive markers.

(ii) We cannot use (C) to insert or delete in the data because that would require the presence in δ_0 of two consecutive markers $\$_{i-1}\$_i$ or $\$_i\$_{i+1}$, which contradicts our assumptions.

(iii) We cannot use (C) to insert in the program because that would require the presence of a circular strand with two markers, - contradiction with our assumptions.

Arguments *(i)* - *(iii)* show that the only possible recombinations are either deletions in the program using (C), which result in the release of circular strands $\bullet\$_ievf$, or insertions/deletions in the data using (A) and (B).

Assuming that the data contains as a subword the left-hand side of a TM rule $i : u \longrightarrow v$, and assuming that the necessary circular strand $\bullet\$_ievf$ has already been excised from the program, the next step is to show that the only possible insertions/deletions in the data are those simulating a rewriting step of TM using rule i.

Indeed, in this situation,

(1) It is not possible to delete in δ_0 using (A), or insert or delete using (B), as all these operations would require a $\$_i$ in δ_0. Therefore only an insertion in δ_0 using (A) is possible. An insertion according to (A) may only take place between a sequence $ceuf$ and a sequence d, where u contains a state symbol, i.e. the read/write head, c and d have length n and e and f are letters. This means that, for the insertion to take place, the linear word has to be of the form

$$\delta_0 \ \pi = xceufdy \ \pi$$

and the intermolecular recombination with the circular strand $\bullet\$_i evf$ inserts $\$_i evf$ between u and f producing the linear strand

$$\delta_1 \ \pi = xceuf\$_i evfdy \ \pi.$$

Note that, as δ_0 contains only one state symbol and no marker $\$_i$, the newly formed word δ_1 contains only two state symbols (read/write heads), one in u and one in v, and only one marker $\$_i$. (Here we use the fact that every rule $u \longrightarrow v$ of the TM has exactly one state symbol on each side.)

(2) Starting now from $\delta_1 \pi$,

(2a) We can delete in δ_1 using (B) and, as there is only one $\$_i$ in δ_1, there is only one position where the deletion can happen. After the release of the strand $\bullet\$_i euf$ as a circular strand, the linear strand produced is

$$\delta_2 \ \pi = xcevfdy \ \pi.$$

(2b) No insertion in δ_1 using (A) may take place, as the marker $\$_i$ "breaks" the contexts necessary for further insertions.

Indeed, the occurrence of another insertion according to (A) requires that the read/write head symbol be both followed and preceded by at least $(n+1)$ letters different from $\$_i$. In δ_1, the first read/write head is in u and the number of letters following it is at most $|u| - 1 + |f| \leq n - 1 + 1 = n$, which is not enough as a right context for insertion using (A). The second read/write head is in v and the number of letters preceding it is at most $|e| + |v| - 1 \leq 1 + n - 1 = n$, which is not enough as a left context for insertion using (A).

(2c) No deletion in δ_1 using (A) may occur, as this would require the presence of a repeat f bordered by a $\$_i ev$ on each side. This would imply that the current strand δ_1 contains two markers $\$_i$, which is not true.

(2d) No insertion in δ_1 using (B) is possible, as that would require the presence of a circular strand containing $\$_i evfd$. The length of such a strand would be at least $1 + |e| + |v| + |f| + |d|$ that is, at least $n+4$, which is more than the length of any initial or intermediate circular strand. Indeed, all the circular strands produced from the program have length $n+3$ and the only circular strands that are released are, as seen in (2a), of the form $\bullet\$_i euf$ and thus also have lengths at most $n+3$.

The arguments above imply that the only possible operations on the data simulate legal rewritings of the TM by tandem recombination steps that necessarily follow each other.

Together with the arguments that the only operations affecting the program are excisions of circular strands encoding TM rules, and that the circular TM rules do not interact with each other, this proves the converse implication.

From the definition of the Turing machine we see that n, the maximum length of a word occurring in a TM rule, equals 4, which completes the proof of the theorem.

Q.E.D.

The preceding theorem implies that if a word $w \in T^*$ is in $L(TM)$, then $\#^6 s_0 w \#^6 \pi$ belongs to $L_a^k(R)$ for some k and therefore it belongs to $L_a^i(R)$ for any $i \geq k$. This means that, in order to simulate a computation of the Turing machine on w, any sufficiently large number of copies of the initial strand will do. The assumption that sufficiently many copies of the input strand are present at the beginning of the computation is in accordance with the fact that there are multiple copies of each strand available during the (polytene chromosome) stage where unscrambling occurs. Note that the preceding result is valid even if we allow interactions between circular strands or within a circular strand, particular cases of which have been formally defined in [**8**].

The proof that a guided recombination system can simulate the computation of a Turing machine suggests that the micronuclear gene, present in multiple copies, consists of a sequence encoding the input data, combined with a sequence encoding a program, i.e., a list of encoded computation instructions. The "computation instructions" can be excised from the micronuclear gene and become circular "rules" that can recombine with the data. The process continues then by multiple intermolecular recombination steps involving the linear strand and circular "rules", as well as intramolecular recombinations within the linear strand itself. The resulting linear strand, which is the functional macronuclear copy of the gene, can then be viewed as the output of the computation performed on the input data following the computation instructions excised as circular strands.

The last step, telomere addition and the excision of the strands between the telomere addition sites, can easily be added to our model as a final step consisting of the deletion of all the markers, rule delimiters and remaining rules from the output of the computation. This would result in a strand that contains only the output of the computation (macronuclear copy of the gene) flanked by end markers (telomere repeats). This also provides a new interpretation for some of the vast quantity of non-encoding DNA found in micronuclear genes.

In conclusion, we have developed a model for the acrobatic process of gene unscrambling in hypotrichous ciliates. While the model is consistent with our limited knowledge of this biological process, it needs to be rigorously tested using molecular genetics. We have shown, however, that the model is capable of universal computation. This both hints at future avenues for exploring biological computation and opens our eyes to the range of complex behaviors that may be possible in ciliates, and potentially available to other evolving genetic systems.

Acknowledgements. Jarkko Kari for essential contribution to the proof of the theorem in its present form. Rani Siromoney, Gilles Brassard for suggestions, Erik Winfree and Gheorghe Păun for comments. Grzegorz Rozenberg, Richard Lipton, David Prescott and Hans Lipps for discussion, Mark Daley for Figure 2.

References

[1] Csuhaj-Varju, E., Freund, R., Kari, L., and G. Paun. 1996. DNA computing based on splicing: universality results. In Hunter, L. and T. Klein (editors). *Proceedings of 1st Pacific Symposium on Biocomputing.* World Scientific Publ., Singapore. Pages 179-190.

[2] Eilenberg, S., 1984. *Automata, Languages and Machines.* Academic Press, New York.

[3] Head, T. 1987. Formal language theory and DNA: an analysis of the generative capacity of specific recombinant behaviors. *Bull. Math. Biology* 49: 737-759.

[4] Head, T. (1991). Splicing schemes and DNA. In *Lindenmayer systems* (Rozenberg, G. and Salomaa, A., Eds.). Springer Verlag, Berlin. Pages 371-383.

[5] Head, T., Paun, G. and Pixton, D. 1997. Language theory and molecular genetics. In *Handbook of Formal Languages* (Rozenberg, G. and Salomaa, A. Eds.), vol 2., Springer Verlag, Berlin. Pages 295-358.

[6] Hoffman, D.C., and D.M. Prescott. 1997. Evolution of internal eliminated segments and scrambling in the micronuclear gene encoding DNA polymerase α in two *Oxytricha* species. *Nucl. Acids Res.* 25: 1883-1889.

[7] Kari, L., and G. Thierrin. 1996. Contextual insertion/deletions and computability. *Information and Computation* 131: 47-61.

[8] Landweber, L.F., Kari, L. 1998. The evolution of cellular computing: nature's solution to a computational problem. Proceedings of 4th DIMACS meeting on DNA based computers, Philadephia. Pages 3-15.

[9] Meyer, E. and Duharcourt, S. 1996. Epigenetic Programming of Developmental Genome Rearrangements in Ciliates. *Cell* 87 : 9-12.

[10] Păun, G. 1995. On the power of the splicing operation. *Int. J. Comp. Math* 59 : 27-35.

[11] Pixton, D., 1995. Linear and circular splicing systems. Proceedings of the First International Symposium on Intelligence in Neural and Biological Systems, IEEE Computer Society Press, Los Alamos. Pages 181-188.

[12] Salomaa, A. 1973. *Formal Languages.* Academic Press, New York.

[13] Siromoney, R., Subramanian, K.G. and Rajkumar Dare, V. 1992. Circular DNA and splicing systems. In Parallel Image Analysis. Lecture Notes in Computer Science 654, Springer Verlag, Berlin. Pages 260-273.

[14] Yokomori, T., Kobayashi, S., and Ferretti, C. 1997. Circular Splicing Systems and DNA Computability. In Proc. of IEEE International Conference on Evolutionary Computation'97. Pages 219-224.

(L. Kari) Department of Computer Science, University of Western Ontario, London, ON, N6A 5B7 Canada

E-mail address: `lila@csd.uwo.ca`

URL: `http://www.csd.uwo.ca/~ lila`

(L. Landweber) Department of Ecology and Evolutionary Biology, Princeton University, NJ 08544-1003 USA

E-mail address: `lfl@princeton.edu`

URL: `http://www.princeton.edu/~ lfl`

DIMACS Series in Discrete Mathematics
and Theoretical Computer Science
Volume **54**, 2000

Membrane Computing Based on Splicing

Gheorghe Păun and Takashi Yokomori

ABSTRACT. Membrane computing devices were recently introduced (under the name of P systems) as distributed parallel computing models of a biochemical type. Multisets of objects are placed in a cell-like hierarchical structure of membranes and they evolve according to given rules, which are applied in a synchronous manner: at each step, all objects which can evolve, from all membranes, must evolve. The result of a computation consists of objects which exit the system. Here we consider objects described by strings, which evolve by means of splicing rules. Both usual P systems based on membrane structures and systems working on planar maps are considered. Several simple variants are shown to be computationally universal: systems with only a few number of membranes (described by a star tree or a linear tree) or working on maps described by asymmetric graphs and without any target indication when communicating strings characterize the recursively enumerable languages.

1. Introduction

The P systems are a class of distributed parallel computing devices of a biochemical type which were recently introduced in [**7**].

In short, in the basic model one considers a *membrane structure* consisting of several cell-membranes which are hierarchically embedded in a main membrane, called the *skin* membrane. Figure 1 illustrates such a structure. The membranes delimit *regions*, where we place *objects*, elements of a finite set (an alphabet).

The objects evolve according to given *evolution rules*, which are associated with the regions. An object can evolve independently of the other objects in the same region of the membrane structure, or in cooperation with other objects. In particular, we can consider *catalysts*, objects which evolve only together with other objects, but are not modified by the evolution (they just assist other objects to evolve). The evolution rules are given in the form of multiset transition rules, they can be the subject of a given priority relation, and in their right hand members contain symbols $a_{here}, a_{out}, a_{in_j}$, where a is an object. The meaning is that one occurrence of the symbol a is produced and remains in the same region, is sent out of the respective membrane, or is sent to membrane j (which should be reachable from the region where the rule is applied), respectively. The membranes can be *dissolved* (by the special action δ). When such an action takes place, all the objects of the dissolved membrane remain free in the membrane placed immediately outside, but the evolution rules of the dissolved membrane are lost. The skin membrane

2000 *Mathematics Subject Classification.* Primary 68Q05, 68Q10, 68Q85; Secondary 92C37, 92C40.

© 2000 American Mathematical Society

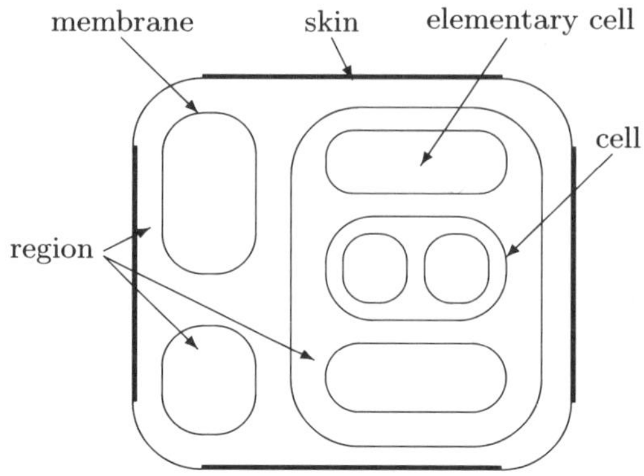

FIGURE 1. A membrane structure.

is never dissolved. Note that the membranes are both separators and channels of communication.

The application of evolution rules is done in a maximally parallel manner: at each step, all objects which can evolve should evolve.

Starting from an initial configuration and using the evolution rules, we get a *computation*. A computation is considered completed when it halts, no further rule can be applied. There are two possible ways of assigning a result to a computation: by considering the multiplicity of objects present in a designated membrane in a halting configuration, or by concatenating the symbols which leave the system, in the order they are sent out of the skin membrane (if several symbols are expelled at the same time, then any ordering of them is accepted). Thus, in the first case we compute vectors of natural numbers, while in the second case we generate a language.

Several variants of P systems were already considered; see the early survey from [**8**] and the references given at the end of the present paper[1]. In most of the cases, we obtain characterizations of recursively enumerable (RE) number relations or representations of permutation closures of RE languages (because we work with multisets, we cannot guarantee the ordering of symbols in strings, even if we read the result outside the system).

When membrane division is possible (we do not use this feature here), NP-complete problems can be solved in polynomial time; this was proved in [**10**] for SAT, which can be solved in linear time.

In [**7**] one also considers the case when the objects are not "atomic" symbols, but they are described by strings. Two variants are investigated: with the evolution rules based on context-free rewriting rules or on the splicing operation. In both cases, characterizations of RE languages are obtained.

The P systems of the type considered in [**7**] have a basic unrealistic feature, the indication of targets when communicating objects, in the form a_{in_j}. For symbol-objects, this has been improved in [**9**] by considering polarized membranes (signs + or − are associated with them) and also assigning an "electrical charge" to objects (also + or −). An object of a given polarity will enter any of the adjacent lower

[1]Answerring a suggestion of a referee, we give a complete bibliography of the domain, at the end of May 1999, even if several titles are not quoted here.

membranes of an opposite polarity. Again computational universality is obtained, but at the price of using membranes of a variable thickness: the evolution rules can have special action-symbols, δ, τ: when introduced, δ decreases the thickness while τ increases it. We can communicate objects only through membranes of thickness one.

A still more unrestricted (hence "realistic") mode of communicating was considered in [**13**]: when an object a must be sent to a lower membrane, just indicate a_{in}; the object will pass to any of the adjacent lower membrane, non-deterministically chosen (we may claim that this is a model of communicating by pressure, by gradient reasons). The indications *here* and *out* are used as in the initial model.

We prove here the same result as in [**7**] – computational universality – for the case of P systems using string-objects evolving by splicing and the non-deterministic way of communicating described above, by indicating a_{in} only, without the address of the target membrane; moreover, we get characterizations of RE languages by systems of a very simple form (as the number of membranes and their arrangement).

It was proposed in [**13**] to consider P systems working on other types of structures, for instance, planar maps described by asymmetric graphs. We also investigate such a case here, again proving universality (this time, the degree of the system is no longer bounded by a small value).

In some sense, these results are not surprising: we already know that the splicing operation is very powerful, weak controls on it entail computational universality, see [**11**] and its references. However, the simple architectures of the P systems considered below show that the combination of the splicing with the P system model is elegant and powerful even in the "minimalistic" framework we work here.

2. Splicing P Systems

Let us first remind the splicing operation introduced in [**5**] as a formal model of the DNA recombination under the influence of restriction enzymes and ligases.

Consider an alphabet V and two symbols $\#, \$$ not in V. A *splicing rule* over V is a string $r = u_1\#u_2\$u_3\#u_4$, where $u_1, u_2, u_3, u_4 \in V^*$ (V^* is the set of all strings over V). For such a rule r and for $x, y, w, z \in V^*$ we define

$$(x, y) \vdash_r (w, z) \quad \text{iff} \quad x = x_1u_1u_2x_2,\ y = y_1u_3u_4y_2,\ w = x_1u_1u_4y_2,\ z = y_1u_3u_2x_2, \\ \text{for some } x_1, x_2, y_1, y_2 \in V^*.$$

(One cuts the strings x, y in between u_1, u_2 and u_3, u_4, respectively, and one recombines the fragments obtained in this way.) When r is understood, we write $\vdash$ instead of $\vdash_r$. For clarity, we usually indicate by a vertical bar the place of splicing: $(x_1u_1|u_2x_2, y_1u_3|u_4y_2) \vdash (x_1u_1u_4y_2, y_1u_3u_2x_2)$.

A pair $\sigma = (V, R)$, where V is an alphabet and R is a set of splicing rules over V, is called an *H scheme*. With respect to an H scheme $\sigma = (V, R)$ and a language $L \subseteq V^*$ we define

$$\begin{aligned} \sigma(L) = \quad & \{w \in V^* \mid (x, y) \vdash_r (w, z) \text{ or } (x, y) \vdash_r (z, w), \\ & \text{for some } x, y \in L, r \in R, z \in V^*\}, \\ \sigma^*(L) = \quad & \bigcup_{i \geq 0} \sigma^i(L), \text{ for} \\ \sigma^0(L) = \quad & L \text{ and } \sigma^{i+1}(L) = \sigma^i(L) \cup \sigma(\sigma^i(L)),\ i \geq 0. \end{aligned}$$

An *extended H system* is a construct $\gamma = (V, T, A, R)$, where V is an alphabet, $T \subseteq V, A \subseteq V^*$, and $R \subseteq V^*\#V^*\$V^*\#V^*$. ($T$ is the *terminal* alphabet, A is the set of *axioms*, and R is the set of *splicing rules*.) When $T = V$, the system is said to be *non-extended*. The pair $\sigma = (V, R)$ is the *underlying H scheme* of γ.

The language generated by γ is defined by $L(\gamma) = \sigma^*(A) \cap T^*$. (We iterate the splicing operation according to rules in R, starting from strings in A, and we keep only the strings composed of terminal symbols.)

It is known that extended H systems with finite sets of axioms and of splicing rules characterize the regular languages, [**1**], [**18**].

Let us now pass to splicing P systems, the object of our investigation.

We identify a membrane structure with a string of correctly matching parentheses, placed in a unique pair of matching parentheses; each pair of matching parentheses corresponds to a membrane. Graphically, a membrane structure is represented by a Venn diagram.

A *P system* is a membrane structure with multisets of *objects* placed in its regions and provided with *evolution rules* for these objects. We define here only the splicing P systems, in the variant we investigate in this paper.

A *splicing P system* (of degree $m, m \geq 1$) is a construct

$$\Pi = (V, T, \mu, L_1, \ldots, L_m, R_1, \ldots, R_m),$$

where:

(i) V is an alphabet; its elements are called *objects*;
(ii) $T \subseteq V$ (the *output* alphabet);
(iii) μ is a membrane structure consisting of m membranes (labeled with $1, 2, \ldots, m$);
(iv) $L_i, 1 \leq i \leq m$, are languages over V associated with the regions $1, 2, \ldots, m$ of μ;
(v) $R_i, 1 \leq i \leq m$, are finite sets of *evolution rules* associated with the regions $1, 2, \ldots, m$ of μ, given in the following form: $(r; tar_1, tar_2)$, where $r = u_1\#u_2\$u_3\#u_4$ is a usual splicing rule over V and $tar_1, tar_2 \in \{here, out\} \cup \{in_j \mid 1 \leq j \leq m\}$.

Note that, as usual in H systems, when a string is present in a region of our system, it is assumed to appear in arbitrarily many copies (any number of copies of a DNA molecule can be obtained by amplification). Thus, we do not use here multisets, as in basic P systems.

Any m-tuple $(M_1, \ldots, M_m)$ of languages over V is called a *configuration* of Π. For two configurations $(M_1, \ldots, M_m), (M'_1, \ldots, M'_m)$ of Π we write $(M_1, \ldots, M_m) \Longrightarrow (M'_1, \ldots, M'_m)$ if we can pass from $(M_1, \ldots, M_m)$ to $(M'_1, \ldots, M'_m)$ by applying the splicing rules from each region of μ, in parallel, to all possible strings from the corresponding regions, and following the target indications associated with the rules. More specifically (but not completely formal), if $x, y \in M_i$ and $(r = u_1\#u_2\$u_3\#u_4, tar_1; tar_2) \in R_i$ such that we can have $(x, y) \vdash_r (w, z)$, then w and z will go to the regions indicated by tar_1, tar_2, respectively. If $tar_j = here$, then the string remains in M_i, if $tar_j = out$, then the string is moved to the region immediately outside membrane i (maybe, in this way the string leaves the system), if $tar_j = in_k$, then the string is moved to the region k, providing that this is immediately below; if not, then the rule cannot be applied. Note that the strings x, y

are still available in region M_i, because we have supposed that they appear in arbitrarily many copies (an arbitrarily large number of them were spliced, arbitrarily many remain), but if a string w, z is sent out of region i, then no copy of it remains here.

A sequence of transitions between configurations of a given P system Π, starting from the initial configuration $(L_1, \ldots, L_m)$, is called a *computation* with respect to Π. The result of a computation consists of all strings over T which are sent out of the system at any time during the computation. We denote by $L(\Pi)$ the language of all strings of this type. We say that $L(\Pi)$ is *generated* by Π.

Note two facts: if a string leaves the system but it is not terminal, then it is ignored; if a string remains in the system, even if it is terminal, then it does not contribute to the language $L(\Pi)$. It is also worth mentioning that we do not consider here halting computations. We leave the process to continue forever and we just observe it from outside and collect the terminal strings which leave it.

We denote by $SPL(tar, m, p)$ the family of languages $L(\Pi)$ generated by splicing P systems as above, of degree at most $m, m \geq 1$, and of depth at most $p, p \geq 1$. (The depth of a P system is equal to the height of the tree describing its membrane structure.)

If all target indications tar_1, tar_2 in the evolution rules of a P system are of the form *here, out, in*, then we say that Π is of the *i/o type*; the strings produced by splicing and having associated the indication *in* are moved into any lower region immediately below the region where the rule is used. The family of languages generated by P systems with this weaker target indication and of degree at most m and depth at most p is denoted by $SPL(i/o, m, p)$.

Remark 1. It is important to emphasize the differences between the systems we consider here and those investigated in [**7**]; in [**7**] the evolution rules are of the form $(r, y)tar$, where r is a usual splicing rule, y is a string, and tar is a target indication; always we splice current strings x with y, $(x, y) \vdash_r (w, z)$ or $(y, x) \vdash_r (w, z)$, and one keeps only the first result of the operation, w, which is sent to the region indicated by tar. Moreover, one accepts as correct only halting computations. So, we cannot directly compare our systems with those in [**7**]. Also, there is a clear difference between our systems and the distributed H systems (see the corresponding chapter of [**11**]). For instance, in the communicating distributed H systems (which are closer to P systems as their architecture), the communication is done on a complete graphs, by broadcasting, with filters at the receivers; none of these features is present here.

3. The Power of Splicing P Systems

We denote by *REG, CF, RE* the families of regular, context-free, and recursively enumerable languages, respectively, and by MAT the family of languages generated by matrix grammars without appearance checking. It is known that $REG \subset CF \subset MAT \subset RE$ and that each one-letter matrix language is regular.

We start by four results showing the computational universality of splicing P systems of rather simple forms.

Theorem 1. $SPL(i/o, 3, 3) = RE$.

Proof. Let $G = (N, T, S, P)$ be a type-0 Chomsky grammar and let B be a new symbol. Assume that $N \cup T \cup \{B\} = \{\alpha_1, \ldots, \alpha_n\}$ and that P contains m

rules, $u_i \to v_i, 1 \le i \le m$. Consider also the rules $u_{m+j} \to v_{m+j}, 1 \le j \le n$, for $u_{m+j} = v_{m+j} = \alpha_j$.

We construct the splicing P system (of degree 3)

$$\begin{aligned}
\Pi &= (V, T, \mu, L_1, L_2, L_3, R_1, R_2, R_3),\\
V &= N \cup T \cup \{B, X, X', Y, Z, Z'\} \cup \{X_i \mid 1 \le i \le n+m\} \cup \{Y_i \mid 0 \le i \le n+m\},\\
\mu &= [_1 [_2 [_3\ ]_3]_2]_1,\\
L_1 &= \{XSBY, Z'\} \cup \{ZY_i \mid 0 \le i \le n+m\},\\
L_2 &= \{X'Z\} \cup \{X_i v_i Z, X_i Z \mid 1 \le i \le n+m\},\\
L_3 &= \{ZY\},\\
R_1 &= \{(\#u_i Y\$Z\#Y_i; in, here),\ (\#Y_i\$Z\#Y_{i-1}; in, here) \mid 1 \le i \le n+m\}\\
&\cup \{(\#BY\$Z'\#; here, here),\ (X\#\$\#Z'; here, out)\},\\
R_2 &= \{(X\#\$X_i v_i \#Z; here, out) \mid 1 \le i \le n+m\}\\
&\cup \{(X_i\#\$X_{i-1}\#Z; here, out) \mid 2 \le i \le n+m\}\\
&\cup \{(X_1\#\$X'\#Z; here, in),\ (X'\#\$X\#Z; here, out)\},\\
R_3 &= \{(\#Y_0\$Z\#Y; out, here)\}.
\end{aligned}$$

The idea of the proof is the "rotate-and-simulate" technique, much used in H systems area.

The sentential forms generated by G are simulated in Π in a circular permutation: Xw_1Bw_2Y, maybe with variants of X, Y, will be present in a region of Π if and only if w_2w_1 is a sentential form of G. Note that we can remove the nonterminal symbol Y only together with B from strings of the form $XwBY$. In this way, we ensure that the string is in the correct permutation.

The simulation of rules in P and the rotation are done in the same way. Assume that some string Xwu_iY is present in region 1, $1 \le i \le n+m$; initially we have here the string $XSBY$. We can perform $(Xw|u_iY, Z|Y_i) \vdash (XwY_i, Zu_iY)$; the string XwY_i is sent to region 2, Zu_iY remains in region 1 (and all possible splicings which involve it produce two strings starting with Z). In region 2 we can only perform a splicing of the form $(X|wY_i, X_jv_j|Z) \vdash (XZ, X_jv_jwY_i)$, for some $1 \le j \le n+m$. The string $X_jv_jwY_i$ is sent back to region 1, XZ remain here and will be used later. Now, in region 1 the only splicing which can involve the string $X_jv_jwY_i$ is $(X_jv_jw|Y_i, Z|Y_{i-1}) \vdash (X_jv_jwY_{i-1}, ZY_i)$. The string $X_jv_jwY_{i-1}$ is sent to region 2, ZY_{i-1} remains in region 1 (it is an axiom). In region 2 we now decrease by one the subscript of X_j, by $(X_j|v_jwY_{i-1}, X_{j-1}|Z) \vdash (X_jZ, X_{j-1}v_jwY_{i-1})$. The process of decreasing the subscripts continues.

If at some moment we reach X_1, hence in region 2 we have a string $X_1v_jwY_k$, then we perform $(X_1|v_jwY_k, X'|Z) \vdash (X_1Z, X'v_jwY_k)$ and $X'v_jwY_k$ is sent to membrane 3. If $k \neq 0$, then nothing can be done, the string is "lost". Otherwise, Y_0 is replaced with Y and the string $X'v_jwY$ is sent to region 2; X' is replaced here by X and the string Xv_jwY is sent to the skin membrane.

If at some moment in region 2 we get a string $X_kv_jwY_0$, for $k \ge 2$, then this string cannot be processed in the skin membrane, hence it is "lost".

Thus, we can correctly continue only when $i = j$, hence we have passed from Xwu_iY to Xv_iwY; in this way we have either correctly simulated a rule from P or we have circularly permuted the string with one symbol.

The process of simulating a rule or of rotating the string with one symbol can be iterated. Therefore, all derivations in G can be simulated in Π and, conversely, all correct computations in Π correspond to correct derivations in G. Because we collect only terminal strings which leave the system, we have the equality $L(G) = L(\Pi)$. □

We were not able to minimize the depth of the used system without paying some price. In the next theorem, this price is the use of target information (for saving space, we omit the proof; it is similar to the previous one, but slightly simpler, because we can make use of the powerful indications of the form in_j):

Theorem 2. $SPL(tar, 3, 2) = RE$.

However, at the price of using two more membranes, we can get rid of the target addresses, still using a system of depth 2:

Theorem 3. $SPL(i/o, 5, 2) = RE$.

Proof. Starting from a type-0 grammar $G = (N, T, S, P)$ as above and with B a new symbol, we construct the system (of degree 5)

$$\begin{aligned}
\Pi &= (V, T, \mu, L_1, L_2, L_3, L_4, L_5, R_1, R_2, R_3, R_4, R_5),\\
V &= N \cup T \cup \{B, X, X', Y, Y', Y'', Z, Z'\} \cup \{X_i \mid 1 \leq i \leq n+m\}\\
&\cup\ \{Y_i \mid 0 \leq i \leq n+m\},\\
\mu &= [_1 [_2\ ]_2 [_3\ ]_3 [_4\ ]_4 [_5\ ]_5]_1,\\
L_1 &= \{XSBY, ZY'', Z'\} \cup \{ZY_i \mid 0 \leq i \leq n+m\},\\
L_2 &= \{X_i v_i Z \mid 1 \leq i \leq n+m\},\\
L_3 &= \{X_i Z \mid 1 \leq i \leq n+m\},\\
L_4 &= \{XZ, ZY\},\\
L_5 &= \{X'Z, ZY'\},\\
R_1 &= \{(\#u_i Y\$Z\#Y_i; in, here),\ (\#Y_i\$Z\#Y_{i-1}; in, here) \mid 1 \leq i \leq n+m\}\\
&\cup\ \{(\#Y'\$Z\#Y''; in, here),\ (\#BY\$Z'\#; here, here),\ (X\#\$\#Z'; here, out)\},\\
R_2 &= \{(X\#\$X_i v_i\#Z; here, out) \mid 1 \leq i \leq n+m\},\\
R_3 &= \{(X_i\#\$X_{i-1}\#Z; here, out) \mid 2 \leq i \leq n+m\},\\
R_4 &= \{(X'\#\$X\#Z; here, out),\ (\#Y''\$Z\#Y; here, here)\},\\
R_5 &= \{(X_1\#\$X'\#Z; here, here),\ (\#Y_0\$Z\#Y'; out, here)\}.
\end{aligned}$$

This system works in a way similar to the systems in the proofs of the previous theorems, but the fact that we can continue the simulation of rules in P or the circular permutation of strings only in the correct manner is much more subtle.

In region 1 we start to simulate the application of rules in P and the circular permutation of a string, by cutting from the right hand end of the string a substring $u_i, 1 \leq i \leq n+m$. In membrane 2 we add to the left end of the string a prefix v_j. These operations are "memorized" also in the subscripts of the end markers X and Y. The subscript of X is decreased in membrane 3 while the subscript of Y is decreased in membrane 1. Only when the two subscripts reach the value 0 at the same time the process should correctly continue. This happens, indeed.

In order to see this, let us consider the two possible cases when in region 1 we get a string with one of the subscripts of X and Y of a minimal value (this string should be send to an inner membrane). When the subscript of X minimal, our string is of the form X_1wY_k, with $k \geq 1$. When the subscript of Y is minimal, the string is of the form X_kwY_0, for some $k \geq 1$. Only the case X_1wY_0 should continue correctly. We consider in detail each case:

1. The string $X_1wY_k, k \geq 1$, can enter no splicing in membranes 2, 3, 4, while the only possible splicing in membrane 5 leads to the string $X'wY_k$ which stops here. No continuation is possible.
2. A string of the form X_1wY_0 can enter no splicing in membranes 2, 3, 4. In membrane 5 we have two possibilities:
   2.1. If we produce the string X_1wY' which is sent to the skin membrane, this strings is transformed here into X_1wY'' and it must again enter an inner membrane. No splicing is possible in membranes 2, 3. In membrane 4 we can produce X_1wY, but the string does not leave the membrane. In membrane 5 we can produce $X'wY''$ and again the string is "lost", because it cannot leave the membrane.
   2.2. If in membrane 5 we apply both productions, then we produce the string $X'wY'$ which is sent to the skin membrane. Here we produce $X'wY''$, which should again enter an inner membrane. In membranes 2, 3, 5 we can apply no splicing rule, but in membrane 4 we have two possibilities. If we produce the string XwY'' which is sent to the skin membrane, then the only possible splicing here is that by the rule $(X\#\$\#Z'; here, out)$. The obtained string leaves the system, but it is not terminal. If in membrane 4 we apply both rules, then we get XwY, which is the correct continuation. The process of simulating rules in P and of circularly permuting the string can be continued.
3. A string of the form $X_kwY_0, k \geq 2$, enters no splicing in membranes 2 and 4. In membrane 3 we can produce $X_{k-1}wY_0$, which is sent to the skin membrane, but no further splicing is possible here. In membrane 5 we can produce the string X_kwY' which is sent to the skin membrane. From membrane 1 we return to the inner membranes the string X_kwY''. This string can enter no splicing in membranes 2 and 5. In membrane 3 we can produce $X_{k-1}wY''$, which is sent to the skin membrane, where no further splicing can involve this string. In membrane 4 we can produce X_kwY, but this string does not leave the membrane.

We conclude that all computations in Π which lead to terminal strings which leave the system correspond to terminal derivations in G. Conversely, it is easy to see that each terminal derivation with respect to G can be simulated in Π. In conclusion, $L(G) = L(\Pi)$. □

Another possibility to obtain a characterization of RE by using systems of depth two is to impose the following natural restriction on the functioning of a splicing P system: *if a string enters a splicing operation, then all its copies are exhausted by this operation*; thus, from a step to the next one we pass only strings resulting from splicing operations and strings which enter no splicing. Note that we do not work with multiplicities associated with strings. We denote by $S'PL(i/o, m, p)$ the family of languages generated by splicing P systems of degree at most m and depth at most p working under this restriction.

Theorem 4. $S'PL(i/o, 3, 2) = RE$.

Proof. We start again from a type-0 grammar $G = (N, T, S, P)$; let B be a new symbol. With the same notations as in the previous proofs (in addition, we denote $U = N \cup T \cup \{B\}$), we construct the splicing P system

$$
\begin{aligned}
\Pi &= (V, T, \mu, L_1, L_2, L_3, R_1, R_2, R_3),\\
V &= N \cup T \cup \{B, H, X, Y, Y', Z, Z'\} \cup \{J_i, X_i, Y_i \mid 1 \leq i \leq n+m\},\\
\mu &= [_1[_2\ ]_2[_3\ ]_3]_1,\\
L_1 &= \{ZY, Z'\} \cup \{ZY_i \mid 1 \leq i \leq n+m\},\\
L_2 &= \{XSBY, ZY, Z'\} \cup \{Xv_iX_iZ \mid 1 \leq i \leq n+m\},\\
L_3 &= \{ZY'\} \cup \{HaX_iJ_i \mid a \in U, 1 \leq i \leq n+m\},\\
R_1 &= \{(\#u_iY\$Z\#Y_i; in, here),\ (Z\#Y_i\$Z\#Y_i; here, here) \mid 1 \leq i \leq n+m\}\\
&\cup\ \{(\#Y'\$Z\#Y; in, here),\ (Z\#Y\$Z\#Y; here, here),\\
&\qquad (\#BY\$Z'\#; out, here),\ (\#Z'\$\#Z'; here, here)\}\\
&\cup\ \{(\#X_iaJ_i\$H\#aX_i; in, here) \mid a \in U, 1 \leq i \leq n+m\},\\
R_2 &= \{(X\#\$Xv_iX_i\#Z; here, here),\ (Xv_iX_i\#Z\$Xv_iX_i\#Z; here, here) \mid\\
&\qquad 1 \leq i \leq n+m\}\\
&\cup\ \{(\#Y\$Z\#Y; out, here),\ (X\#\$\#Z'; here, out)\},\\
R_3 &= \{(X_ia\#\$HaX_i\#J_i; out, out),\ (HaX_i\#J_i\$HaX_i\#J_i; here, here) \mid\\
&\qquad a \in U, 1 \leq i \leq n+m\}\\
&\cup\ \{(\#X_iY_i\$Z\#Y'; out, here) \mid 1 \leq i \leq n+m\} \cup \{(Z\#Y'\$Z\#Y'; here, here)\}.
\end{aligned}
$$

The construction is based on a combination of the rotate-and-simulate technique and the way of checking the correctness of using these operations as introduced in [**3**].

Assume that we have a string of the form XwY in membrane 2; initially, $w = SB$. The string can be sent out changed or not changed. In the first case, a prefix v_iX_i is added to w, for some $1 \leq i \leq n+m$. In the second case, no terminal string can be produced: a string from the skin membrane can be sent to membrane 2 only if it ends with Y_j, for some $1 \leq j \leq n+m$, and such a string can exit membrane 2 only by removing the left marker X; then, the symbol Y_j can never be removed. Thus, we have to leave membrane 2 with a string of the form Xv_iX_izY. In the skin membrane we can use the rule $(\#BY\$Z'\#; out, here)$ only if B is adjacent to Y, but the string sent out of the system is not terminal. Thus, we have to replace a suffix u_jY with Y_j, for some $1 \leq j \leq n+m$. The obtained string $Xv_iX_izY_j$ is sent to lower membranes, but if it enters membrane 2, then it cannot leave it again (otherwise than by removing X, and this will not lead to a terminal string: even if we introduce Y' in membrane 3 and then we introduce the string again in a lower membrane, the string goes never out). In membrane 3 we start moving the symbol X_i to the right, step by step, by using the rule $(X_ia\#\$HaX_i\#J_i; out, out)$. For instance, in a general case we perform the operation: $(Xv_ix_1X_ia|x_2Y_j, HaX_i|J_i) \vdash (Xv_ix_1X_iaJ_i, HaX_ix_2Y_j)$. The resulting strings are sent to the skin membrane, where the only possible splicing is $(Xv_ix_1|X_iaJ_i, H|aX_ix_2Y_j) \vdash (Xv_ix_1aX_ix_2Y_j, HX_iaJ_i)$. The symbols X_i and a were interchanged. The process of moving X_i to the right continues until X_i reaches Y_j. The only way to continue is by using the rule $(\#X_iY_i\$Z\#Y'; out, here)$. In this way we check the equality of i and j. In the positive case we send to the skin membrane the string Xv_izY' and from membrane 1 we send to membrane 2 the string Xv_izY. This means that either a rule of G was correctly simulated, or a symbol was correctly moved from the right end of the string to the left end.

The correctness of this process essentially relies upon the fact that from a step to the next one we do not pass strings which were already spliced. In this way, we do not mix strings of different "generations". This is crucial, for instance, in the skin membrane, when two strings $Xv_ix_1X_iaJ_i, HaX_ix_2Y_j$ are spliced in such a way to "reconstruct" the string which was cut in parts in membrane 3.

We can continue as long as we want. In any moment, the symbol X is removed in membrane 2. The resulting string can lead to a string which can be sent out of the system only if Y will not be replaced again by some Y_j and only if B is adjacent to Y (hence the string is in the same circular permutation as the sentential form of G whose derivation was simulation in Π). If the string sent out of the system is terminal, then it will be included in $L(\Pi)$.

All the axioms different from $XSBY$ are passed from a step to the next one by the rules of the form $(x_1\#x_2\$x_1\#x_2; here, here)$, present in each region.

We conclude that $L(G) = L(\Pi)$. □

We do not know whether or not these results can be improved. Here are some information about the power of systems of a lower degree.

It is obvious that for degrees 1 (no communication) and 2 (no choice, because we have only one inner membrane), we have $SPL(tar, j, j) = SPL(i/o, j, j), j = 1, 2$.

Theorem 5. $SPL(i/o, 1, 1) = REG$.

Proof. The inclusion $\supseteq$ can be directly proved (see, for instance, the construction in the proof of Lemma 7.18 from [**11**]; add the indication (*here, here*) to all splicing rules not producing a terminal string and (*out, here*) to the rules producing a terminal string).

Conversely, let us consider a P system $\Pi = (V, T, [_1\ ]_1, L_1, R_1)$ and associate with it the non-extended H system $\gamma = (V, L_1, R_1)$. All the strings ever present in Π form the (regular) language $L(\gamma)$. Consider now all the strings which can be obtained by one splicing according to the rules in R_1 starting from the strings in $L(G)$; from each such splicing we collect the first resulting string if using a rule $(r; out, here)$, the second one if using a rule $(r; here, out)$, and both the resulting strings when the rule is of the form $(r; out, out)$. The obtained language, let us denote it by $splic(L(\gamma))$, is still regular (this is easy to be seen, because we consider only one-step splicing). The intersection of $splic(L(\gamma))$ with T^* equals $L(\Pi)$ and is a regular language. □

As expected, the increase of the degree increases the power of P systems:

Theorem 6. $SPL(i/o, 2, 2) - MAT \neq \emptyset$.

Proof. Consider the system

$$\begin{aligned} P &= (\{a, b, c, c', d, d', e\}, \{a, b, c, d\}, [_1[_2\ ]_2]_1, \\ &\quad \{cabd, ce, ed, ed'\}, \{caae, c'be, ed\}, R_1, R_2), \\ R_1 &= \{(\#ad\$e\#d; in, here), (c'\#\$c\#e; here, here), (\#bd\$e\#d'; in, here), \\ &\quad (bd\#\$bd\#; out, out)\}, \\ R_2 &= \{(c\#\$caa\#e; here, out), (c\#\$c'b\#e; here, here), (\#d'\$e\#d; out, here)\}. \end{aligned}$$

We obtain the language $L(\Pi) = \{ca^{2^n}bd \mid n \geq 0\}$, which is not in MAT (this family is closed under morphisms – restricted morphisms, here – and we know that the one-letter languages in MAT are regular; this is not the case with $\{a^{2^n} \mid n \geq 0\}$).

The idea of the work of this system is that we repeatedly cut one occurrence of a from the right end of a string and we append a^2 in its left end; similarly, we cut b from the right end and we append b in the left end. The two operations are not mixed.

For instance, take a string ca^iba^jad. We perform $(ca^iba^j|ad, e|d) \vdash (ca^iba^jd, ead)$ in membrane 1 and ca^iba^jd is sent to membrane 2. If we append here $c'b$ in the left end, then the string cannot leave the membrane. If in membrane 2 we add caa in the left end, then this is the desired operation. If in a string ca^ibd we replace bd with d' in membrane 1, then in membrane 2 we can perform $(c|a^id', caa|e) \vdash (ce, caaa^id')$, but nothing can be done in membrane 1 with the string $caaa^id'$. If in membrane 2 we replace d' with d and send the string out, then we can never add a symbol b to the string, so the string cannot be sent out of the system, hence it does not participate to the language $L(\Pi)$. Thus, in membrane 2 we have first to add a copy of b in the left end of the string and then to replace d' by d; we get cba^id, which is what we want.

Thus, the number of copies of a is doubled at each phase ended with the move of b from an end of the string to the other end. At any time, any string of the form ca^ibd can be sent out of the system. It is clear that for each such a string, i is a power of 2. □

4. P Systems on Asymmetric Graphs

The observation that a membrane structure can be described by a tree implies the possibility of considering a P system as a computing device working on a tree (this loses the biochemical intuition, but can be mathematically appealling) and this suggests a generalization, already examined in [**13**] for the case of symbol-objects: consider a similar computing machinery working on a graph of a form different from that of a tree. One class of graphs particularly interesting from the point of view of molecular computing is that of asymmetric planar graphs (for each two nodes i, j, at most one of $(i, j), (j, i)$ is an edge). Such a graph is a representation of a planar map such that each border segment can be passed in one direction only. (From a practical point of view, this remark is attractive, because we can devise a support for the computation consisting of several adjacent "rooms" whose "walls" permit the passing of objects in one direction only.) When formally giving such a system, instead of a membrane structure we indicate the underlying graph.

Because on a planar map we do not have levels of regions, when moving objects we do not use the indications *here, out, in*, but *here, out, go*; *here* indicates that an object remains in the same region, *out* refers to objects which have to leave the system (through a wall which permits communication to the region outside the system; if no such a wall exists for the region where we work, then a rule which introduced *out* cannot be used); *go* indicates that the object must go to another room, a neighboring one, non-deterministically choosing it among the rooms to which we can move through a wall which permits communication.

Modulo these remarks, the functioning of a P system (actually, we say P′ system, in order to point out that we do not work on a "pure" membrane structure) using an asymmetric graph is identical to the functioning of an usual P system. We denote by $SP'L(go, m)$ the family of languages generated by splicing P′ systems of degree (the number of vertices in the underlying graph) at most $m, m \geq 1$ (the depth makes no sense in this case; other graph complexity parameters can be considered – for instance, the maximal in- or out-degree – but we do not start such a study here). The union of all these families is denoted by $SP'L(go, *)$.

We start investigating the power of splicing P′ systems with two results similar to Theorems 5 – 6.

Theorem 7. $SP'L(go, 1) = SP'L(go, 2) = REG$.

Proof. If a system has only two rooms, then they can communicate in only one direction, or they do not communicate at all. Each room produces a regular language (which can be generated by a non-extended H system – see again the informal proof of Theorem 5). If the two rooms do not cooperate, then outside the system we get the union of languages generated by two one-room systems, which is a regular language (the same argument as in the proof of Theorem 5). If we pass from one room to another one, then a regular language (that produced in the first room) is iteratively spliced in the second one, according to a finite set of rules. This leads to a regular language, too, see [**1**], [**18**]. Again what we get out is a regular language. Thus, we obtain $SP'L(go, 2) \subseteq REG$.

The inclusions $REG \subseteq SP'L(go, 1) \subseteq SP'L(go, 2)$ are obvious. □

Theorem 8. $SP'L(go, 3) - MAT \neq \emptyset$.

Proof. We take the P system Π from the proof of Theorem 6 and we construct a P′ system Π′ with three rooms, described by the ring $(\{1, 2, 3\}, \{(1, 2), (2, 3), (3, 1)\})$. In room 1 we put the strings and the rules from membrane 1 of Π (with *in* replaced by *go*), in room 2 we put the strings and the rules from membrane 2 (with *out* replaced by *go*), and in room 3 we only put rules which pass the strings unchanged to room 1. (The details are left to the reader.) It is obvious that the system Π′ is equivalent with the system Π. □

As expected, also splicing P′ systems are computationally universal. However, our proof of this assertion does not give a (small) bound on the degree of the necessary systems.

Theorem 9. $SP'L(go, *) = RE$.

Proof. Let $G = (N, T, S, P)$ be a type-0 Chomsky grammar, B a new symbol, $u_i \to v_i, 1 \leq i \leq m$, the rules in P and $u_{m+j} \to v_{m+j}, 1 \leq j \leq n$, new rules, with $u_{m+j} = v_{m+j} = \alpha_j$, for $N \cup T \cup \{B\} = \{\alpha_1, \dots, \alpha_n\}$. We construct the P′ system (of degree $2(n + m + 1)$)

$$\begin{aligned}
\Pi &= (V, T, g, L_0, L_{0'}, L_1, L_{1'}, \dots, L_{(n+m+1)}, L_{(n+m+1)'}, R_0, R_{0'}, R_1, R_{1'},\\
&\quad \dots, R_{(n+m+1)}, R_{(n+m+1)'}),\\
V &= N \cup T \cup \{B, X, Y, Z, Z'\} \cup \{X_i, Y_i \mid 0 \leq i \leq n + m\},\\
g &= (\{0, 0', 1, 1', \dots, (n + m + 1), (n + m + 1)'\},\\
&\quad \{(i, i') \mid 0 \leq i \leq n + m + 1\}\\
&\quad \cup \{(i', i + 1),\ (i', n + m + 1) \mid 0 \leq i \leq n + m\} \cup \{((n + m + 1)', 0)\},
\end{aligned}$$

$$
\begin{aligned}
L_0 &= \{XSBY, Z'\} \cup \{ZY_i \mid 1 \le i \le n+m\},\\
L_{0'} &= \{X_i v_i Z \mid 1 \le i \le n+m\},\\
L_i &= \{ZY_j \mid 0 \le j \le n+m\},\ 1 \le i \le n+m,\\
L_{i'} &= \{X_j Z \mid 0 \le j \le n+m\},\ 1 \le i \le n+m,\\
L_{(n+m+1)} &= \{XZ\},\\
L_{(n+m+1)'} &= \{ZY\},\\
R_0 &= \{(\#u_iY\$Z\#Y_i; go, here) \mid 1 \le i \le n+m)\}\\
&\cup \{(\#BY\$Z'\#; here, here),\ (X\#\$\#Z'; here, out)\},\\
R_{0'} &= \{(X\#\$X_iv_i\#Z; here, go) \mid 1 \le i \le n+m\},\\
R_i &= \{(\#Y_j\$Z\#Y_{j-1}; go, here) \mid 1 \le j \le n+m\},\ 1 \le i \le n+m,\\
R_{i'} &= \{(X_j\#\$X_{j-1}\#Z; here, go) \mid 1 \le j \le n+m\},\ 1 \le i \le n+m,\\
R_{(n+m+1)} &= \{(X_0\#\$X\#Z; here, go)\},\\
R_{(n+m+1)'} &= \{(\#Y_0\$Z\#Y; go, here)\}.
\end{aligned}
$$

The graph in this system and a planar map associated with it are given in Figures 2 and 3, respectively.

The system works as follows. Any string of the form Xwu_iY from region 0 (initially we have $XSBY$) enters here a splicing $(Xw|u_iY, Z|Y_i) \vdash (XwY_i, Zu_iY)$, for some $1 \le i \le n+m$; the string XwY_i passes to region $0'$, where we perform $(X|wY_i, X_jv_j|Z) \vdash (XZ, X_jv_jwY_i)$, for some $1 \le j \le n+m$. The string $X_jv_jwY_i$ passes now from region to region, clockwise, and in regions k the subscript of Y is decreased by one, while in regions k' the subscript of X is decreased by one. If in a region k we receive a string of the form X_rzY_0, then this string cannot be procesed. Similarly, if a string of the form X_0zY_s arrives in a region k', then it cannot be processed. In any moment, from a region k' we can pass to region $(n+m+1)$. Only strings of the form X_0zY_s can be spliced here: $(X_0|zY_s, X|Z) \vdash (X_0Z, XzY_s)$. The string XzY_s is passed to region $(n+m+1)'$, where we can process this string only if $s=0$. This means that the subscripts of X and Y have reached the value 0 at the same time, that is we have $i=j$. The simulation of rules in P and the circular permutation of symbols is done in a correct manner.

Thus, any derivation in G can be simulated in Π, with the sentential form circularly permuted, and, conversely, each computation in Π corresponds to a correct derivation in G.

A string can leave the system only from region 0; if no nonterminal is present (hence also B and Y were removed), this means that the symbol B, Y were adjacent when removing them, that is, the obtained string is in the same permutation as in G. We conclude that $L(G) = L(\Pi)$. □

Remark 2. If we start the construction in the previous proof from a universal type-0 grammar, then we get a universal (hence a *programmable*) P′ system, of a fixed degree, so the hierarchy on the degree collapses also in this case. Finding a reasonably small value for m such that $RE = SP'L(go, m)$ remains as an open problem (the previous proof gives a huge value for m).

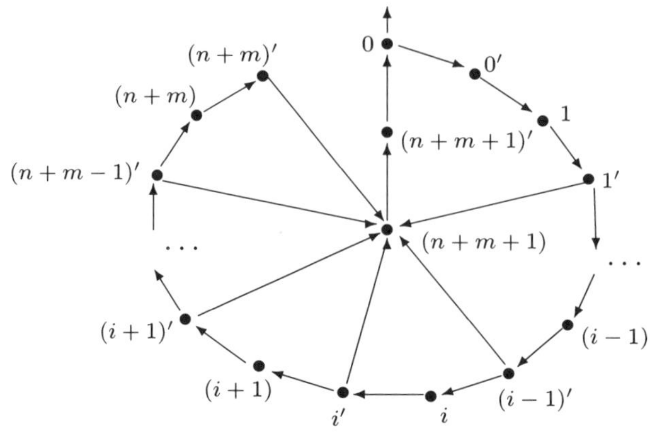

FIGURE 2. The graph of the system in the proof of Theorem 9.

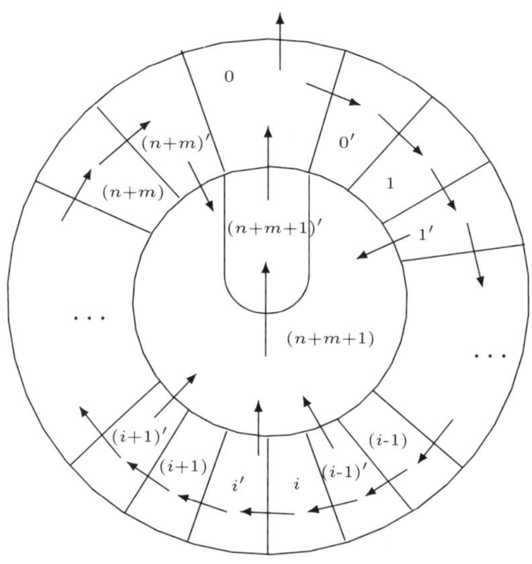

FIGURE 3. The planar map of the system in the proof of Theorem 9.

5. Final Remarks

We have considered splicing P systems of a rather simple form (in what concerns the underlying tree and the way of communicating strings from a region to another one), as well as variants of such systems working on planar maps described by asymmetric graphs. In all cases, characterizations of recursively enumerable languages were found. Because the proofs are constructive and start from type-0 grammars, the existence of universal P systems of these types is obtained as a corollary of the proofs.

We stress the naturalness of splicing P systems, which are intrinsically bio-chemical-like : both the membrane computing and the splicing operation are bio-chemically inspired. Of course, this means nothing in what concerns the possibility of implementing such computing devices. This is a question (of a definite importance) far beyond the scope of this paper. (As far as we know, up to now no implementation attempt has been made.)

Acknowledgements

Research supported by "Research for Future" Program no. JSPS-RFTF 96I00101, from the Japan Society for the Promotion of Science.

References

[1] K. Culik II, T. Harju, Splicing semigroups of dominoes and DNA, *Discrete Appl. Math.*, 31 (1991), 261–277.

[2] J. Dassow, Gh. Păun, On the power of membrane computing, *J. Univ. Computer Sci.*, 5, 2 (1999), 33–49 (www.iicm.edu/jucs).

[3] C. Ferretti, S. Kobayashi, G. Mauri, T. Yokomori, On the universality of Post and splicing systems, *Theoretical Computer Sci.*, to appear.

[4] F. Freund, Generalized P systems, *Proc. of FCT'99 Conf.*, Iaşi, Romania (to appear in *LNCS*, Springer-Verlag, 1999).

[5] T. Head, Formal language theory and DNA: an analysis of the generative capacity of specific recombinant behaviors, *Bull. Math. Biology*, 49 (1987), 737–759.

[6] M. Maliţa, Membrane computing in Prolog, manuscript, 1999.

[7] Gh. Păun, Computing with membranes, submitted, 1998 (see also *TUCS Research Report* No. 208, November 1998, www.tucs.fi).

[8] Gh. Păun, Computing with membranes. An introduction, *Bulletin of the EATCS*, 67 (Febr. 1999), 139–152.

[9] Gh. Păun, Computing with membranes. A variant, submitted, 1999 (see also *CDMTCS Research Report* No. 098, 1999, Auckland Univ., New Zealand, www.cs.auckland.ac.nz/CDMTCS).

[10] Gh. Păun, P systems with active membranes: Attacking NP complete problems, submitted 1999 (see also *CDMTCS Research Report* No. 102, 1999, Auckland Univ., New Zealand, www.cs.auckland.ac.nz/CDMTCS).

[11] Gh. Păun, G. Rozenberg, A. Salomaa, *DNA Computing. New Computing Paradigms*, Springer-Verlag, Heidelberg, 1998.

[12] Gh. Păun, G. Rozenberg, A. Salomaa, Membrane computing with external output, submitted, 1999 (see also *TUCS Research Report* No. 218, December 1998, www.tucs.fi).

[13] Gh. Păun, Y. Sakakibara, T. Yokomori, P systems on graphs of restricted forms, submitted, 1999.

[14] Gh. Păun, G. Thierrin, Multiset processing by means of systems of finite state transducers, *Workshop on Implementing Automata* WIA99, Potsdam, August 1999 (see also *CDMTCS Research Report* No. 101, 1999, Auckland Univ., New Zealand, www.cs.auckland.ac.nz/CDMTCS)..

[15] Gh. Păun, T. Yokomori, Simulating H systems by P systems, *Journal of Universal Computer Science*, 5 (1999), to appear (www.iicm.edu/jucs).

[16] Gh. Păun, S. Yu, On synchronization in P systems, *Fundamenta Informaticae*, 34 (1999) (see also *University of Western Ontario Report* TR 539, 1999, www.csd.uwo.ca/faculty/syu/TR539.html).

[17] I. Petre, A normal form for P systems, *Bulletin of the EATCS*, 67 (Febr. 1999), 165 - 172.

[18] D. Pixton, Regularity of splicing languages, *Discrete Appl. Math.*, 69 (1996), 101–124.

(G. Păun) Institute of Mathematics of the Romanian Academy, PO Box 1 - 764, 70700 Bucureşti, Romania

E-mail address: `gpaun@@imar.ro`

(T. Yokomori) Department of Mathematics, School of Education, Waseda University, 1-6-1 Nishi-waseda, Shinjuku-ku, Tokyo 169-8050, Japan

E-mail address: `yokomori@@mn.waseda.ac.jp`

DIMACS Series in Discrete Mathematics
and Theoretical Computer Science
Volume **54**, 2000

DNA-based Cryptography

Ashish Gehani, Thomas H. LaBean, and John H. Reif

ABSTRACT. Recent research has considered DNA as a medium for ultra-scale computation and for ultra-compact information storage. One potential key application is DNA-based, molecular cryptography systems. We present some procedures for DNA-based cryptography based on one-time-pads that are in principle unbreakable. Practical applications of cryptographic systems based on one-time-pads are limited in conventional electronic media, by the size of the one-time-pad; however DNA provides a much more compact storage media, and an extremely small amount of DNA suffices even for huge one-time-pads. We detail procedures for two DNA one-time-pad encryption schemes: (i) a substitution method using libraries of distinct pads, each of which defines a specific, randomly generated, pair-wise mapping; and (ii) an XOR scheme utilizing molecular computation and indexed, random key strings. These methods can be applied either for the encryption of (appropriately recoded) natural DNA, and also for the encryption of DNA encoding binary data. In the latter case, we also present a novel use of chip-based DNA micro-array technology for 2D data input and output. Finally, we examine a class of DNA steganography systems, which secretly tag the input DNA and then disguise it (without further modifications) within collections of other DNA. We consider potential limitations of these steganography methods, showing that with some assumptions on the information theoretic entropy of the plaintext messages, certain DNA steganography systems may not be cryptographically secure, and can be broken. We also discuss various modified DNA steganography systems which appear to have improved security.

1. Introduction

1.1. Biomolecular Computation. Biotechnological methods (e.g., recombinant DNA) have been developed for a wide class of operations on DNA and RNA strands, including site-specific edits and splicing operations. There has recently arisen a new area of research known as DNA computing, which makes use of recombinant DNA techniques for doing computation. More generally, biomolecular computing (BMC) makes use of such biotechnological methods for doing computation (see the comprehensive survey of Reif [R98]). Various recombinant DNA operations have been shown sufficient to allow for universal computation [H92]. BMC methods have been proposed to solve difficult combinatorial search problems such as the Hamiltonian path problem [A94] and the Data Encryption Standard (DES) (see Boneh, et al [BDL95] and Adleman, et al [ARRW96]), using the vast parallelism available via recombinant DNA to do the combinatorial search among a large number of possible solutions represented by DNA strands. While these methods for solving hard combinatorial search problems may succeed for fixed sized problems, they are ultimately limited by their volume requirements, which may

2000 *Mathematics Subject Classification.* Primary 92C40, 92B05, 68Q22, 68Q25, 94A60.

© 2000 American Mathematical Society

grow exponentially with input size. However, BMC has many exciting further applications beyond pure combinatorial search. For example, DNA and RNA are appealing media for data storage due to the very large amounts of data that can be stored in compact volume. They vastly exceed the storage capacities of conventional electronic, magnetic, optical media. A gram of DNA contains about 1021 DNA bases, or about 108 tera-bytes. Hence, a few grams of DNA may have the potential of storing all the data stored in the world. Most recombinant DNA techniques can be applied at concentrations of 5 grams of DNA per one liter of water. The DNA strands may constitute (a) a "wet" data base of biological data, or (b) data obtained from more conventional binary storage media. In the former case (a), the natural DNA obtained from biological sources may be recoded using nonstandard bases as described in Landweber and Lipton [LL97], to allow for subsequent BMC processing. In the later case (b), the DNA data may be moved to conventional binary storage media via a variety of techniques including DNA chip arrays (see Section 3), and the binary data may be encoded in DNA strands by use of an alphabet of short oligonucleotide sequences. Baum [B95] has discussed methods for fast associative searches within DNA databases using hybridization. Other BMC techniques (see Reif [R95]) might perform more sophisticated data base operations on DNA data such as the data base join operations and various massively parallel operations on the DNA data.

1.2. Cryptography. Data security and cryptography are critical aspects of conventional computing and may be also important to possible DNA data base applications. Here we provide basic terminology used in cryptography (see Schneier [S96]). Suppose a sender wishes to send a message to a receiver so that no one else can read the message. The message is initially written in plaintext (i.e., in non-encrypted form) over a finite alphabet. Encryption (also known as encipherment) is the process of scrambling the plaintext message, transforming it into an encrypted message known as cipher text. For example, a fixed codebook may be used to provide an initial mapping from characters in the finite plaintext alphabet to a finite alphabet of codewords, and then a sophisticated algorithm depending on a key may be applied to further encrypt the message. Decryption (also known as decipherment) is the reverse process of transforming the encrypted message back to the original plaintext message. A cryptosystem (also known as a cipher) is a method for both encryption and decryption of data. Hence, the processes of encryption and decryption require: an informational message to be transmitted; a physical system for message read-in, transformation, and read-out; and agreed upon encryption and decryption algorithms. An unbreakable cryptosystem is one for which successful cryptanalysis is not possible.

1.3. Our Results: BMC methods for Cryptography. This paper investigates a variety of biomolecular methods for encrypting and decrypting data. We assume the plaintext message data is encoded in DNA strands. For example, the DNA strands in solution in a test tube may form a "wet" data base of biological data (e.g., the DNA of personnel of an organization) which might need to be secured (this secrecy might be needed for privacy reasons, or perhaps in the context of our current administration, for legal or political reasons). Alternatively, a test tube of DNA might contain encoded data obtained from more conventional binary or analog storage media. We briefly discuss input and output of the DNA data to conventional binary storage media via (photo- sensitive and/or photo-emitting) DNA chip arrays. In this case we assume the plaintext message data is encoded in DNA strands by use of a (publicly known) alphabet of short oligonucleotide sequences. In this latter case, one important potential advantage of using DNA for storage is its compactness. First we present a class of DNA cryptography techniques that are in principle unbreakable. We initially secretly assemble a library of one-time-pads in the form of DNA strands. Then we propose a number of methods whereby a large number of short message sequences can be encrypted using these one-time-pads. The aspect that restricts use one-time-pad encryption systems in practice for conventional electronic media is the large amount of one-time-pad data which must be created and transmitted securely. Here, DNA shines because it is an extremely compact form of representing data and significant amounts of information can be carried in a limited amount of physical space. Further, we present an interesting concrete example of a DNA cryptosystem technique in which the input is a two-dimensional image. By use of photo- sensitive DNA-on-a-chip technology, this image is converted to a solution of DNA strands encoding the image, these strands can then be encrypted via DNA cryptography techniques. The reverse process of decryption of the image is also described. The decrypted image can be output as a two-dimensional image, by conventional (fluorescent) DNA-on-a-chip technology. Finally, we discuss a general class of methods, known as steganography, where the original plaintext is not actually encrypted but instead disguised within other data. In DNA steganography methods, the DNA plaintext messages are appended with one or more secret keys and the resulting appended DNA strands are then hidden by mixing them within many other irrelevant DNA strands (e.g., randomly constructed DNA strands). For example, Clelland, Risca, and Bancroft [CRB99] recently proposed genomic steganography: techniques using amplifiable microdots. As in conventional steganography methods, the plaintext messages are not otherwise preprocessed, and so the methods are very appealing due to their simplicity. Unfortunately, we show, with some assumptions on the bounded information theoretic entropy of the plaintext messages found in practice, that a class of such DNA steganography systems can be broken. We also discuss some improvements that increase the security of DNA steganography systems.

1.4. Organization. In this Section 1, we have introduced BMC and cryptography terminology, and also have discussed our results. In Section 2, we describe some unbreakable DNA crptosystems using randomly assembled one-time pads. In Section 3, we give a concrete, detailed example of a DNA cryptosystem for two dimensional images, using a DNA chip for I/O and also using a randomly assembled one-time pad. In Section 4, we discuss a class of DNA steganography techniques and show that they can be broken with some modest assumptions on the entropy of the plaintext, even if they employ perfectly random one-time pads. We conclude our paper in Section 5.

2. DNA Cryptosystems Using Random One-Time Pads

The input plaintext messages. As mentioned above, we assume input to our DNA cyptosystems are short segments of distinct plaintext messages. See below for a discussion of the design of word libraries and see Section 3 for details of a possible method (a chip-based DNA micro-array) for input and output of messages from conventional storage media. One-time-pad encryption uses a random codebook to convert short segments of plaintext messages to encrypted text. Two points critical to security issues should be made about codebooks; they must be truly random, and they must be used only once. This class of cryptosystems using a secret random one-time pad are the only cryptosystems known to be absolutely unbreakable; see

Schneier[S96]. (In contrast, the security of encryption using pseudo-random encryption pads is uncertain; while certain such systems have been shown to be attractive targets for cryptanalysis since their inherent periodicity supplies information to potential code breakers, Yao [Y82] has proven other systems to be unbreakable in polynomial time, given some (as yet unproven) certain complexity theoretic assumptions.) The reuse of pads, likewise, increases the risk of message decryption by an undesired interceptor. These two principles dictate certain features of the DNA sequences and of sequence libraries, which will be discussed more below. We will first assemble a large one-time-pad in the form of a DNA strand, which is randomly assembled from short oligonucleotide sequences, and then isolated and cloned. These one-time-pads will be assumed to be constructed in secret, and we further assume that specific one-time-pads are shared in advance by both the sender and receiver of the secret message. This assumption requires initial communication of the one-time-pad between sender and receiver, which is facilitated by the very compact nature of DNA in solution. We will provide methods for constructing such a DNA one-time-pad. We propose two methods whereby a large number of short message sequences can be encrypted, to ensure the original plaintext message cannot be determined from the resulting DNA. In these encryption methods, the one-time-pad is used in effect as a random codebook (as opposed to a fixed codebook, which cannot be assumed to be secret in an unrestricted cryptosystem) for mapping short message sequences into encrypted sequences. The mapping methods used by these encryption methods include: (1) the use of substitution, where we encrypt each message sequence by associated matching with corresponding sections of the one-time DNA pad, or alternatively, (2) the use of bit-wise XOR computation via biomolecular computing techniques. The decryption is done by similar methods.

2.1. Our DNA Cryptosystem Using Substitution. Substitution One-time-pad Cryptosystems. The input to a substitution one-time- pad system, is a plaintext binary message of length n, partitioned into plaintext words of fixed length, and a substitution one-time-pad consisting of a table randomly mapping all possible strings of plaintext words into cipher words of fixed length, such that there is a unique reverse mapping. The plaintext is encrypted by substituting each ith block of the plaintext with the cipher word given by the table, and is decrypted by reversing these substitutions. DNA Implementation. In the case of encryption by substitution, we wish to convert one test tube of short DNA strands (the plaintext messages) into another set of entirely different strands (the encrypted messages) in a random yet reversible way. Our DNA encoded messages are manipulated such that the plaintext are converted to cipher strands and the plaintext strands are removed. Our substitution method requires one-time-pad DNA sequences to accomplish this conversion. The overall scheme involves long DNA pads containing many segments, where each segment contains a cipher word followed by a plaintext word. The cipher word acts as a hybridization site for binding of a primer, which is then specifically appended with a plaintext word to produce word-pairs. The word-pair DNA strands can be used as a type of lookup table in the first step of the conversion of plaintext into cipher text. Additional manipulations, described in later sections of this paper, accomplish the remaining steps to complete encryption and decryption. DNA one-time pads. An ideal library of one-time pads would contain a huge number of pads and each pad would provide a perfectly unique, random mapping of plaintext to cipher word pairs. Our construction procedure

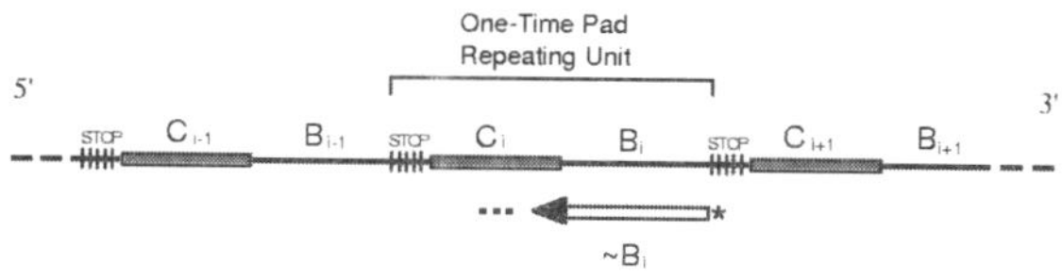

FIGURE 1. One-time-pad Codebook DNA Sequences.

approaches these goals. The structure of an example pad is given in Figure 1. The repeating unit is made up of: one sequence word, Bi, from the set of cipher or codebook-matching words; one sequence word, Ci, from the set of plaintext words; and a polymerase "stopper" sequence. Each sequence pair i, uniquely associates a plaintext word with a cipher word. Oligo with sequence ~Bi, corresponding to the Watson-Crick complement of cipher word Bi, can be used as polymerase primer and be extended by specific attachment of plaintext word Ci. The stopper sequence prohibits extension of the growing DNA strand beyond the boundary of the paired plaintext word. A library of unique codebook strands is constructed using this theme. Each individual strand from this codebook library specifies a particular, unique set of word pairings. The one-time-pad consists of a DNA strand of length n containing $d = n/(L_1 + L_2 + L_3)$ copies of the repeating pattern: a cipher word of length L_2, a plaintext word of length L_1, and stopper sequence of length L_3. (We set $L1 = c_1 \log_2 n, L_2 = c_2 log_2 n$, and $L_3 = c_3$, for fixed integer constants $c_1, c_2, c_3 > 1$.) Each repeat unit specifies a single mapping pair, and no codebook word or plaintext word will be used more than once on any pad. Therefore, given a codebook word B_i we are assured that it maps to only a single plaintext word C_i and vice versa. The stopper sequence acts as "punctuation" between repeat units so that DNA polymerase will not be able to continue copying down the template (pad) strand. Stopper sequences consist of a run of identical nucleotides which act to halt strand copying by DNA polymerase given a lack of complementary nucleotide triphosphate in the test tube. For example, the sequence TTTT will act as a stop point if the polymerization mixture lacks its base-pairing complement, A. Stopper sequences of this variety have been prototyped previously by Hagiya, et al [HA+97]. Given this structure, we can anneal primers and extend with polymerase in order to generate a set of oligonucleotides corresponding to the plaintext/cipher lexical pairings (or word-pair strands). The set of word-pair strands is essentially a lookup table for a random codebook. Therefore, the feasibility depends upon: the size of the lexicon; the number of possible pads available; and the size, complexity, and frequency of message transmissions. Lexicon size d gives the total number of words available for use within a single codebook pad (and, if we use the $2D$ chip grid for I/O (see below Section 3), the number of pixels available on the $2D$ chip grid). Word sizes L_1, L_2 determine the total number of possible DNA sequences available for a lexicon, however lexicon size is reduced by elimination of complementary sequences and division of the possible sequences into two lexicons. Pad diversity stands for the total number of random pads generated during a single pad construction experiment. Multiple construction cycles can be used to prepare greater numbers of unique pads. The estimate of message size relative to lexicon size assumes that the random codebooks are used in conjunction with the DNA chip I/O outlined in Section 3 below.

Codebook Libraries. Construction of the libraries of codebook pads can be approached using segmental assembly or build-up procedures used successfully in previous gene library construction projects [LK93, LB97] and DNA word encoding methods used in DNA computation [DMGFS96, DMGFS98, DMRGF+97, FTCSC97, GDNMF97, GFBCL+96, HGL98, M96]. The design of sequence words for the plaintext and cipher lexicons represents the first technical challenge. We would like the lexicons to be distinct, or disjoint, in the mathematical sense. We also need essentially complete coverage of the lexicon on each pad, as well as unique word mapping, so a single plaintext word pairs with a single cipher word and vice versa. (While perfectly disjoint lexicons are ideal, the system may function with biased lexicons containing a small degree of potential overlap.)
Construction of DNA one-time pads. There are a number of possible methodologies for construction of the plaintext and cipher pairs used for the pad. One methodology is the random assembly of one-time pads in solution (e.g. on a synthesis column). We view such methods less favorably due to the difficulty of achieving both full coverage and yet still avoiding possible conflicts by repetition of plaintext and/or cipher words. (We can set the constants c1 and c2 large enough so that the probability of getting repeated words on a pad of length n is very small, but then the coverage may be reduced.) The methodology, which we favor since it is very controllable, is to employ a DNA chip (see [FRP+91, PSS+94. CYH+96, BKH96. S98] for basic DNA chip technology). Such DNA chips are currently commercially available and chemical methods for construction of custom variants are well developed. The DNA chip has an array of immobilized DNA strands, so that multiple copies of a single sequence are grouped together in a microscopic pixel. The microscopic arrays of the DNA chip are optically addressable, and there is known technology for growing distinct DNA sequences at each (optically addressable) site of the array. Light-directed synthesis allows the chemistry of DNA synthesis to be conducted in parallel at thousands of locations (i.e. it is a combinatorial synthesis). Therefore, the number of sequences prepared far exceeds the number of chemical reactions required. For preparation of oligonucleotides of length L, the 4L sequences are synthesized in 4n chemical reactions. For example, the ~65,000 sequences of length 8 require 32 synthesis cycles, and the 1.67x107 sequences of length 10 require only 48 cycles. The plaintext and cipher pairs can be constructed so that there is nearly complete coverage of the lexicon on each pad, as well as a nearly unique word mapping between plaintext and cipher pairs. These resulting cipher word, plaintext word pairs can be assembled together in random order (with possible repetitions) on a long DNA strand by a number of known methods (e.g., simply blunt end ligation, or for higher efficiency, the use of a technique of hybridization assembly with complemented pairs as done in Adleman's original DNA experiment [A97]). Cloning or PCR may be used to amplify the resulting one-time pad.

2.2. Our DNA XOR One-time-pad Cryptosystem. *Vernam Cipher Cryptosystems:* In classical cryptography (see Kahn[K67]), the Vernam cipher (now known as the XOR one-time-pad cryptosystem) is deployed by generating a sequence, S, of R independently distributed random bits known as a one-time-pad, replicating the one-time-pad, and storing one copy at the source and one at the destination. Let L be the number of bits of S that remain unused, where initially $L = R$. Recall that XOR is the operation that given two Boolean inputs, yields 0 if the inputs are the same, and otherwise is 1. When a plaintext binary message M which is $n < L$ bits long needs to be sent, each bit M_i is XOR'ed with the bit $K_i = SR - L + i$ to produce encrypted bits $Ci = M_i\text{XOR}K_i$ for $i = 1, \ldots, n$. The n bits of S that have been consumed are then destroyed at the source and the encrypted sequence $C = (C_1, C_2, ..., C_n)$ is despatched to the destination. At the destination the identical process is repeated - that is the sequence C is used in the place of M, performing bitwise XOR with bits from S, destroying the bits of S after they are consumed. The commutative property of the XOR results in the initial message being reproduced since $C_iXORK_i = M_i$ since $C_i = M_iXORK_i$ and Mi XOR K_i XOR$K_i = M_i$. (Note that for efficient DNA encoding, we may alternatively do this in quaternary, that is over modular base 4, rather than in binary, since DNA has four nucleotides. In this case, we assume the input plaintext and one-time-pad are in quaternary, and for encryption we use addition modulo 4 instead of XOR, and for decryption we subtract the one-time-pad elements modulo 4. For simplicity, we will continue the discussion for the binary case.) We wish again to convert one test tube of short DNA strands (the plaintext messages) into another set of entirely different strands (the encrypted messages) in a random yet reversible way. Each of the plaintext messages are assume to have appended unique prefix index tags of fixed length L0 indexing them. Each of the one-time-pad DNA sequences are also assumed to have appended unique prefix index tags of the same length L0, which form the complements of the plaintext message tags. By use of known recombinant DNA techniques (e.g., annealing and ligation), each corresponding pair of a plaintext message and a one-time-pad sequence, with the same tag, can be concatenated into a single DNA strand. Our DNA encoded messages are modified in this case by a bit-wise XOR computation, so that fragments of the plaintext are converted to cipher strands using the one-time-pad DNA sequences, and the plaintext strands are removed. The reverse decryption is similar, using the commutative property of the bit-wise XOR operation. Clearly, as the remaining basic building block, we need to describe a method to effect bit-wise XOR on vectors. Methods developed to effect binary arithmetic operations such as addition are similar to the bit-wise XOR computation, but have the additional constraint of needing to deal with a carry bit. If we have a method to execute addition, we could use it, ignoring the carry bit and slightly altering the mapping to reflect the rules of bit-wise XOR instead of those of bit addition (which takes into account also carry bits). The required bit-wise XOR computation can thus be done by such modification (which involves a considerable simplification) of at least two previously known biomolecular computing techniques for integer addition:

(a) Guarnieri, Fliss, and Bancroft prototyped [GFB96] the first BMC addition operations (on single bits) in recombinant DNA. This experimental work was very significant. However, it suffered from some limitations: (i) only two numbers where added, so it did not take advantage of the massive parallel processing capabilities of BMC and (ii) the outputs were encoded distinctly from the inputs, so it did not allow for repeated operations. Subsequent proposed methods [OGB97,LKSR97,GPZ97] for basic operations such as arithmetic (addition and subtraction) permit chaining of the output of these operations into the inputs to further operations, and to allow operations to be executed in massive parallel fashion. In particular, Rubin el al [RKL98] gave an experimental demonstration of a BMC method for chained integer arithmetic. This work also gave one of the first demonstrations in BMC of logically reversible computation, by way of the conditional XOR operation. This method, which O(n) recombinant DNA operations for

A: Message **B: Encrypted Message** **C:Decrypted Message**

FIGURE 2. Simulated Read-Outs from DNA Chip Input/Output.

vectors of lengh n, can be used directly for our required XOR vector operations (i.e., by simply disabling the logic used for carry-sums from previous bits).

(b) Another method being developed for effecting general binary addition, and XOR in particular as well, is by the use of DNA tiles (LaBean, et al [LWR99]). Their idea is based on the DNA self assembly work by Winfree, et al [W95, W96, Win98a, WLW+98, WYS96] and the compact assembly methods of Reif [R97]. Tiles with specific uncomplemented pads at the corners were constructed, with the purpose of effecting self-assembly. The tiles that have been prototyped in the laboratory by LaBean have been designed so that self-assembly of an output of a computation can occur based on an initial assembly created from tiles that represent the input. More specifically, given a binary input string, each bit is represented by a single tile. The tiles are designed such that they assemble linearly to represent the binary string. The use of a special corner tile allows two such linear tile assemblies representing two binary input strings respectively, to come together and create a closed framework within which specially designed output tiles can fit in. This process allows for unmediated parallelized binary addition or XORing. At the end of the process, there exists a single strand that runs through the entire assembly which will contain the two inputs and the output, as a result of the special design of these tiles. By using this property we are able to effect the Vernam cipher in DNA. For details of this XOR computation, using self-assembly of DNA tiles, see LaBean, et al [LYR+98,LWR99].

3. A DNA Cryptosystem for 2D Images using a DNA Chip and a Randomly Assembled One-Time Pad

3.1. Overview of Method. In this section, we outline an example system capable of encryption and decryption of input and output data in the form of 2D images recorded on the microscopic arrays of a DNA chip. (Again, see [FRP+91, PSS+94. CYH+96, BKH96. S98] for basic DNA chip technology and also see Mills, et al [MYP98] for a previous use of DNA chips for I/O.) The system we describe here consists of: a data set to be encrypted; a chip bearing immobilized DNA strands; and a library of one-time pads encoded on long DNA strands. The data set for encryption in this specific example is a 2-dimensional image, but variations on the method may be useful for encoding and encrypting other forms of data or types of information. Recall the DNA chip contains an addressable array of nucleotide sequences immobilized such that multiple copies of a single sequence are grouped together in a microscopic pixel. Again, we note that such DNA chips are currently commercially available and chemical methods for construction of custom variants are well developed. Further chip details will be given below. The library of one-time pads was described in Section 2.1 above. Figure 2 gives a coarse grained outline of the I/O method. Fluorescent-labeled, word-pair DNA strands are prepared from a substitution pad codebook as described in Section 2.1. These are annealed specifically to their sequence complements at unique sites (pixels) on

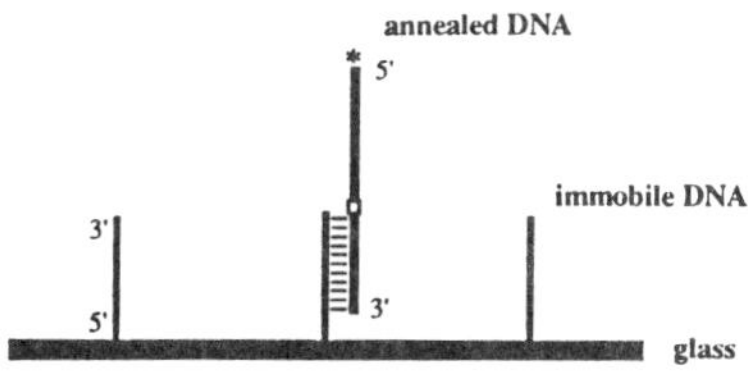

FIGURE 3. Components and Organization of the DNA Chip.

the DNA chip. The message information (Panel A) is transferred to a photo mask with transparent (white) and opaque (black) regions. Following a light-flash of the mask-protected chip, the annealed oligonucleotides beneath the transparent mask pixels are cleaved at a photo-labile position; and their 5′ sections dissociated from the annealed 3′ section and are collected in solution. This test tube of strands is the encrypted message. Annealed oligos beneath opaque mask are unaffected by the light-flash and can be subsequently washed off the chip and discarded. If the encrypted message oligos are reannealed onto a (washed) DNA chip, they would now anneal to new specifc locations and the message information would be unreadable (Panel B). (Note: if the sequence lexicons for 5′ segment (cipher word) and 3′ segment (plaintext word) are disjoint, no binding would occur and the chip in Panel B would be completely black.) Decrypting is accomplished by using the fluorescent labeled oligos as primers in one-way (lopsided) PCR with the same one-time codebook which was used to prepare the initial word pair oligos. When the word-pair PCR product is bound to the same DNA chip, the decrypted message is revealed (Panel C). Input/Output as observed by fluorescence microscopy of the DNA I/O chip. presents a coarse grained outline of the I/O method. The annealed DNA in Figure 3 corresponds to the word-pair strands prepared from a random substitution pad as described in Section 2.1 above. Immobile DNA strands are located on the glass substrate of the chip in a sequence addressable grid according to currently used techniques. The annealed strand contains: a fluorescent label on its 5′ end (asterisk); a codebook-matching sequence word (not base-paired on the chip); a photo-labile base (white square) capable of cleaving the DNA backbone; and a chip-matching word (base-paired to immobile strand). The 5′ (unannealed) end carries a cipher word while the 3' (annealed) end carries a plaintext word. These word-pair strands are special in that they contain a photo-cleavable base analog between the two sequence words. The photo-cleavable base analog can be chemically added to the 3′ end of the cipher word during oligo synthesis. Figure 4 gives step by step procedures for encryption and decryption. We start with a DNA chip displaying the sequences complementary (in the Watson-Crick sense) to the plaintext lexicon. In step one, the fluorescent-labeled word-pair strands prepared from a one-time-pad are annealed to the chip at the pixel bearing the complement to their plaintext 3′ end. In the next step, the mask (heavy black bar) protects some pixels from a light-flash. At unprotected regions, the DNA backbone is cleaved at a site between the plaintext and cipher words. In the final step, the cipher word strands, still labeled with fluorophore at their 5′ ends, are collected and transmitted as the encrypted message. A message can be decrypted only by using a one-time-pad and DNA chip identical to those used in the encrypting process. First, the word-pair strands must be reconstructed by appending the cipher word with the proper plaintext word. This is accomplished by polymerase extension or lop-sided PCR using cipher words as primer and one-time-pad as template. The cipher strands bind to their specific locations on the pad and are appended with their proper plaintext

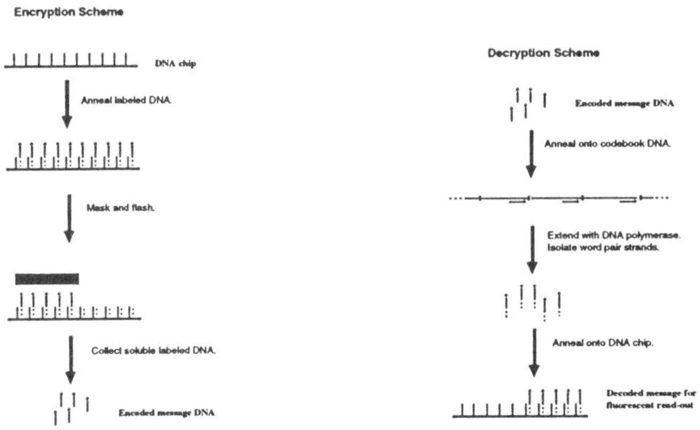

FIGURE 4. Step by step procedures for encryption and decryption.

partner. Note that in the decrypting process the fluorescent label is still required, but the photo-labile base is unnecessary and not present. The final step of decryption involves binding the reformed word-pair strands to the DNA chip and reading the message by fluorescent microscopy.

4. DNA Steganography Techniques

4.1. Steganography. Steganography is a class of techniques that hide secret messages within other messages. In a steganography system, the original plaintext is not actually encrypted but is instead disguised or hidden within other data. Historical examples of steganography systems are the use of grills that mask out all of an image except the secret message, micro-photographs placed within larger images, invisible inks, etc. The cryptography literature (see Schneier[S96]) generally consider conventional steganography methods to have low security (in fact there is disagreement whether steganography is actually encryption, since the plaintext is not actually encrypted but instead disguised within other media) and there are numerous cases where steganography methods have been broken in practice (e.g., see Kahn[K67] and Schneier[S96]). However, it is very appealing due to it's simplicity. There are a number of techniques for applying steganography in the context of biomolecular computation. One method is to take one or more input DNA strands (considered to be the plaintext message) and append to them one or more randomly constructed "secret key" strands. The resulting "tagged plaintext" DNA strands are then hidden by mixing them within many other additional "distracter" DNA strands which might also be constructed by random assembly. Given knowledge of the "secret key" strands, the resulting solution of DNA strands can be decrypted by a number of possible known recombinant DNA separation methods. For example, the plaintext message strands may be separated out by hybridization with the complements of the "secret key" strands might be placed in solid support on magnetic beads or on a prepared surface. These separation steps may combined with amplification steps and/or PCR (see Barnes [B94] and Roberts [R94]).

4.2. Cryptanalysis of DNA Steganography Systems. In a true cryptographic cipher the security is dependent on a secret key. The security of the above system is expected to derive from – as is the case with all steganographic systems – the fact that the adversary is unaware of the existence of the message in the medium of transmission, and/or can not distinguish the plaintext message from the medium. As soon as this assumption is no longer valid, the system can generally be compromised. In particular, this DNA steganography system's security is entirely dependent on the degree that the message DNA strands are indistinguishable from the "distracter" DNA strands. It is reasonable to assume the secret tags are indistinguishable from the "distracter" DNA strands. However, if (a) the plaintext is not initially compressed, and comes from a natural source such as English or natural DNA, and (b) the "distracter" DNA strands are constructed by random assembly, then the original plaintext portion of the "tagged plaintext" DNA strands are distinguishable from the other additional "distracter" DNA strands. We now consider cryptanalysis of such a steganography system, without knowledge of the "secret key" strands, and with the assumption that the plaintext is generated from a source that is distinguishable from the additional "distracter" DNA strands that are mixed with them. In particular, we assume that we know both (i) the probability distribution of the source generating the set of plaintext message DNA strands and we also know (ii) the probability distribution of the source generating the set of "distracter" DNA strands, and we further assume these probability distributions are distinct. The Shannon information theoretic entropy E_S (see [CT 91]) provides a measure of the factor that a source can be compressed without loss of information. For example, many images have entropy nearly 4, English text has entropy about 3, and computer programs have entropy about 5. The overall entropy of most classes of DNA, (including human DNA) is in the range of 1.2 to 2 (see [GT94], [LY97],[NW99]), and is larger on non-coding sequences than on the protein coding sequences. As a simple example, let us for the moment assume: (i) the plaintext DNA source has Shannon entropy E_S bounded by a constant above 1, (ii) the probability distribution of the plaintext source is known (e.g., generated by an ergodic stationary process), and (iii) the "distracter" DNA strands have a random uniform distribution. We also assume for simplicity that both the plaintext and "distractor" DNA strands are of some large length n. Let $L = E_S \log_2 n$. Then, following certain known lossless data compression schemes such as Lempel-Ziv [ZL77] (also see [CT 91]), we form a dictionary D of the $d = n/L$ most frequently occurring subsequences of length at least L in the known plaintext source distribution. We will separate out the plaintext message strands by repeated rounds of hybridization with the complements of the elements of this dictionary. On each round of separation, we assume a test tube T which contains a mixture of "tagged plaintext" DNA strands mixed with a high concentration of "distracter" DNA strands. Let $r(T)$ be the ratio of concentration of "distracter" DNA strands to "tagged plaintext" DNA strands. We can assume that we have constructed a dictionary D whose size is is much larger than $\log r(T)$. On each round we form a new test tube $F(T)$ with expected $r(F(T))$ considerably reduced from the previous ratio $r(T)$, by the following steps:

[1] Pour a fraction s (e..g, let $s = 1/2$) of the volume of the current test tube T into a test tube T_1 and pour the remaining fraction $1 - s$ of the volume T into a test tube T_2.

[2] Choose a random text phrase x in D (not previously considered in a prior trial), and using the Watson-Crick complement of that phrase x, do a separation on test tube T_2 with respect to phrase x, yielding a new test tube T_3 whose contents are only DNA strands containing the phrase x.

[3] Pour the contents of test tubes T_1 and T_3 into a new test tube $F(T)$.

(Note that the series of separations, used to eliminate the distracters, will shape the distracter population into one whose distribution more closely matches that of the plaintext. Nevertheless, our proof given below will remain correct on subsequent trails, since on each trail we choose a random text phrase from the dictionary D that had not previously been considered in a prior trial. Furthermore, the distracters are assumed to have a random uniform distribution. Hence we can assume independent probabilities from trial to trial.) It will be useful to note that for large n, we can approximate $(1-1/n)^n \sim 1/e$. By known results used in lossless data compression methods (see Cover and Thomas [CT 91]), each subsequence of length L of the plaintext DNA strand whose source was the same as used in formation of the dictionary D, the probability of containing a given text phrase x of length at least L from the dictionary D is expected to be $1/d$. Since the plaintext DNA strand is assumed to be of length n, and contains at least d= n/L distinct phrases of length L, it follows that it has probability at most $(1-1/d)^d \sim 1/e$ of not containing dictionary text phrase x. (Note that since this event is not a certainty, the above protocal uses re- pooling of test tubes T_1 and T_3.) On the other hand, each subsequence of length L of a "distracter" DNA strand is random, so has probability $p = 1/4^L = 1/4^{(\log n)E_S} = 1/n^{2E_s}$ of containing a given text phrase of length at least L. Since the "distracter" DNA strands are assumed to be of length n, it follows that each of them has probability at least $(1-p)^n$ of not containing a given text phrase of length at least L. But this probability is $(1-p)^n = (1-p)^{(1/p)(np)}$ $\sim (1/e)^{np} = 1/exp(np) = 1/exp(n^{1-2E_S}) \sim 1$ for large n and entropy $E_S > 1$. Therefor, we have shown that it is very unlikely that the "distracter" DNA strands contain a given random text phrase x of length at least L from the dictionary D. Hence, we have shown: (1) with probability at most $1 - 1/exp(n^{1-2E_S}) \sim 0$, any given "distracter" DNA strand originally within test tube T_2 is added to test tube T_3 via the separation step, and (2) in contrast, with probability at least $1-1/e$, any plaintext DNA strand originally within test tube T_2 is added to test tube T_3 via the separation step. Since test tubes T_1, T_2 consist of fractions $s, 1-s$ of the volume test tube T, and $F(T)$ is constructed by pouring together test tubes T_1, T_3 we have that the relative concentration of DNA strands in $F(T)$ as compared to the original test tube T expects to decrease by: (1') a factor of $s + (1-s)/exp(n^{1-2E_S}) \sim s$ for the "distracter" DNA strands, and (2') a factor of $s + (1-s)/e$ for any plaintext DNA. Let $c = 1/(1-1/e+1/s) < 1$. It follows that the ratio $r(F(T))$ of "distracter" DNA strands to plaintext DNA will expect to decrease from the original ratio $r(T)$ by at least this constant multiplier c factor: $(s + (1-s)/exp(n^{1-2E_S}))/(s + (1-s)/e)$ $\sim s/(s + (1-s)/e) = 1/(1 + (1/s - 1)/e) = c < 1$, for $0 < s < 1$. Hence, for any given $r' < r$, after $\log(r/r')/\log(1/c) = O(\log(r/r'))$ repeated rounds of this process, the ratio of concentration in the test tube T will expect to decrease from initially $r = r(T)$ to the given smaller ratio r'. Cryptanalysis using hints. Another cryptanalysis technique for breaking steganographic systems is to use a number of "hints" that disambiguate the plaintext. As a concrete example, consider the case where we wish to make secret the DNA of an individual (say, the President), and we do this using such an improved steganography system where the other additional "distracter" DNA strands (that are mixed with the DNA of an individual) are DNA from a similar but not identical genetic pool. Then the steganography system may often be broken by use of sufficiently many distinguishing "hints" concerning the DNA of the individual, such as the fact that the individual might have a particular set of observable expressed gene sequences (e.g., for baldness, etc.). These hints may allow for the subsequent identification of the full secret DNA by use of a series of separation steps using the complement of portions of the known gene sequences.

4.3. Improved DNA Steganography Systems. We may enhance the security of DNA steganography systems by a number of methods:

A. Mimicking Distribution of "Distracter" DNA. The first technique is an improved construction of the set of "distracter" DNA strands, so that their distribution better mimics the plaintext source distribution. In particular, we can construct the "distracter" DNA strands by random assembly from elements of the dictionary D, so it is more difficult to distinguish the probability distribution of the plaintext source from that of the "distracter" DNA strands. (Also note that we may alternatively apply a series of separations, used to eliminate the distracters in our above cyptoanalysis of steganography systems, to shape the distracter population into one whose distribution more closely matches that of the plaintext.) This difference between these two distributions (that is, between the source distribution and the distribution resulting from random assembly from elements of the dictionary D) can be formally given by the relative entropy (see definition in Cover and Thomas [CT 91]). It is possible that such an improved steganography system may be in turn susceptible to a modified cryptanalysis attack - there may remain enough special properties of realistic source distributions such as English text and natural DNA that are not reflected in the fixed size dictionary D, so that the resulting relative entropy might still be sufficient to break the resulting DNA steganographic system. In particular, the resulting enhanced DNA steganographic system might still be broken by simply using a similar cryptanalysis technique as given above, but with instead a somewhat larger dictionary D', providing a better model of the source plaintext distribution.

B. Compression of the Plaintext. Another method for an improved DNA steganography system is to recode the plaintext using a universal lossless compression algorithm method such as Lempel-Ziv [ZL77]. In this case the resulting distribution of the recoded plaintext approximates a universal distribution, so uniformly random assembled distracter sequences may suffice to provide improved security. The drawback of this system is that unlike conventional steganography methods, the plaintext messages needs to be preprocessed. This recoding of plaintext messages can easily be done electronically; but the recoding would have to be done by BMC means in the case where the plaintext is derived from natural DNA. In the latter case, we can do the recoding by a method similar to that discussed in Section 2 for substitution one-time-pads (using a Lempel-Ziv dictionary D for the substitution mapping). There is no known proof of the security of the resulting enhanced DNA steganography system; in contrast to the absolute security provided by the use of one-time-pads for equivalent effort.

5. Conclusion

This paper has presented an initial investigation into the use of DNA-based methods for cryptosystems. We have discussed two classes of methods: (i) DNA cyptography methods based on DNA one-time-pads, and (ii) DNA steganography methods. Our DNA substitution and XOR methods based on one-time-pads are in principle unbreakable. Furthermore, we have provided an example of a concrete implementation of our DNA cyptography methods which includes 2D input and output. In contrast, it remains to be seen if DNA steganography systems with natural DNA plaintext input can or cannot be made to be unbreakable. We have shown that a certain class of DNA steganography methods offer only limited security, and can be broken with some reasonable assumptions on the entropy of the plaintext messages. We also considered a number of modified DNA steganography systems with apparently improved security.

Acknowledgements. The authors wish to sincerely thank the referees for their excellent suggestions for improvements to this paper, and also wish to thank Jeremy De Bonet for helpful editorial comments. This work was supported in part by Grants NSF/DARPA CCR-9725021, CCR-96-33567, NSF IRI- 9619647, ARO contract DAAH-04-96-1-0448, and ONR contract N00014-99-1-0406.

References

[A94] Adleman, L., Molecular Computation of Solution to Combinatorial Problems, Science, 266, 1021, (1994).

[A95] Adleman, L., On Constructing a Molecular Computer, Dept of CS, U.S.C., (1995). Available via anonymous ftp from ftp.usc.edu/pub/csinfo/papers/adleman/molecular_computer.ps.

[ARRW96] Adleman, L.M., P.W.K. Rothemund, S. Roweis, E. Winfree, On Applying Molecular Computation To The Data Encryption Standard, 2nd Annual DIMACS Meeting on DNA Based Computers, Princeton, June, 1996

[BCGT96] Bach, E., A. Condon, E. Glaser, and C. Tanguay, Improved Models and Algorithms for DNA Computation, Proc. 11th Annual IEEE Conference on Computational Complexity, J. Computer and System Sciences, to appear.

[B94] Barnes, W.M., PCR amplification of up to 35-kb DNA with high fidelity and high yield from bacteriophage templates, Proc. Natl. Acad. Sci., 91, 2216–2220, (1994).

[B95] Baum, E. B., How to build an associative memory vastly larger than the brain, Science, April 28, 1995.

[B96] Baum, E. B., DNA Sequences Useful for Computation, 2nd Annual DIMACS Meeting on DNA Based Computers, Princeton University, June 1996.

[BKH96] Blanchard, A. P., R. J. Kaiser and L. E. Hood, High-density oligonucleotide arrays, Biosens. Bioelec., Vol. 11, 687-690, (1996).

[BDL95] Boneh, D., C. Dunworth, R. Lipton, Breaking DES Using a Molecular Computer, Princeton CS Tech-Report number CS-TR-489-95, (1995).

[CYH+96] Chee, M. ,R. Yang, E. Hubbell, A. Berno, X. C. Huang, D. Stern, J. Winkler, D. J. Lockhart, M. S. Morris and S. P. A. Fodor, Accessing genetic information with high-density DNA arrays, Science, Vol. 274, 610-614, (1996).

[CT 91] Cover, T. M. and J. A. Thomas, Elements of Information Theory, John Wiley, NY, (1991).

[DMGFS96] Deaton, R., R.C. Murphy, M. Garzon, D.R. Franceschetti, and S.E. Stevens, Jr., Good encodings for DNA-based solutions to combinatorial problems, Proceedings of the 2nd Annual DIMACS Meeting on DNA Based Computers, June 1996.

[DMGFS98] Deaton, R., R.C. Murphy, M. Garzon, D.R. Franceschetti, and S.E. Stevens, Jr., Reliability and efficiency of a DNA-based computation, Phys. Rev. Lett. 80, 417-420 (1998).

[DMRGF+97] Deaton, R., R.C. Murphy, J.A. Rose, M. Garzon, D.R. Franceschetti, and S.E. Stevens, Jr., A DNA Based Implementation of an Evolutionary Search for Good Encodings for DNA Computation, ICEC'97 Special Session on DNA Based Computation, Indiana, April, 1997.

[CRB99] Clelland, C.T., Risca, V., and C. Bancroft. Genomic Steganography: Amplifiable Microdots. To appear in Nature, 1999.

[FRP+91] Fodor, S. P. A., J. L. Read, C. Pirrung, L. Stryer, A. T. Lu and D. Solas, Light-directed spatially addressable parallel chemical synthesis, Science, Vol. 251, 767-773, (1991).

[FTCSC97] Frutos, A.G., A.J. Thiel, A.E. Condon, L.M. Smith, R.M. Corn, DNA Computing at Surfaces: 4 Base Mismatch Word Design, 3rd DIMACS Meeting on DNA Based Computers, Univ. of Penns., (June, 1997).

[GDNMF97] Garzon, M., R. Deaton, P. Neathery, R.C. Murphy, D.R. Franceschetti, S.E. Stevens Jr., On the Encoding Problem for DNA Computing, 3rd DIMACS Meeting on DNA Based Computers, Univ. of Penns., (June, 1997).

[GFBCL+96] Gray, J. M. T. G. Frutos, A.M. Berman, A.E. Condon, M.G. Lagally, L.M. Smith, R.M. Corn, Reducing Errors in DNA Computing by Appropriate Word Design, University of Wisconsin, Department of Chemistry, October 9, 1996.

[GT94] Grumbach, S., and F. Tahi, Compression of DNA Sequences, Proceedings of the IEEE Data Compression Conference (DCC'94), Snowbird, UT, 72–82, March 1994.

[GB96] Guarnieri, F., and C. Bancroft, Use of a Horizontal Chain Reaction for DNA-Based Addition, Proceedings of the 2nd Annual DIMACS Meeting on DNA Based Computers, June 10-12, 1996, American Mathematical Society, Providence, RI (in press), (1996).

[GFB96] Guarnieri, F., Fliss, M., and C. Bancroft, Making DNA Add, Science, 273, 220–223, (1996).

[GPZ97] Gupta, V., S. Parthasarathy, M.J. Zaki, Arithmetic and Logic Operations with DNA, 3rd DIMACS Meeting on DNA Based Computers, Univ. of Penns., (June, 1997).

[HGL98] Hartemink, A., D. Gifford, J. Khodor, Automated constraint-based nucleotide sequence selection for DNA computation, 4th Int. Meeting on DNA-Based Computing, Baltimore, Penns., (1998).

[HA+97] Hagiya, M., M. Arita, D. Kiga, K. Sakamoto, and S. Yokoyama, Towards Parallel Evaluation and Learning of Boolean mu Formulas with Molecules, 3rd DIMACS Meeting on DNA Based Computers, Univ. of Penns., (June, 1997).

[H92] Head, T., Splicing schemes and DNA, In: Lindenmayer Systems: Impacts on Theoretical Computer Science, Computer Graphics, and Developmental Biology, Ed. by G.Rozenberg and A.Salomaa, Springer-Verlag, 371-383, (1992). Also appears in: Nanobiology 1, 335–342, (1992).

[HH92] Henikoff, S. and Henikoff, J.G, Amino acid substitution matrices from protein blocks, Proc Natl Acad Sci, Nov 15;89(22):10915-9 (1992).

[K67] Kahn, D. The Codebreakers: The Story of Secret Writing, New York: Macmillan Pub. Comp., (1967).

[K98] J. Kelsey, B. Schneier, D. Wagner and C. Hall, Cryptoanalytic Attacks on Pseudorandom Number Generators, Lecture Notes in Computer Science, 1998, p 168.

[KBDL98] Kotera,M., A.-G. Bourdat, E. Defrancq and J. Lhomme, A highly efficient synthesis of oligodeoxyribonucleotides containing the 2'-deoxyribonolactone lesion, J. Am. Chem. Soc., 120, 11810-11811, (1998).

[LK93] LaBean, T.H. and S.A. Kauffman, Design of synthetic gene libraries encoding random sequence proteins with desired ensemble characteristics. Protein Science 2, 1249-1254. (1993).

[LB97] LaBean, T.H. and T.R. Butt, (U.S. Patent # 5,656,467), Methods and materials for producing gene libraries. Date of issue 20-August-1997.

[LYR+98] LaBean, T. H., H. Yan, J. H. Reif and N. Seeman, Construction and Analysis of a DNA Triple Crossover Molecule, (November. 1998).

[LWR99] LaBean, T. H., E. Winfree, J. H. Reif, Experimental Progress in Computation by Self-Assembly, 5th Annual DIMACS Meeting on DNA Based Computers, MIT, Cambridge, MA (June 1999).

[LL97] Landweber, L.F. and R. Lipton, DNA 2 DNA Computations: A Potential 'Killer App'?, 3nd Annual DIMACS Meeting on DNA Based Computers, University of Pens., (June 1997).

[LY97] Loewenstern, D. and Yainilos, P., Significantly lower entropy estimates for natural DNA sequences, J.A Storer and M Cohn (Eds.), IEEE Data Compression Conference, Snowbird, UT, pp. 151-161, (March, 1997).

[MYP98] Mills, A., B. Yurke, P. Platzman, Error-tolerant massive DNA neural- network computation, 4th Int. Meeting on DNA-Based Computing, Baltimore, Penns., (June, 1998).
[M96] Mir, K.U., A Restricted Genetic Alphabet for DNA Computing, 2nd Annual DIMACS Meeting on DNA Based Computers, Princeton University, (June 1996).
[NW99] Nevill-Manning, C.G. and I.H. Witten, Protein is Incompressible, J.A Storer and M Cohn (Eds.), IEEE Data Compression Conference, Snowbird, UT, pp. 257-266, (March, 1999).
[PSS+94] Pease, A. C. , D. Solas, E. J. Sullivan, M. T. Cronin, C. P. Holmes and S. P. Fodor, Light-generated oligonucleotide arrays for rapid DNA sequence analysis, Proc. Natl Acad. Sci. USA, Vol. 91, 5022-5026, (1994).
[P99] Pirrung, M., personal communication, (1999).
[R95] Reif, J.H., Parallel Molecular Computation: Models and Simulations, Seventh ACM Symp. on Parallel Algorithms and Architectures (SPAA95), ACM, Santa Barbara, 213–223, June 1995. Algorithmica, special issue on Computational Biology, 1999. (http://www.cs.duke.edu/~reif/paper/ paper.html)
[R97] Reif, J.H., Local Parallel Biomolecular Computation, 3rd DIMACS Meeting on DNA Based Computers, Univ. of Penns., (June, 1997). DIMACS Series in Discrete Mathematics and Theoretical Computer Science, ed. H. Rubin, (1999). (http://www.cs.duke.edu/~reif/ paper/Assembly.ps and /Assembly.fig.ps)
[R98] Reif, J.H., Paradigms for Biomolecular Computation, First International Conference on Unconventional Models of Computation, Auckland, New Zealand, January 1998. Unconventional Models of Computation, edited by C.S. Calude, J. Casti, and M.J. Dinneen, Springer Pub., Jan. 1998, pp 72-93. (http://www.cs.duke.edu/~reif/paper/paradigm.ps)
[R94] Roberts, S.S., Turbocharged PCR, Jour. of N.I.H. Research, 6, 46–82, (1994).
[RKL98] Rubin, H., J. Klein, T. Leete, A biomolecular implementation of logically reversible computation with minimal energy dissipation, 4th Int. Meeting on DNA-Based Computing, Baltimore, Penns., (June, 1998).
[S96] Schneier, B., Applied Cryptography, 2nd Edition, John Wiley, (1996).
[S98] Suyama, A., DNA chips - Integrated Chemical Circuits for DNA Diagnosis and DNA computers, To appear,(1998).
[WQF+98] Wang, L., Q. Liu, A. Frutos, S. Gillmor, A. Thiel, T. Strother, A. Condon, R. Corn, M. Lagally, L. Smith, Surface-based DNA computing operations: DESTROY and READOUT, 4th Int. Meeting on DNA-Based Computing, Baltimore, Penns., (June, 1998).
[W95] Winfree, E., Complexity of Restricted and Unrestricted Models of Molecular Computation, Princeton DIMACS Technical Report workshop on DNA-based computers, April 4, 1995.
[W96] Winfree, E., On the computational power of DNA annealing and ligation, DNA based computers, Lipton, R.J. and Baum, E.B. eds., Am. Math. Soc., Providence, RI, (1996).
[Win98a] Winfree, E., Simulations of computing by self-assembly, 4th Int. Meeting on DNA-Based Computing, Baltimore, Penns., (June, 1998).
[WLW+98] E. Winfree, F. Liu, Lisa A. Wenzler, N. C. Seeman, Design and Self- Assembly of Two Dimensional DNA Crystals, Nature 394: 539–544, 1998. (1998).
[WYS96] Winfree, E., X. Yang, N.C. Seeman, Universal Computation via Self- assembly of DNA: Some Theory and Experiments, 2nd DIMACS Meeting on DNA Based Computers, Princeton, June, 1996.
[Y82] Yao, A., Theory and Applications of Trapdoor Functions, Proc. of the 23rd IEEE Symp. On Foundations of Computer Science, pp. 80-91, (1982).
[ZS92] Zhang, Y., and N.C. Seeman, A Solid-Support Methodology for the Construction of Geometrical Objects from DNA, J. Am. Chem. Soc., 114, 2656– 2663, (1992).

[ZL77] J. Ziv and A. Lempel, A Universal Algorithm for Sequential Data Compression, IEEE Transactions on Information Theory, 23:3, 337–343,(1977).

(Ashish Gehani) D223 LSRC Building, Research Drive, Department of Computer Science, Duke University, Box 90129 Durham, NC 27708-0129
E-mail address: geha@cs.duke.edu

(Thomas LaBean) D223 LSRC Building, Research Drive, Department of Computer Science, Duke University, Box 90129 Durham, NC 27708-0129
E-mail address: thl@cs.duke.edu

(John Reif) D223 LSRC Building, Research Drive, Department of Computer Science, Duke University, Box 90129 Durham, NC 27708-0129
E-mail address: reif@cs.duke.edu